# HIGH PRESSURE BOILERS

## Sixth Edition

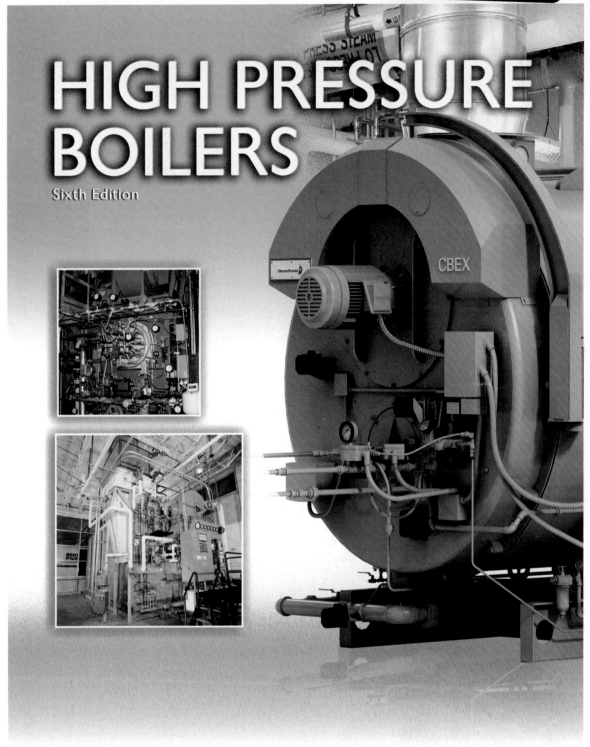

**Frederick M. Steingress**

**Harold J. Frost**

**Daryl R. Walker**

**AMERICAN TECHNICAL PUBLISHERS**
Orland Park, Illinois

American Technical Publishers Editorial Staff

Editor in Chief:
    Jonathan F. Gosse
Vice President—Editorial
    Peter A. Zurlis
Assistant Production Manager:
    Nicole D. Bigos
Technical Editor:
    Eric F. Borreson
Supervising Copy Editor:
    Catherine A. Mini
Copy Editor:
    Jeana M. Platz
Editorial Assistant:
    Erin E. Magee
Cover Design:
    Richard O. Davis

Art Supervisor:
    Sarah E. Kaducak
Illustration/Layout:
    Nick G. Doornbos
    Nick W. Basham
    Bethany J. Fisher
    Christopher S. Gaddie
    Steven E. Gibbs
    Ashlee A. Stevens
Digital Media Manager:
    Adam T. Schuldt
Digital Resources:
    Robert E. Stickley
    Mark A. Passine
    Tim A. Miller

6 7 8 9 – 18 – 9 8 7 6 5 4 3 2 1

Printed in the United States of America

ISBN 978-0-8269-4331-6

 This book is printed on recycled paper.

# ACKNOWLEDGMENTS

*The authors and publisher are grateful for the photographs and technical information provided by the following companies and organizations:*

ABB Inc., Drives & Power Electronics
ASI Robicon
Aurora Pump
Babcock and Wilcox Co.
CH20
Cleaver-Brooks
Dwyer Instruments, Inc.
Eastern Illinois University Renewable Energy Center
Fireye, Inc.
Fluke Corporation
Fred Hutchinson Cancer Research Center
Hach Company, USA

Jenkins Bros.
Lab Safety Supply, Inc.
McDonnell & Miller
The Permutit Co., Inc.
Port of Seattle
Quad/Graphics, Inc.
Rosemount Analytical
Saftronics Inc.
Sarco Co., Inc.
Teledyne Farris Engineering
TSI Incorporated

*Technical review assistance:*

Stephen F. Connor
      Cleaver-Brooks
Byron D. Nichols
      President, Thermal System Training, LLC
      American Society of Power Engineers, Board of Directors

# CONTENTS

# CONTENTS

# CONTENTS

# INTRODUCTION

*High Pressure Boilers* provides information on the safe and efficient operation of high pressure boilers and related equipment. The content and the format of the textbook are specifically designed for use in preparation for boiler operator or facility operator engineer licensing. The textbook can also be used as an introduction to stationary engineering or as a reference for upgrading skills. The extensive illustrations show equipment details and sequential operating procedures for common boiler operator duties. Photographs from industry and leading equipment manufacturers are included to illustrate what a boiler operator may see on the job.

This new edition has been thoroughly updated with many energy efficiency and environmental topics emphasized throughout. Additional content on heat exchanger principles, boiler emissions requirements, and blowdown temperature control and heat recovery has been added. A new section on environmental regulations provides a summary of operating requirements.

Tech tips and case studies are located throughout the textbook to enhance the content. Sample exams following each chapter are used to test for understanding of information presented in the chapter. Also, an appendix, glossary, and index are provided for easy reference. Questions to develop critical thinking skills are provided at the end of each section.

# FEATURES

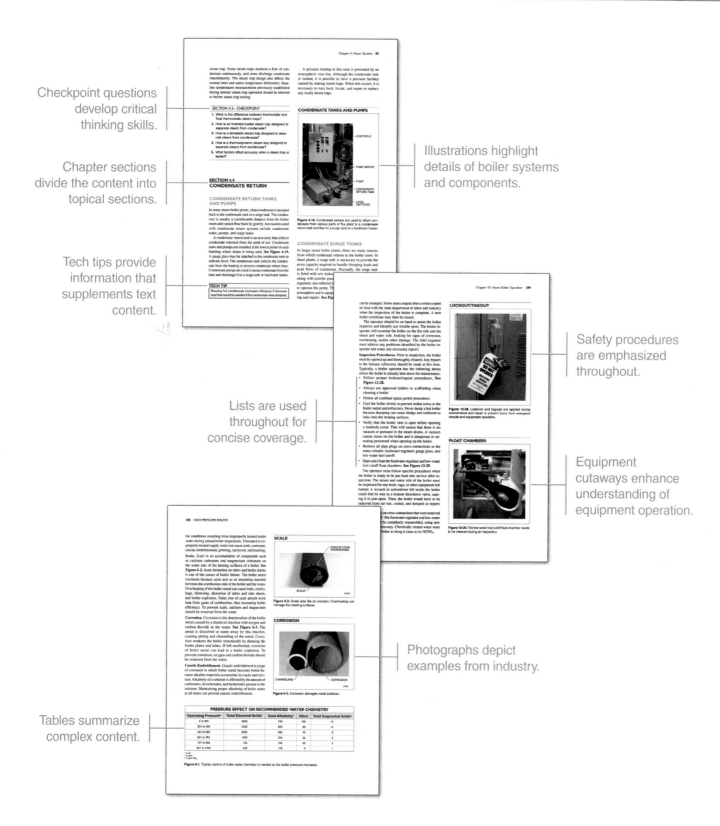

Checkpoint questions develop critical thinking skills.

Chapter sections divide the content into topical sections.

Tech tips provide information that supplements text content.

Lists are used throughout for concise coverage.

Illustrations highlight details of boiler systems and components.

Safety procedures are emphasized throughout.

Equipment cutaways enhance understanding of equipment operation.

Photographs depict examples from industry.

Tables summarize complex content.

# LEARNER RESOURCES

*High Pressure Boilers* includes access to online learner resources that reinforce content and enhance learning. These online resources can be accessed using either of the following methods:

- Key ATPeResources.com/quicklinks into a web browser and enter QuickLinks™ code **472658**.
- Use a Quick Response (QR) reader app to scan the QR code with a mobile device.

The online learner resources include the following:

- **Quick Quizzes**™ that provide interactive questions for each section, with embedded links to highlighted content within the textbook and to the Illustrated Glossary
- **Illustrated Glossary (English/Español)** that serves as a helpful reference to commonly used terms in both English and Spanish, with selected terms linked to illustrations
- **Flash Cards** that provide a self-study/review of common terms and their definitions as well as provide recognition practice for ASME symbol stamps
- **Master Math**™ **Applications** that help learners prepare for a more advanced license
- **Sample Licensing Exams** provide opportunities to practice taking a licensing exam
- **Boiler Operator Resources** provide the learner with printable tables of supplemental information
- **Media Library** that consists of videos and animations that reinforce textbook content
- **ATPeResources.com,** which provides access to additional online resources that support continued learning

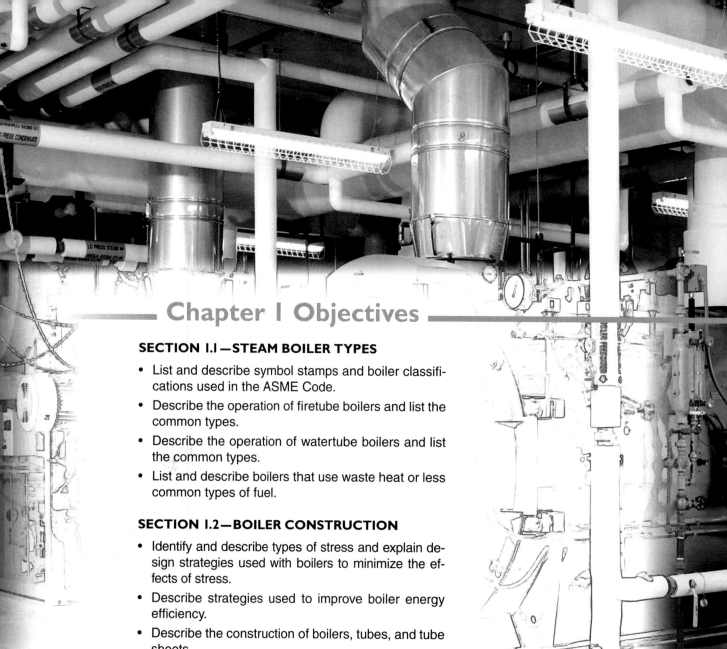

# Chapter 1 Objectives

### SECTION 1.1—STEAM BOILER TYPES

- List and describe symbol stamps and boiler classifications used in the ASME Code.
- Describe the operation of firetube boilers and list the common types.
- Describe the operation of watertube boilers and list the common types.
- List and describe boilers that use waste heat or less common types of fuel.

### SECTION 1.2—BOILER CONSTRUCTION

- Identify and describe types of stress and explain design strategies used with boilers to minimize the effects of stress.
- Describe strategies used to improve boiler energy efficiency.
- Describe the construction of boilers, tubes, and tube sheets.

### SECTION 1.3—USES OF STEAM

- Describe common industrial, institutional, and commercial uses of steam.
- Describe the main features of power generation.

## ARROW SYMBOLS

| AIR | GAS | WATER | STEAM | FUEL OIL | CONDENSATE | AIR TO ATMOSPHERE | GASES OF COMBUSTION |
|-----|-----|-------|-------|----------|------------|-------------------|---------------------|

**Digital Resources**
ATPeResources.com/QuickLinks
Access Code: **472658**

2

# CHAPTER 1

# STEAM BOILERS

## INTRODUCTION

Modern society depends on steam for many uses. Steam generated by boilers is used to supply electricity, water, and gas. Steam is also used for various industrial and commercial processes and to heat water and buildings. A steam boiler is a closed vessel containing water. Water in a steam boiler is pressurized and turned into steam when heat is added. When a fuel is burned, the chemical energy (ability of the chemical to do work) in the fuel is transformed into heat. This heat, which is a form of energy, is contained in the steam. The total heat in steam from a specific point (usually 32°F) is called enthalpy.

## SECTION 1.1
## STEAM BOILER TYPES

### ASME CODE

Most states have adopted the *ASME Boiler and Pressure Vessel Code* (ASME Code). The *ASME Code* is a code written by the American Society of Mechanical Engineers (ASME) that governs and controls the types of material, methods of construction, and procedures used in the installation of boilers. A local pressure vessel code can be consulted to determine whether there are any differences from the ASME definitions. ASME symbol stamps indicate that parts within the boiler system conform to ASME standards. **See Figure 1-1.**

In the ASME Code, steam boilers are generally classified as either low pressure steam boilers or high pressure steam boilers. A *boiler* is a closed vessel used for heating water to generate steam by direct application of heat from combustion fuels or electricity. Steam boilers, however, can be further classified as firetube steam boilers or watertube steam boilers, as well as other ways. The type of steam boiler selected for a particular application depends on the pressure, temperature, and amount of steam required.

### TECH TIP

*The ASME Code has been the accepted standard for many years for the construction of pressure vessels. It is legally enforceable in most jurisdictions.*

## AMERICAN SOCIETY OF MECHANICAL ENGINEERS (ASME) SYMBOL STAMPS

| Symbol | Before 2013 | 2013 and After |
|---|---|---|
| Power boiler | S | S |
| Boiler assembly | A | A |
| Electric boiler | E | E |
| Miniature boiler | M | M |
| Pressure piping | PP | PP |
| Safety valve | V | V |
| Heating boiler | H | H |

**Figure 1-1.** Both the older ASME symbol stamps and the 2013 designs indicate that boiler components conform to ASME standards.

## Boiler Classifications

Steam and hot water boilers may be further classified by the ASME Code, state codes, local codes, and ordinances. **See Appendix.** For example, boilers are classified by the ASME Code and many local jurisdictions as follows:

- Hot water supply boiler—a low pressure hot water heating boiler having a volume exceeding 120 gal., a heat input exceeding 200,000 Btu/hr, or an operating temperature exceeding 200°F that provides hot water to be used externally to itself
- Low pressure hot water heating boiler—a boiler in which water is heated for the purpose of supplying heated water at pressures not exceeding 160 psi (pounds per square inch) and temperatures not exceeding 250°F
- Low pressure steam heating boiler—a boiler operated at pressures not exceeding 15 psi for steam
- Power hot water boiler (high temperature water boiler)—a boiler used for heating water or liquid to a pressure exceeding 160 psi or to a temperature exceeding 250°F
- Power steam boiler—a boiler in which steam or other vapor is generated at pressures exceeding 15 psi
- Small power boiler—a boiler with pressures exceeding 15 psi but not exceeding 100 psi and having less than 440,000 Btu/hr heat input

## Low Pressure Steam Boilers

A *low pressure steam boiler* is a boiler that operates at a maximum allowable working pressure (MAWP) of not more than 15 psi. This may vary in some jurisdictions. Low pressure steam boilers are used primarily for heating buildings, such as schools, apartments, warehouses, and factories, and for heating domestic water. These boilers can be firetube, watertube, or cast-iron sectional designs. Boiler size will vary based on the quantity of steam required.

### TECH TIP

*Manufacturers of pressure vessels that comply with the ASME Code must have an agreement with an inspection agency to certify that each vessel is manufactured according to the ASME Code.*

## High Pressure Steam Boilers

A *high pressure steam boiler* is a boiler that operates at a maximum allowable working pressure (MAWP) of more than 15 psi. High pressure boilers are also known as power boilers. *Boiler horsepower (BHP)* is the amount of energy equal to the evaporation of 34.5 lb of water/hr from and at a feedwater temperature of 212°F.

High pressure steam boilers are used in generating electricity and in industrial and commercial locations where steam is used for processes other than heating. Paper mills use steam for dryers. Breweries use steam in equipment such as brew kettles and mash tubs. Pasteurizing and sterilizing facilities use steam generated by high pressure boilers in their processes.

## FIRETUBE STEAM BOILERS

A *firetube steam boiler* is a boiler in which hot gases of combustion pass through tubes that are surrounded by water. **See Figure 1-2.** Firetube steam boilers may be either high pressure or low pressure boilers. All firetube boilers have the same basic operating principle. The heat produced by gases of combustion passes through the tubes while water surrounds the tubes. However, firetube boilers have different designs based on application and installation considerations. Firetube boilers are used where moderate pressures are needed and large quantities of steam are not required. Three common types of firetube steam boilers are the horizontal return tubular boiler, scotch marine boiler, and vertical firetube boiler.

## FIRETUBE STEAM BOILERS

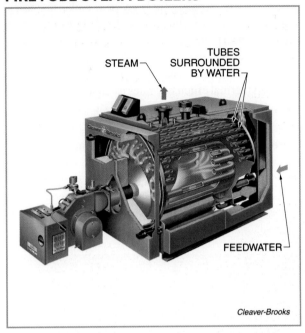

Figure 1-2. In firetube steam boilers, heat and gases of combustion pass through tubes surrounded by water.

### Advantages and Disadvantages of Firetube Boilers

Firetube boilers are designed for pressure up to a maximum of 350 psi and approximately 2000 BHP. The advantages of a firetube boiler are as follows:

- can be factory assembled, thus giving better quality control
- initial cost is less than a watertube boiler
- requires little or no setting (brickwork)
- contains larger volume of water for a given size compared to a watertube boiler
- requires less headroom

Some of these advantages can be disadvantages. Because of the large volume of water these boilers contain, serious damage may occur. **See Figure 1-3.** Knowledge of the following basic principles of boiler operation can prevent serious accidents:

- Water will boil and turn to steam when it reaches 212°F at atmospheric pressure at sea level.
- The higher the steam pressure, the higher the boiling point of water in the boiler.
- As steam pressure in a boiler increases, there is a corresponding increase in temperature.

## BOILER EXPLOSIONS

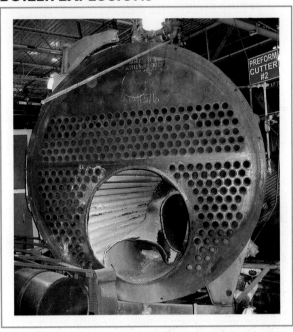

Figure 1-3. A boiler explosion can cause property damage and loss of life.

When a steam boiler is operating at 100 pounds per square inch gauge pressure (psig), the temperature of the water and steam will be approximately 338°F. **See Figure 1-4.** If there is a sudden drop in pressure from 100 psig to 0 psig without a corresponding drop in water temperature, the water at 338°F immediately flashes into steam. The volume increases by a factor of about 1600 when water flashes into steam. This can result in a disastrous explosion. At this temperature and pressure, approximately 13% of the water flashes into steam. The remaining water cools to 212°F at the corresponding atmospheric pressure. Specific steam pressure and temperature relationships are identified in the Properties of Saturated Steam tables. **See Appendix.**

### Principles of Firetube Boiler Operation

When water is heated, it increases in volume and becomes less dense. This warmer water, now less dense, rises and the cooler water drops to take its place. **See Figure 1-5.** As steam bubbles begin to form, this steam/water mixture becomes less dense, which greatly increases the natural circulation in the boiler. The steam bubbles break through the surface and enter the steam space.

| STEAM PRESSURE AND TEMPERATURE RELATIONSHIP | |
|---|---|
| **Gauge Pressure*** | **Temperature†** |
| 20 | 259 |
| 30 | 274 |
| 40 | 287 |
| 50 | 298 |
| 60 | 307 |
| 70 | 316 |
| 80 | 324 |
| 90 | 331 |
| 100 | 338 |
| 150 | 366 |
| 200 | 388 |
| 250 | 406 |
| 300 | 422 |

* in psi
† in °F
**NOTE:** *gauge pressure + atmospheric pressure (14.7) = absolute pressure*
100 psig + 14.7 = 114.7 psia

**Figure 1-4.** The boiling point of water increases as steam pressure increases.

**Heating Surfaces.** A *heating surface* is the part of a boiler with water on one side and heat and gases of combustion on the other. By increasing the heating surface, more heat is transferred from the gases of combustion. **See Figure 1-6.** This results in more rapid water circulation and faster formation of steam bubbles. Adding tubes inside the shell increases the heating surface. Tube sheets at each end support the tubes.

**Thermal Efficiency.** When larger quantities of steam are generated, the thermal efficiency of a boiler increases. *Thermal efficiency* is the ratio of the heat absorbed by a boiler (output) to the heat available in the fuel (input) including radiation and convection losses. Modern firetube boilers with improved design and heat transfer rates have achieved thermal efficiency rates as high as 90% to 95%.

**Furnaces.** A *furnace* is a location where the combustion process takes place. A furnace is also known as a firebox or combustion chamber. Placing an internal furnace within the boiler shell greatly increases the heating surface. **See Figure 1-7.** This also increases heat absorption through radiant heat transfer. Also, the boiler creates steam more quickly.

*Firetube boilers are typically available as package boilers.*

## FIRETUBE BOILERS—WATER CIRCULATION

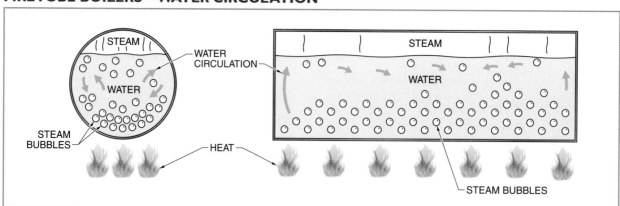

**Figure 1-5.** Water becomes less dense when heated. This warmer water rises, releasing steam bubbles, as cooler, denser water drops.

## FIRETUBE BOILERS—HEATING SURFACES

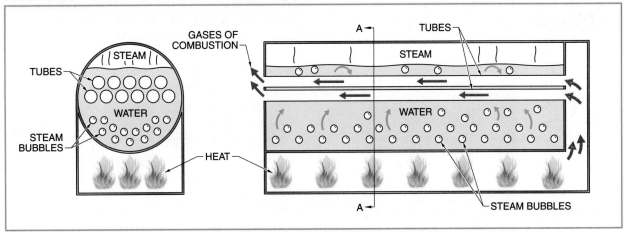

**Figure 1-6.** Boiler efficiency increases as the number of tubes in a boiler increases.

## FIRETUBE BOILERS—INTERNAL FURNACES

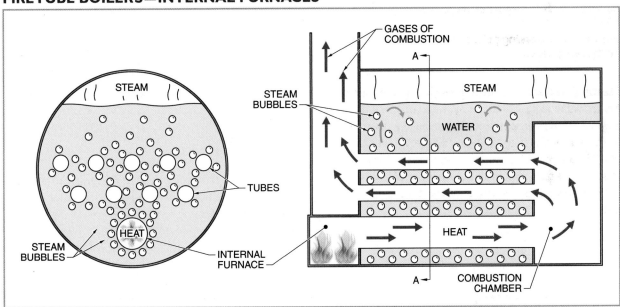

**Figure 1-7.** Adding an internal furnace to a boiler increases the heating surface, which increases boiler efficiency.

**Dry-Back and Wet-Back Boilers.** Firetube boilers are designed as dry-back or wet-back boilers. A *dry-back boiler* is a firetube boiler with a refractory-lined door or chamber at the rear or top of the boiler that directs the gases of combustion from the furnace to the first pass of tubes or from one section of tubes to another. A *wet-back boiler* is a firetube boiler with three tube sheets and a water-cooled turnaround chamber, with a water leg formed between the rear tube sheet and the chamber. This water leg directs the gases of combustion from the furnace to the first pass of tubes and then through succeeding passes.

By inserting baffles on the fire side, the boiler can be made into a two-, three-, or four-pass boiler with a larger number of fire tubes. **See Figure 1-8.** By increasing the number of passes and the number of fire tubes, the water can absorb more heat. This increases thermal efficiency.

## FIRE-SIDE BAFFLES

*Cleaver-Brooks*

**Figure 1-8.** Baffles can be placed on the fire side to create a multipass boiler.

### Horizontal Return Tubular Boilers

A *horizontal return tubular (HRT) boiler* is a firetube boiler that consists of a shell that contains tube sheets and fire tubes mounted over a firebox or furnace. **See Figure 1-9.** A modern design of an HRT boiler is a package boiler with a welded steel firebox that can be equipped with a gas or oil burner. For many years, HRT boilers were the workhorse in industry. However, HRT boilers are less common now. HRT boilers are fired with fuel oil, gas, wood, or coal. Some HRT boilers are also equipped with a dry pipe to ensure a higher quality of steam.

For older styles of HRT boilers, the two methods used to support the drum are columns with a suspension sling or supporting brackets resting on metal plates in a brick setting. **See Figure 1-10.** The tube sheets of an HRT boiler are supported with through stays (braces) and diagonal stays. The feedwater pipe extends approximately three-fifths of the length of the drum. It enters through the front tube sheet. This

helps to heat the feedwater and direct it to a cooler location in the boiler, which increases water circulation in the boiler.

The drum is riveted on older HRT boilers. Since the bottom of the drum is in the firebox, leaks often develop around the rivets. The bottom blowdown line also must pass through the firebox. The blowdown line is protected by a sleeve and refractory or by a wrapping of protective insulation. Modern boiler drums are welded.

### Scotch Marine Boilers

A *scotch marine boiler* is a firetube boiler with an internal furnace. **See Figure 1-11.** Scotch marine boilers were used on ships for many years and have a corrugated or plain furnace, combustion chamber, and tubes passing through the boiler to the front tube sheet. Scotch marine boilers used in industry today have been modified to meet the demands of stationary plants. The furnace is completely surrounded by water. This increases the boiler heating surface, which in turn increases boiler efficiency.

## MODERN HORIZONTAL RETURN TUBULAR (HRT) BOILERS

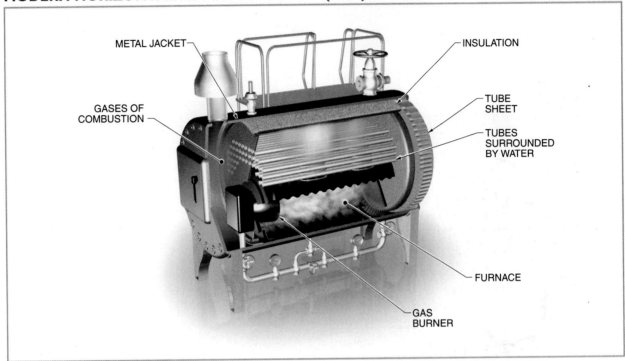

**Figure 1-9.** A modern HRT boiler consists of a shell that contains tube sheets and fire tubes mounted over a firebox or furnace.

## TRADITIONAL HRT BOILERS

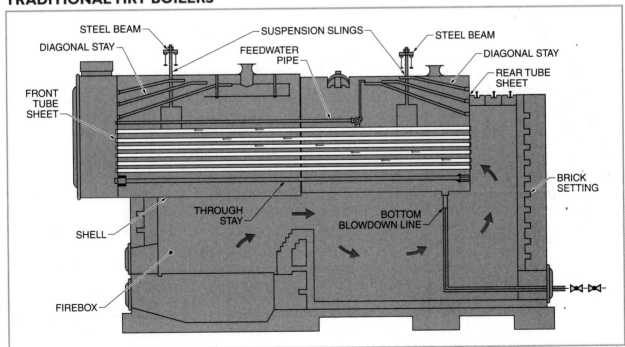

**Figure 1-10.** On an older HRT boiler, the drum is suspended over the firebox. Stays are used to support the flat surfaces.

## SCOTCH MARINE BOILERS

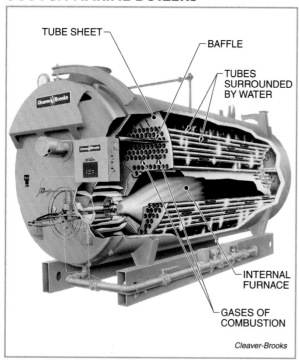

TUBE SHEET
BAFFLE
TUBES SURROUNDED BY WATER

INTERNAL FURNACE

GASES OF COMBUSTION

*Cleaver-Brooks*

**Figure 1-11.** In a scotch marine boiler, the furnace is inside the drum.

### Vertical Firetube Boilers

A *vertical firetube boiler* is a one-pass firetube boiler that has tubes in a vertical position. The gases of combustion cannot be retained in the boiler by the use of baffles. **See Figure 1-12.** The combustion chamber, or firebox, of a vertical firetube boiler is supported by staybolts.

A water leg is formed by the construction of the inner and outer wrapper sheets. The water leg is subject to overheating if deposits of sludge and sediment accumulate. *Sludge* is the soft, muddy residue produced from impurities in water that accumulates in low spots in a boiler. *Sediment* is the hard, sandy particles of foreign matter that have precipitated out of water.

An advantage of vertical firetube boilers is that the floor space required is minimized. This makes vertical firetube boilers popular in the dry-cleaning industry. A disadvantage of vertical firetube boilers is that they often require a high ceiling.

A *dry-top boiler* is a vertical firetube boiler with an upper tube sheet that is dry. A dry-top boiler produces steam that is slightly superheated, which can cause tubes in the upper tube sheet to leak. Dry-top boilers are rarely used.

## VERTICAL FIRETUBE BOILERS

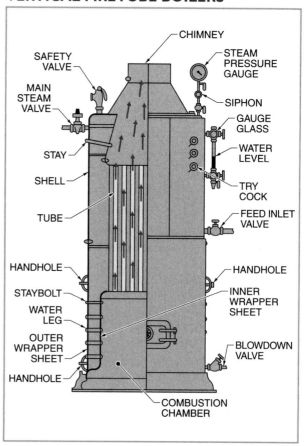

CHIMNEY
SAFETY VALVE
STEAM PRESSURE GAUGE
MAIN STEAM VALVE
SIPHON
GAUGE GLASS
WATER LEVEL
STAY
SHELL
TRY COCK
TUBE
FEED INLET VALVE
HANDHOLE
HANDHOLE
STAYBOLT
INNER WRAPPER SHEET
WATER LEG
OUTER WRAPPER SHEET
BLOWDOWN VALVE
HANDHOLE
COMBUSTION CHAMBER

**Figure 1-12.** Vertical firetube boilers require less floor space than HRT boilers.

## WATERTUBE STEAM BOILERS

A *watertube steam boiler* is a boiler that has water inside the tubes with heat and gases of combustion around the tubes. Watertube boilers were developed as the rapid growth of industry prompted a need for steam at higher pressures. As boilers evolved, the demand for the higher pressures and temperatures limited the use of firetube boilers because they are not practical under these conditions. Firetube boilers need large drum diameters and thicker plate construction to meet the demand for higher pressures and capacities.

Watertube boilers were designed to operate at pressures as high as 3206 psi, a point known as the critical pressure of steam. *Critical pressure* is the pressure at which the density of water and steam are the same. Boilers operating at these extremely high pressures are called critical pressure boilers and are typically used for power generation.

## TECH TIP

*Heat is transferred from the gases of combustion to water to produce steam. This heat transfer occurs through the metal tubes to the water using gases of combustion inside the tubes in a firetube boiler or using gases of combustion outside the tubes in a watertube boiler.*

Watertube boilers carry a smaller volume of water per unit of output than firetube boilers. They are also capable of handling large steam loads and respond quickly to fluctuating steam loads. Because of their construction, watertube boilers can carry extremely high steam pressures and temperatures.

### Principles of Watertube Boiler Operation

Unlike a firetube boiler, a watertube boiler has the heat and gases of combustion surrounding the tubes. **See Figure 1-13.** The heated water inside the tubes becomes less dense, causing it to rise. The steam bubbles that form at the heating surfaces rise and finally break through the water surface in the steam drum. The cooler water drops to the bottom. A mud drum is the lowest part of the water side of a watertube boiler. Factors that cause water circulation in a boiler include the following:

- the difference in the density between the denser liquid in downcomer tubes and the less-dense mixture of steam bubbles and liquid in hotter "steaming" riser tubes
- the difference in temperature between the fluid in the downcomer tubes and riser tubes
- the buoyancy of steam bubbles, which causes the bubbles to rise faster than the water surrounding them

### Straight-Tube Boilers

A *straight-tube boiler* is a watertube boiler with box headers connected together by straight, inclined water tubes. **See Figure 1-14.** In a straight-tube boiler, the water enters a box header and circulates through the inclined tubes to another box header where the steam separates from the water. Downcomers allow for circulation between the lower drum and the box headers.

When heated, the water expands in volume, becomes less dense, and moves toward the rear header where the steam bubbles begin to form. The steam bubbles rise up the rear header to the steam drum. Here, they break through the surface of the water, entering the steam space. Generally, the tubes in straight-tube watertube boilers are expanded into the drums and headers and are slightly flared to prevent leakage.

## WATERTUBE BOILERS— WATER CIRCULATION

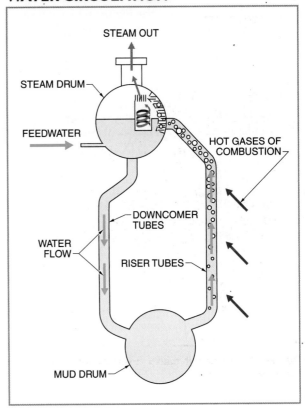

**Figure 1-13.** Heated water becomes less dense and rises to the steam drum of watertube boilers and gives up heat. Cooled water becomes denser and sinks down to the mud drum.

## STRAIGHT-TUBE WATERTUBE BOILERS

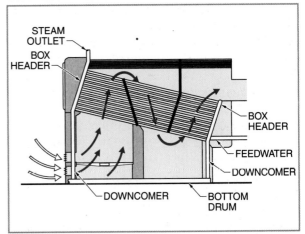

**Figure 1-14.** Straight-tube watertube boilers have inclined tubes to increase circulation. Water is introduced to the boiler through an internal feedwater line.

## Bent-Tube Boilers

A *bent-tube boiler* is a watertube boiler with multiple drums connected by shaped tubes. **See Figure 1-15.** Circulation becomes more complex in a bent-tube boiler. The rear tube bank acts as a downcomer to the mud drum. Most of the steaming occurs in the front tube banks. The steam and water mixture enters the steam drum below the water surface. Bent-tube boilers operate at higher pressures and higher steam capacities than straight-tube boilers. The design eliminates the need for headers. Common bent-tube boiler configurations include A, D, and O styles.

## BENT-TUBE WATERTUBE BOILERS

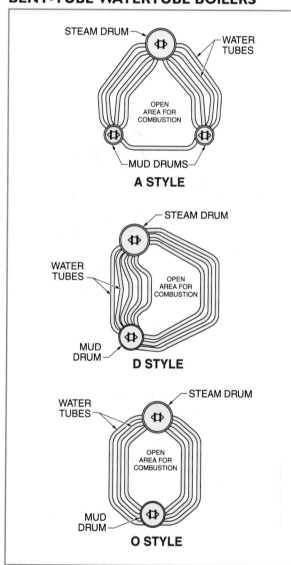

**Figure 1-15.** Tube shape and the location of drums determine the type of watertube boiler.

**TECH TIP**

*To reduce the cost of site assembly, large boilers are prefabricated to the greatest degree feasible.*

## Membrane Boilers

A *membrane boiler* is a watertube boiler that uses strips of metal alloy welded between the tubes to form a seal. **See Figure 1-16.** This eliminates the need for a seal-welded inner casing. The gases of combustion pass across the tubes to heat water. The boiler vessel is gastight, insulated on the outside, and covered by a steel casing. Membrane boilers are a recent design and are used where space is limited. They can be used for heating water or when steam is required.

## MEMBRANE BOILERS

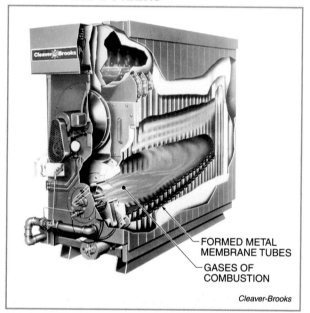

*Cleaver-Brooks*

**Figure 1-16.** In membrane boilers, the gases of combustion pass between the formed membranes to heat water.

## Flex-Tube Boilers

A *flex-tube boiler* is a watertube boiler in which replaceable serpentine tubes are connected to the upper and lower drums and surround the firebox. **See Figure 1-17.** The tubes have a tapered bushing welded on the ends, which is fitted into the headers by a press fit. This allows the tubes to be replaced without welding. The serpentine shape allows the tubes to expand and contract. A flex-tube boiler is normally used where floor space is limited.

## FLEX-TUBE BOILERS

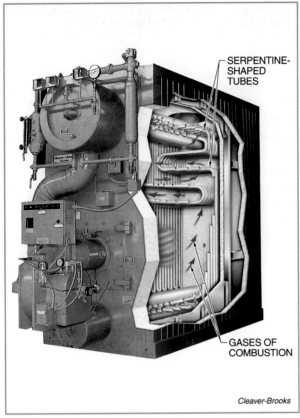

*Cleaver-Brooks*

**Figure 1-17.** The serpentine shape of the tubes in flex-tube boilers allows them to expand and contract in a vertical direction.

## OTHER BOILER TYPES

There are several other types of boilers commonly used. These boilers use alternative fuels or heat that otherwise might be wasted. Refuse boilers and waste boilers are very similar. They may burn similar fuels and the name used varies from one industry to another.

### Electric Boilers

An *electric boiler* is a boiler that produces heat using electrical resistance coils or electrodes instead of burning a fuel. **See Figure 1-18.** Electric boilers operate similarly to boilers fired by fuel oil, gas, or coal, except that electric boilers do not burn a fuel. In resistance-coil electric boilers, electricity flows through a coiled conductor. Resistance created by the coiled conductor generates heat. Resistance-coil electric boilers are used as low capacity boilers and are more common than electrode electric boilers.

## ELECTRIC BOILERS

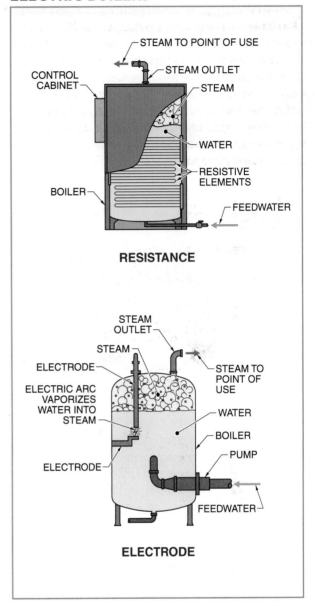

**Figure 1-18.** In electric boilers, heat is generated by a resistive element or electrical arc.

In electrode electric boilers, heat is generated by electric current flowing from one electrode to another electrode through the boiler water. Conductivity of the boiler water will affect the flow of electricity and the amount of heat generated. Conditions that affect boiler water conductivity must be carefully monitored to ensure safe and efficient operation. For optimum efficiency, the manufacturer recommendations for boiler water conditioning should be followed.

## Refuse Boilers

A *refuse boiler* is a boiler that uses the municipal solid waste that would normally go to landfills. Using solid waste in refuse boilers helps minimize the impact of landfills on the environment. These boilers are designed with sophisticated controls to handle varying load swings. Most refuse boilers use an inclined grate design or a rotary-kiln combustor. The boiler is classified as a refuse-derived fuel (RDF) boiler if the solid waste is burned outside the boiler furnace and the gases are burned in the furnace. The boiler is classified as a mass-burning boiler if the municipal solid waste is not sorted or only minimally sorted before it is introduced into the boiler.

Most refuse boilers are watertube boilers using special tube and furnace designs to handle the slagging and corrosive nature of the fuel. Care must be taken to design the boilers to handle the bottom ash residue and the fly ash created from the solid fuel.

## Waste Boilers

A *waste boiler* is a boiler that uses fuel from an industrial process that would normally be wasted. Waste boilers can use the heat from the incineration of industrial waste such as wood, bark, sawdust, garbage, tires, used oil, industrial by-products, and black liquor residue from the paper industry. **See Figure 1-19.** These boilers are usually conventional firetube or watertube boilers fitted with special burners and burner accessories. In some instances, there are special equipment requirements to ensure that boiler emissions meet environmental regulations.

## WASTE BOILER FUEL

CONVEYOR — WOOD CHIPS —

**Figure 1-19.** Waste boilers can use the heat from the incineration of industrial waste.

*The fuel costs of operating a large boiler over its lifetime can far exceed the initial cost of a boiler.*

## Heat Recovery Steam Generators

A *heat recovery steam generator (HRSG)* is a boiler that uses heat recovered from a hot gas stream, such as hot exhaust gases from a kiln or from a gas turbine. **See Figure 1-20.** It does not require a burner to burn a fuel because the hot gases already exist as waste from another process.

HRSGs have four major sections: the economizer, evaporator, superheater, and preheater. The economizer section is a heat exchanger that heats the feedwater before it enters the steam drum. The evaporator section is a heat exchanger that vaporizes the water to produce saturated steam. The superheater section is a heat exchanger that heats the saturated steam to a temperature above the saturated temperature to produce dry steam. The preheater section is a heat exchanger that heats another medium (water, glycol, air, etc.) to be used in the plant.

HRSGs can be modular, packaged, or once-through steam generators. Modular HRSGs are larger units and the sections are built off site, shipped in sections, and then field erected on site. Packaged HRSGs are smaller units that are shipped as a unit that was fully assembled at the factory. Once-through steam generators are a specialized type of HRSG without boiler drums. In this design, the inlet feedwater follows a continuous path without segmented sections for economizers, evaporators, and superheaters to produce superheated steam.

## HEAT RECOVERY STEAM GENERATORS (HRSGs)

**Figure 1-20.** HRSGs use heat recovered from a hot gas stream.

### SECTION 1.1–CHECKPOINT

1. What is the ASME Code?
2. What are the operating pressures of low pressure steam boilers and high pressure steam boilers?
3. What are some advantages of firetube boilers?
4. What are the three common types of firetube boilers?
5. What is a heating surface of a boiler?

## SECTION 1.2
# BOILER CONSTRUCTION

### BOILER STRESS

*Stress* is pressure or tension applied to an object. Boilers and components can be damaged when too much stress is applied. The types of stress commonly found in boilers include compressive, tensile, and shear stress. **See Figure 1-21.** The materials used in boiler construction must withstand various boiler temperatures and pressures. Boiler drums and shells, braces, stays, and tubes are subjected to continuous stress from forces created when the boiler is in operation. The ASME Code ensures that boilers and fittings are strong enough and made of suitable materials that can withstand this stress. Sinuous headers and staybolts are design features that are incorporated to improve boiler integrity and help resist the stress in a boiler.

### STRESS

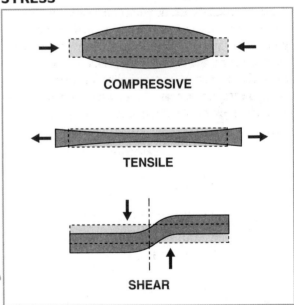

**Figure 1-21.** Boiler materials and fittings are subjected to compressive, tensile, and shear stress.

### Compressive Stress

*Compressive stress* is stress that occurs when two forces of equal intensity act from opposite directions and push toward the center of an object. Fire tubes in a firetube boiler are subjected to compressive stress. Pressure in firetube boilers acts as a force that pushes toward the center of the tubes. The tubes will be crushed if the compressive stress on the tubes exceeds the strength of the tubes. Compressive stress also occurs when a vacuum develops within the boiler.

## Tensile Stress

*Tensile stress* is stress that occurs when two forces of equal intensity act on an object, which pulls the object in opposite directions. Boiler plates and stays are subjected to tensile stress. Pressure in the boiler pushes tube sheets apart. Stays are pulled from each end. The stays will break if the force pulling on the ends exceeds the strength of the stays.

## Shear Stress

*Shear stress* is stress that occurs when two forces of equal intensity act parallel to each other but in opposite directions. Shear pins are used in couplings between a motor and a pump. Shear pins are cut by shear forces if something jams the pump. This allows the motor to turn freely without damage.

*Piping should be color coded to indicate the contents of the pipes.*

## Sinuous Headers and Stays

Some of the early watertube boilers had box headers that required the use of staybolts to prevent bulging. The introduction of sinuous headers eliminated the need for staybolts and made for safer, more efficient boilers. **See Figure 1-22.** The steam drum is dished (concave) and therefore requires no braces or stays.

Through stays are used to prevent bulging in boilers. Staybolts hold the inner and outer wrapper sheets of vertical firetube boilers together. Diagonal stays may be needed in the upper steam space of modified HRT boilers to hold flat and curved surfaces together.

## STAYBOLTS AND SINUOUS HEADERS

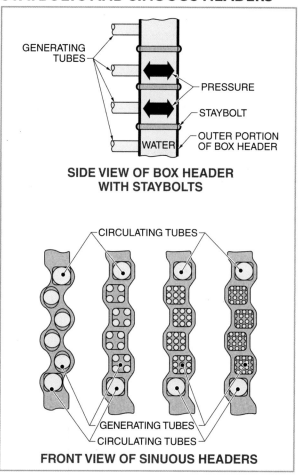

**Figure 1-22.** Sinuous headers eliminate the need for staybolts.

## ENERGY EFFICIENCY

There are several design features of boilers that are important parts of efficient operation of a boiler. Boiler efficiency varies considerably from one design to another. Standard boilers are about 80% efficient, and near-condensing boilers are about 86% efficient. Baffles, waterwalls, and special tube designs are used to improve heat transfer.

## Baffles

It is important that a heating surface absorbs as much heat as possible from the gases of combustion before they enter the stack. Baffles are used to direct the gases of combustion so that they come in close contact with the boiler heating surface. **See Figure 1-23.** Baffles are constructed of steel plates or refractory (firebrick) material and must be maintained for optimum efficiency. If broken baffles are not replaced, the gases of combustion take a direct path to the stack. This results in a sudden increase in the temperature of the flue gas going up the stack and a decrease in boiler efficiency and capacity.

## BAFFLES

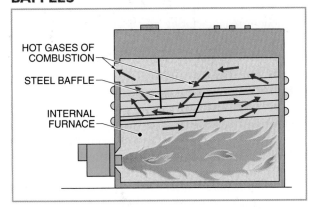

**Figure 1-23.** Baffles are used to direct the gases of combustion so that they come in close contact with the boiler heating surface.

**TECH TIP**

*A variation in ambient temperature of 40°F can affect boiler efficiency by about 1%. Since most boiler rooms are relatively warm, efficiency calculations are commonly based on an 80°F ambient temperature.*

### Waterwalls

Some watertube boilers are equipped with waterwalls to increase the heating surface and the life of the furnace refractory. **See Figure 1-24.** A *waterwall* is a set of tubes that is placed in the furnace area of watertube boilers and used to increase the heating surface of boilers and the service life of refractory. Increased heating surface increases the steaming capacity of the boiler. Additional benefits of waterwalls include smaller furnace volume per unit of output, higher heat absorption in the furnace area, and increased firing rate per unit of furnace volume.

In most watertube boilers that have waterwalls, one side of the furnace has a waterwall that is adjacent to the second pass of boiler tubes. This section of waterwall then acts as a baffle to direct the gases of combustion through the furnace and back through the second pass of boiler tubes.

## WATERWALLS

**Figure 1-24.** Waterwalls extend refractory life and increase the steaming capacity of boilers.

The amount of steam generated using waterwalls is expressed in pounds of steam per hour per square foot (lb/hr/sq ft) of waterwall heating surface. On a hand-fired coal boiler, the amount of steam generated is about 8 lb/hr/sq ft of waterwall heating surface. When using a stoker-fired boiler, the amount of steam generated is about 12 lb/hr/sq ft of waterwall heating surface. When burning fuel oil, gas, or pulverized coal, the amount of steam generated is about 16 lb/hr/sq ft of waterwall heating surface. Some furnaces require a certain amount of heat for combustion efficiency. Because excessive cooling may reduce stability of ignition and combustion efficiency, some furnaces are partially water cooled, or the waterwall is partially insulated.

## Boiler Tubes

Historically, most boiler tubes have been manufactured with smooth internal surfaces. There are new tube designs, however, with extended internal surfaces designed for better heat transfer from the gases of combustion to the boiler water. **See Figure 1-25.** Rifled boiler tubes are available that create turbulence in the fluid flow for better heat transfer efficiency and are typically used for the waterwall tubes in large power boilers. Finned boiler tubes are also available that increase the heat transfer surface.

## HIGH-EFFICIENCY BOILER TUBES

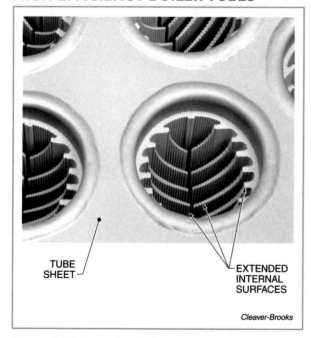

TUBE
SHEET

EXTENDED
INTERNAL
SURFACES

*Cleaver-Brooks*

**Figure 1-25.** New tube designs are made with extended internal surfaces to improve heat transfer.

## BOILER ASSEMBLY

Smaller boilers can be assembled in a factory to help minimize construction costs. Boilers that are too large to be assembled in a factory must be assembled on site. Tubes and tube sheets are critical components of a boiler. A boiler operator must understand the design of tubes and tube sheets so that the boilers can be properly maintained.

### Package Boilers

As modern manufacturing procedures and technology were developed, package boilers became popular. A *package boiler* is a boiler that comes completely assembled with its own pressure vessel, burner, draft fans, and fuel train. **See Figure 1-26.** Package boilers are shipped on skids and arrive ready for operation. A package boiler only requires connecting the feedwater lines, steam lines, fuel lines, electrical connections, and a tie into the breeching and stack. This installation takes a relatively short time, and then the boiler is ready for firing.

Package boilers can be either firetube or watertube boilers. Package boilers are shipped by rail or truck or overseas on ships. The only limitations of package boilers are the pounds of steam per hour capability and the size of the unit. Package boilers are designed as a unit to optimize energy efficiency by ensuring that all components work together. Package boilers enable steam boiler manufacturers to assemble boilers in their plants, and the quality of construction can be closely monitored.

### Field-Erected Boilers

A *field-erected boiler* is a boiler that is assembled at the final site because of size and complexity. **See Figure 1-27.** Field-erected boilers are used to generate electricity or large amounts of steam. Some industrial boilers are six or more stories high. Utility boilers can be up to 70 stories high. Field-erected boilers require very careful quality control during construction.

It is much easier, however, to build a package boiler in a plant using jigs and automated construction techniques than to have a boiler field erected. Field-erected boilers are checked by field inspectors to ensure their compliance with ASME Code standards.

### Tubes and Tube Sheets

The construction of a firetube boiler limits the amount of steam the boiler can produce and its operating pressure. Flat tube sheets must be supported and require careful use of stays to ensure the shell can withstand the pressure. **See Figure 1-28.**

## PACKAGE BOILERS

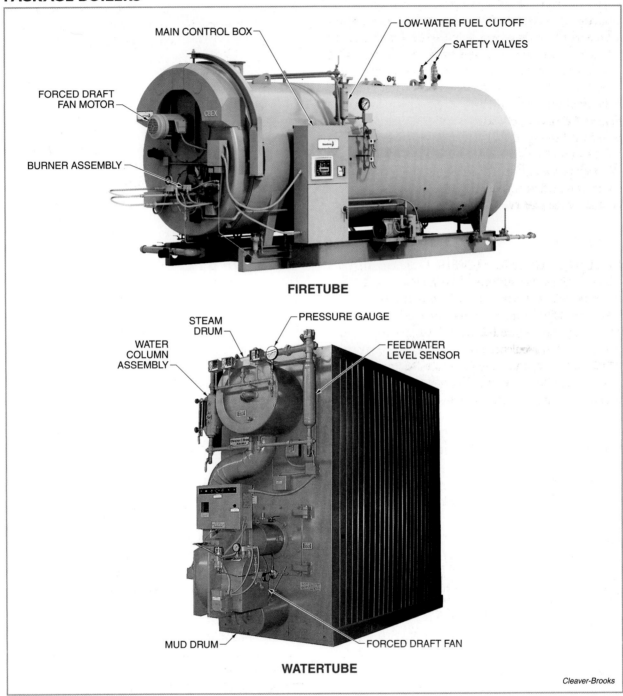

**FIRETUBE**

**WATERTUBE**

*Cleaver-Brooks*

**Figure 1-26.** Package boilers are preassembled at the factory and allow for quick installation.

For both firetube and watertube boilers, the tubes are measured by their outside diameters (OD). In firetube boilers, the tubes are expanded in both the front and rear tube sheets. The protruding ends of the tubes are then rolled to ensure a tight fit and prevent water leakage. Then, the tubes are beaded over and seam welded to prevent the tube ends from burning.

## FIELD-ERECTED BOILERS

*Cleaver-Brooks*

**Figure 1-27.** Field-erected boilers are assembled on the final site because of size and complexity.

## TUBE SHEET SUPPORT

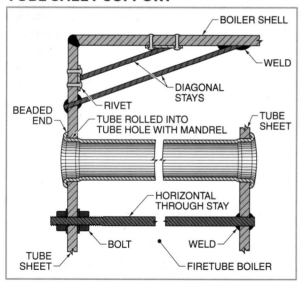

**Figure 1-28.** Stays in firetube boiler construction are used to support flat surfaces and prevent bulging.

SECTION 1.2 – CHECKPOINT

1. What are the different types of stress?
2. Describe design strategies that minimize stress in boilers.
3. What are common strategies for boiler design that improve efficiency?
4. Describe the difference between package boilers and field-erected boilers.
5. Describe the construction of tubes and tube sheets.

### SECTION 1.3
# USES OF STEAM

## USES OF STEAM

Boilers are usually used to generate steam. Steam can be used for many applications throughout a facility. In addition, steam is often used to generate electricity. In this type of application, the steam is used to turn a turbine to power a generator.

### Industrial, Institutional, and Commercial Applications

Boilers and their related equipment provide steam and/or hot water for industrial, institutional, and commercial applications. In industrial applications, steam is used to produce force, heat, or mechanical action for a production process. For example, food processing plants use steam for cooking, pasteurization, and sterilization. Paper mills use steam extensively in pulp processing and drying operations. Oil refineries use steam for distillation of petroleum products in the refining process. **See Figure 1-29.**

Steam is also used in steam jacketing of vessels and steam tracing of piping to maintain uniform temperatures. Steam is used in the process of wood bending, to increase the plasticity of wood products, and in steam kilns for drying lumber. Steam is used to help dry concrete. Steam is also used for soil sterilization in agriculture to replace harmful chemical agents.

Steam is used for heating and hot water needs in institutional facilities such as hospitals, nursing homes, rehabilitation facilities, and correctional facilities. Steam is used in autoclaves to sterilize instruments. It is also used in the laundering and cooking needs of the facility.

## OIL REFINERIES

*ASI Robicon*

**Figure 1-29.** Oil refineries are major users of steam.

Steam's capacity to transfer heat is used in residential and commercial applications for cooking and steam cleaning of fabrics and carpets. Hot water produced in a boiler is primarily used to heat residential and commercial facilities.

### Power Production

High-pressure steam is often used as a prime mover that is used to rotate turbines to generate electricity. A large fraction of the electricity produced uses steam to power turbines that turn electric generators.

**Turbines.** A *turbine* is a machine that converts the energy from the expansion of high-pressure steam into the mechanical energy of rotation. Turbines require superheated steam for steam expansion without condensation. Water droplets from saturated steam cause damage to a turbine. The turbine shaft is connected to a generator shaft to produce electrical power.

**Generators.** A *generator* is a device that converts mechanical energy, usually from a rotating shaft, into electrical energy by means of electromagnetic induction. **See Figure 1-30.** *Electromagnetic induction* is the production of electricity as a result of a conductor passing through a magnetic field. Alternating current (AC) generators convert mechanical energy into AC voltage and current. AC generators are the most common generators used to produce electrical power. Direct current (DC) generators convert mechanical energy into DC voltage and current.

## GENERATORS

**Figure 1-30.** A generator converts mechanical energy into electrical energy.

**Cogeneration.** *Cogeneration* is the process of generating electricity and then using the waste heat from the generating process for heating buildings, providing process heat, or further electrical generation. Interest in cogeneration as a way to reduce total energy use has increased. The development of low-cost gas turbines for smaller power plants have made cogeneration more attractive. Exhaust gases from engines and turbines pass through a heat exchanger to produce hot water or steam for use in plant heating or processes. These plants are also called combined heat and power (CHP) plants.

By using cogeneration, it is possible for one plant to supply most or all of the required electricity, process steam, heated water, and air conditioning needed for a facility. Electricity produced by a cogeneration plant can be used on site to reduce the amount of electric power purchased, or the power can be returned to the electrical grid for credit. The additional capital investment and maintenance costs for cogeneration must be balanced with the reduction of the costs of electricity purchased.

---

SECTION 1.3—CHECKPOINT

1. How does a food processing plant use steam?
2. How can hot water produced in a boiler be used in residential and commercial facilities?
3. How does a generator convert mechanical energy into electrical energy?
4. What is cogeneration?

---

---

## CASE STUDY—WATER LEVEL CONTROL IN AN ELECTRIC BOILER

### Situation

An operating engineer at a correctional facility was responsible for starting, stopping, operating, maintaining, and repairing the boilers, chillers, cooling towers, air compressors, and air handlers throughout the facility. This correctional facility had two boilers. One was a gas-fired scotch marine boiler, and the other was a resistance-coil electric boiler. Each boiler was designed to carry 80% of the designed load.

The gas-fired boiler was fired throughout the year to supply the required steam to try to meet the load requirements of the facility. Because the gas boiler was designed to carry 80% of the designed load, the electric boiler was needed to furnish the remaining 20% of the designed load in the winter months. The electric boiler was also used as a backup boiler when the gas boiler was down for inspections or repair.

The electric boiler had 80 resistance coils arranged in 20 stages of 4 coils per stage. It was a vertical-shell boiler that was about 7′ tall and about 5′ in diameter. Half of the electric coils entered the shell from one side, and the remaining half entered from the other side. The contactors for the coils on one side were in the front of the boiler, and the contactors for the coils on the other side were in the back of the boiler.

Each coil was about 3′ long and covered an area about three-fourths of the diameter of the boiler. The coils were interlaced up both sides of the boiler. The first stage consisted of the lowest coils, which were about 6″ from the bottom. The 20th stage consisted of the top coils, which were located about 20″ from the top of the boiler shell. There was about 15″ of steam space in the top of the boiler.

### Problem

When the electric boiler started producing steam, the water level was erratic and could not be controlled. The water level would continue to rise for several minutes and then suddenly fall. When it was rising, it would go out of sight in the top of the gauge glass. When it was falling, it would go out of sight in the bottom of the gauge glass.

To the feedwater regulator control mechanism, this appeared as an increase in water level. Because of the high level indication, the regulator would stop feeding water into the boiler. Eventually the water level would fall as the steam bubbles broke through the tension at the water surface, collapsing the volume of water above it. The regulator would sense the drop in water level and open to add water to the boiler. Then, the cycle would start over as further steaming occurred. This would continue as the first 10 or 12 stages were energized, but decreased in intensity as more stages turned on. As additional stages were energized, the boiler pressure increased, compressing the size of the steam bubbles.

### Solution

The manufacturer's representative was called to evaluate the situation. The representative determined that large steam bubbles were formed as the first stages of the electric coils were energized, lifting the water above it as it grew in size and rose toward the surface. The representative determined the problem was overfiring the boiler in relation to the demand. This adversely impacted the proper pressure/density in the boiler water, causing the shrink and swell condition.

The service technician rewired the stages according to direction from the manufacturer. The rewiring staged the coils to energize up one side of the boiler first, starting at the bottom, and then up the other side of the boiler. This increased the firing rate modulation ratio, applying heat over a wider range.

The facility's engineering staff also changed the operating procedures of the boilers. In the winter months, the electric boiler was base loaded so that the first four stages (20% capacity) would stay on all the time. The gas boiler would modulate to maintain the steam pressure for the remaining load. Base-loading the electric boiler completely stabilized the water level.

**Name** _____ **Date** _____

_____ 1. A low pressure steam boiler operates at a pressure of not more than ___ psi.
    A. 15
    B. 35
    C. 50
    D. 75

_____ 2. The higher the steam pressure, the higher the ___ of the water in the boiler.
    A. impurities
    B. boiling point
    C. volume
    D. atmospheric pressure

_____ 3. ___ is the amount of energy equal to the evaporation of 34.5 lb of water/hr from and at a feedwater temperature of 212°F.
    A. Factor of evaporation
    B. Boiler horsepower
    C. Latent heat of fusion
    D. Evaporation energy

_____ 4. When a steam boiler is operating at 100 pounds per square inch gauge pressure (psig), the temperature of the water and steam will be approximately ___°F.
    A. 115
    B. 225
    C. 338
    D. 550

_____ 5. A horizontal return tubular (HRT) boiler is a ___ boiler.
    A. watertube
    B. firetube
    C. scotch marine
    D. condensing

_____ 6. As steam pressure in a boiler increases, there is a corresponding increase in the ___.
    A. superheat
    B. volume of steam
    C. boiler horsepower
    D. temperature

_____ 　**7.** A sudden drop in pressure from 100 psig to 0 psig without a corresponding drop in temperature could result in a(n) ___.
　　A. explosion
　　B. flooding condition
　　C. loss of efficiency
　　D. increase in surface tension

_____ 　**8.** Water will boil and turn to steam when it reaches ___°F at atmospheric pressure at sea level.
　　A. 78
　　B. 100
　　C. 212
　　D. 338

_____ 　**9.** ___ efficiency is the ratio of the heat absorbed by a boiler (output) to the heat available in the fuel (input).
　　A. Combustion
　　B. Watertube
　　C. Thermal
　　D. Overall plant

_____ 　**10.** Baffles are used to direct gases so that they come in close contact with the ___.
　　A. feedwater
　　B. combustion chamber
　　C. superheater
　　D. boiler heating surface

_____ 　**11.** ___ stress occurs when two forces of equal intensity act parallel to each other but in opposite directions.
　　A. Compressive
　　B. Shear
　　C. Tensile
　　D. Cutting

_____ 　**12.** Some watertube boilers are equipped with ___ to increase the heating surface and the life of the furnace refractory.
　　A. brickwork
　　B. baffles
　　C. waterwalls
　　D. stays

_____ 　**13.** A disadvantage of a vertical firetube boiler is that it requires ___.
　　A. a high ceiling
　　B. the use of staybolts
　　C. a lot of floor space
　　D. special alloyed tubes

_____   **14.** In a firetube boiler, the heat produced by gases of combustion ___.
    A. passes through the tubes
    B. surrounds the tubes
    C. bypasses the combustion chamber
    D. surrounds the combustion chamber

_____   **15.** Watertube boilers were designed to operate at pressures as high as ___ psi, a point known as the critical pressure of steam.
    A. 512
    B. 1015
    C. 3206
    D. 5000

_____   **16.** A mud drum, if used, is the lowest part of the water side of a(n) ___ boiler.
    A. HRT
    B. watertube
    C. firetube
    D. electric

_____   **17.** A ___ boiler is a watertube boiler in which replaceable serpentine tubes are connected to the upper and lower drums and surround the firebox.
    A. straight-tube
    B. bent-tube
    C. membrane
    D. flex-tube

_____   **18.** In electrode electric boilers, heat is generated by electric current flowing from one electrode to another electrode through the ___.
    A. metal plates
    B. stainless steel tubes
    C. coiled conductor
    D. boiler water

_____   **19.** According to the ASME Code, a high pressure steam boiler operates at a steam pressure higher than ___ psi.
    A. 1
    B. 15
    C. 212
    D. 970

_____   **20.** A condensing economizer recovers ___ Btu/lb from the heat in water vapor.
    A. 32
    B. 212
    C. 970
    D. 1150

_____ **21.** A(n) ___ does not require a burner because it uses hot waste heat gases from another process.
    A. waste boiler
    B. electric boiler
    C. heat recovery steam generator
    D. scotch marine boiler

_____ **22.** In a watertube boiler, the heated water inside the tubes becomes ___, causing it to rise.
    A. less dense
    B. more dense
    C. less viscous
    D. more viscous

_____ **23.** Cogeneration produces ___ and then uses the waste heat in a process or for heating.
    A. hot air
    B. hot water
    C. electricity
    D. gas

_____ **24.** A ___ is a device that converts mechanical energy into electrical energy by means of electromagnetic induction.
    A. generator
    B. transfer switch
    C. transformer
    D. turbine

_____ **25.** A(n) ___ uses municipal solid waste that would normally go to landfills.
    A. electric boiler
    B. coal boiler
    C. refuse boiler
    D. heat recovery steam generator

# Chapter 2 Objectives

### SECTION 2.1—STEAM SYSTEMS
- Define "steam."
- Describe the components of a steam system used with a steam boiler.

### SECTION 2.2—FEEDWATER SYSTEMS
- Define "feedwater."
- Describe the components of a feedwater system used with a steam boiler.

### SECTION 2.3—FUEL SYSTEMS
- List and describe the components of a natural gas system used with a steam boiler.
- List and describe the components of a fuel oil system used with a steam boiler.
- List and describe the components of a coal system used with a steam boiler.

### SECTION 2.4—DRAFT SYSTEMS
- List and describe the components of a draft system used with a steam boiler.
- Explain the difference between forced draft, induced draft, and combination draft systems.

### SECTION 2.5—INSTRUMENTATION AND CONTROL SYSTEMS
- Describe the functions of instrumentation used with a steam boiler.
- Explain the operation of common control systems used with a steam boiler.

## ARROW SYMBOLS

| AIR | GAS | WATER | STEAM | FUEL OIL | CONDENSATE | AIR TO ATMOSPHERE | GASES OF COMBUSTION |
|-----|-----|-------|-------|----------|------------|-------------------|---------------------|
| ⇨ | ⇨ | ➭ | ➡ | ⇨ | ➡ | ⇨ | ➡ |

**Digital Resources**
ATPeResources.com/QuickLinks
Access Code: **472658**

# CHAPTER 2

# BOILER SYSTEMS

## INTRODUCTION

Four separate systems must function in order for a steam boiler to operate: the steam system, feedwater system, fuel system, and draft system. In addition, the functions of the four separate systems are integrated through an instrumentation and control system. The failure of any one of these systems results in a plant shutdown.

The steam system controls and directs the steam to the point where its energy is used for process operations, heating, or the generation of electricity. The feedwater system supplies suitable water to the boiler at the temperature and pressure needed to sustain the steaming cycle. The fuel system supplies fuel in the proper condition for the combustion process. The draft system provides air for combustion and discharges the gases of combustion into the atmosphere. The instrumentation and control system provides control for a boiler or battery of boilers.

## SECTION 2.1
## STEAM SYSTEMS

A steam system has many components that are used to control high-temperature steam to maintain safe and efficient boiler operation. **See Figure 2-1.** Feedwater flows into the boiler to be heated by the gases of combustion. *Steam* is the vapor that forms when water is heated to its boiling point. Steam boilers generate steam to be used in loads to do work.

As steam is produced in the steam drum (1), it creates pressure. This pressure must be limited by pressure controls. If the pressure controls fail, a safety valve (2) is a last resort to control the pressure. Steam may go through steam scrubbers, separators, and/or superheaters (3), depending on the requirements of the steam loads. Scrubbers, separators, and superheaters are used for loads that require dry steam, such as turbines. Superheated steam contains more heat energy than saturated steam at the same pressure. Most types of loads do not need superheated steam.

The steam exits the boiler through piping attached to the boiler drum. Automatic nonreturn valves (4) and main steam stop valves (5) can be used to isolate a boiler from the steam piping. Steam flows through a main steam line (6) to where it is distributed to the rest of the plant. A steam header is used when multiple boilers are in battery. Each distribution line (7) is isolated from the header by a valve, usually a gate valve.

## STEAM SYSTEMS

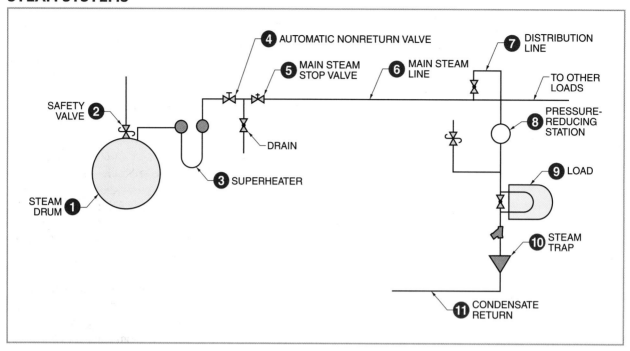

**Figure 2-1.** The steam system consists of all the fittings and accessories used to deliver steam from a boiler to a load. The steam system also includes the condensate return system.

### TECH TIP

*Boilers are inspected periodically. The inspectors are usually an excellent source of knowledge regarding steam system safety, efficiency, and troubleshooting tips.*

Each distribution line conveys steam to a location in the plant where it is needed. A pressure-reducing station (8) can be installed at the point of use to reduce the steam pressure to the pressure needed for the load. Any time there is a reduction in steam pressure through a pressure-reducing station, there must be a safety valve to protect the devices designed to operate at the reduced pressure.

Typical loads (9) include steam turbines, process loads, heating coils, and hot water heaters. As steam gives up its heat, it condenses back to water. Each load device has one or more isolation valves that allow the load device to be taken out of service for repairs. Steam traps (10) are used to send the condensate back to the condensate return (11). In the interest of efficiency, much of the used steam is reclaimed after it gives up its heat and condenses. It is returned to the boiler through the condensate system. During the steam cycle of a boiler, steam and condensate can be transferred many times between different parts of the boiler system.

### SECTION 2.1–CHECKPOINT

1. What is steam?
2. What is the role of a safety valve if the pressure controls fail?
3. What happens to steam as it loses its heat?
4. Describe the difference between a main steam line and a distribution line?

## SECTION 2.2
# FEEDWATER SYSTEMS

*Feedwater* is water supplied to a boiler. The boiler operator must understand the various ways of getting feedwater to the boilers. A steaming boiler can have the water in the gauge glass go from half full to empty in a matter of minutes if water is not added to the boiler. A boiler that is on-line and loses its water can overheat and damage the tubes or even explode, causing injury or loss of life to the operator and damage to the facility and equipment.

## TECH TIP

*Deaerators protect boilers by preheating the feedwater to remove most dissolved gases.*

One pound of water is needed to produce each pound of steam. If the boiler is generating 20,000 lb/hr of steam, it must be supplied with 20,000 lb/hr of feedwater. Water weighs approximately 8.3 lb/gal., so the boiler needs about 2410 gal./hr of feedwater (20,000 ÷ 8.3 = 2410 gal./hr of water).

Condensate returns from the steam system through the condensate return line (1) to the surge tank (2) and is pumped to the deaerator tank (3). **See Figure 2-2.** Oxygen and other noncondensable gases are separated and directed to the atmosphere through the vent (4). Some steam is lost because of leaks in glands and valves or lost in process work. A float (5) opens a feed valve (6) located on a water line (7) to add makeup water to compensate for the condensate lost. To prevent the deaerator from becoming waterlogged, an internal overflow line (8) discharges to waste (9).

The deaerator tank is located above the feedwater pumps so that the water is supplied to the pumps through the suction line (10) at a slight positive pressure. The feedwater pump (11) can be a centrifugal pump and the second feedwater pump (12) may be a steam-driven turbine centrifugal pump. This allows the plant to be flexible. Each feedwater pump has its own set of suction valves (13) and discharge valves (14). This allows either pump to be taken out of service for repairs. Each feedwater pump also has its own check valve (15) to prevent hot water from the boiler from backing up if the check valve located close to the boiler fails.

The feedwater leaves the pump through a feedwater discharge line (16) and enters the closed feedwater heater (17) where it is heated to a relatively high temperature by the gases of combustion discharged from the boiler. The closed feedwater heater is equipped with an inlet valve (18) and an outlet valve (19). A bypass line and valve (20) are needed so the heater can be taken out of service without a shutdown. In some cases, a closed feedwater heater is not used, depending on the size of the plant and whether it is economically feasible.

## FEEDWATER SYSTEM TO FEEDWATER HEATER

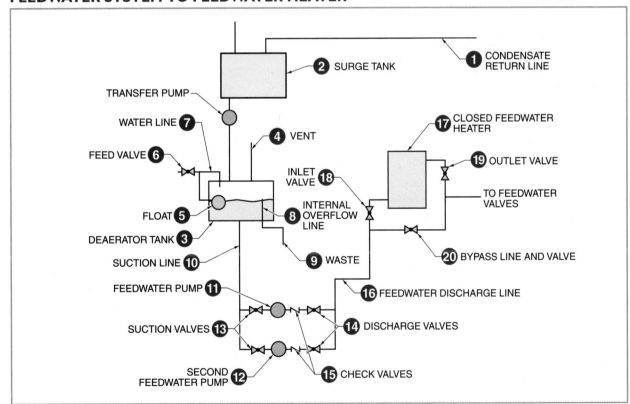

**Figure 2-2.** The feedwater system supplies and prepares feedwater for use in a boiler.

After the feedwater passes through the closed feedwater heater (1), the feedwater flows through a series of valves to the boiler. **See Figure 2-3.** The closed feedwater heater is equipped with an outlet valve (2) and a bypass line and valve (3). The outlet line from the feedwater heater to the boiler has a main feedwater stop valve (4). To maintain a proper level of water in a boiler (5), an automatic feedwater regulator is used.

The control element (feedwater regulator) (6) is located at the normal operating water level. The top of the control element is connected by a steam line (7) that goes to the highest part of the steam side of the boiler. There is a shutoff valve (8) so that the regulator may be taken out of service for repairs. The bottom of the control element is connected to the boiler by a control element line (9) that is well below the normal operating water level. The shutoff valve (10) is used when making repairs. The control element is connected to a feedwater regulating valve (11) located in the feedwater discharge line.

The feedwater regulating valve has two shutoff valves (12) so the regulator may be taken out of service for maintenance. If the regulator is out of service, the water goes through the bypass valve (13) so the boiler does not have to be taken off-line.

A stop valve (14) and a check valve (15) are located on the feedwater line. The stop valve is located closest to the shell of the boiler so that the check valve can be repaired without draining the boiler. The feedwater regulator is equipped with a blowdown valve (16) to ensure that the water and steam lines are free of sludge and sediment. Accessories in a feedwater system vary depending on the size and function of the plant. For example, on larger watertube boilers, an economizer (17) can heat the feedwater by using gases of combustion before the feedwater enters the boiler.

## SECTION 2.2 – CHECKPOINT

1. What is feedwater?
2. What can happen if a boiler loses its water?
3. How many pounds of water are needed to produce a pound of steam?
4. What is the purpose of the float and the feed valve?
5. Why is the deaerator tank typically located above the feedwater pumps?

## FEEDWATER SYSTEM FROM FEEDWATER HEATER TO BOILER

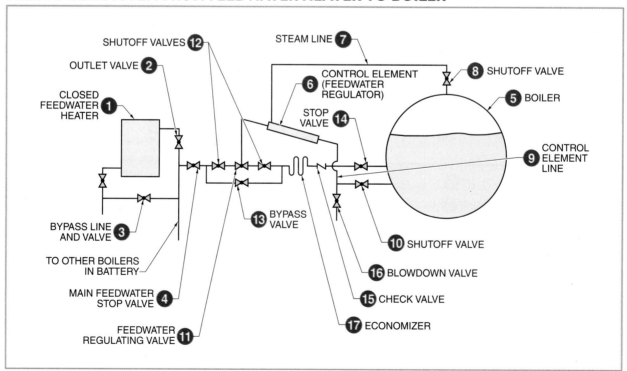

**Figure 2-3.** The feedwater system must supply the amount of feedwater needed to maintain the normal operating water level.

# FUEL SYSTEMS

## NATURAL GAS SYSTEMS

A *fuel system* is a boiler system that provides fuel for combustion to produce the heat needed to evaporate water into steam. Fuel systems commonly used in steam boilers include natural gas, fuel oil, a combination of natural gas and fuel oil, and coal. Other fuels may be used, such as municipal solid waste (MSW), wood chips (hog fuel), coke, or many types of biomass. The fuel used is governed by its cost, availability, and federal and local air pollution standards. Fuel costs reflect purchase price, storage facilities, required auxiliary equipment, and boiler maintenance costs resulting from the use of a specific fuel. For plant flexibility, it is advantageous to be able to burn more than one type of fuel in a boiler.

Natural gas is a clean, minimally polluting fuel that requires no storage and is readily available in most areas. It is very combustible and must be carefully controlled. Because of the possible danger of an explosion, gas leaks cannot be tolerated. Any indication of a gas leak must be promptly addressed.

Natural gas leaks must be located by qualified in-plant personnel or a gas company representative. Gas company officials should be notified immediately at the sign of any gas leak. A leak detector is used to locate the leak, the line is secured, and the repair is made. Personnel in the general area must be notified when repairs are made. The pressure of the gas available at a given location can play an important part in determining whether a low-pressure gas system or a high-pressure gas system should be used.

### Low-Pressure Gas Systems

A low-pressure gas system is safer than a high-pressure gas system because the possibility of gas leakage is minimized due to the lower pressures involved. **See Figure 2-4.** The gas line (1) is fitted with a gas cock (2) that allows the boiler operator to close the gas off from the system when making repairs. The pilot solenoid valve (3) controls gas to the pilot (4). The manual reset valve (5) is an electric valve that cannot be opened until the gas pilot is lit. The zero gas governor (6) reduces the pressure of the gas to 0 psi.

There is a safety switch between the manual reset valve and zero gas governor called a vaporstat (7). The vaporstat is energized by the gas pressure in the line or deenergized when there is insufficient gas pressure. The

main gas solenoid valve (8) opens at the proper time, allowing gas to be drawn down to the injector (9) where it is mixed with air.

## LOW-PRESSURE GAS SYSTEMS

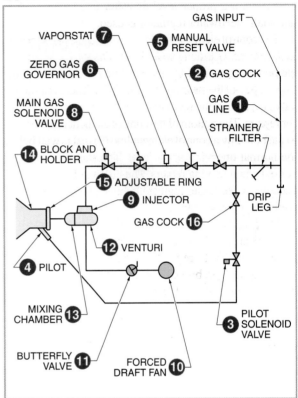

**Figure 2-4.** In a low-pressure gas system, the rate of combustion is controlled by the amount of air supplied to the venturi.

The forced draft fan (10) sends air through the butterfly valve (11). The air passes through a venturi (12) and draws the gas with it to the mixing chamber (13). The block and holder (14) is a method of mounting the burner assembly on the boiler front. As the air and gas mixture passes into the furnace, it is ignited by the pilot. Secondary air is controlled by an adjustable ring (15) so complete combustion occurs. A gas cock (16) found on the gas line to the pilot can be used to secure the gas to the pilot.

### High-Pressure Gas Systems

A high-pressure gas system supplies gas to the burner at a set pressure. **See Figure 2-5.** A typical high-pressure gas system is sized at about 2,000,000 Btu/hr to

84,000,000 Btu/hr. The gas line (1) has a capped drip leg (2) and a gas strainer/filter (3) to catch any debris in the line. Most installations have a pressure gauge (4) before every control device to determine the gas pressure from the main gas service. The main gas cock (5) is used to isolate the main gas line for repairs or testing. A gas-pressure regulator (6) controls the gas pressure supplied to the burner.

Another pressure gauge (7) shows the gas pressure on the downstream side of the gas-pressure regulator. Some systems have a pressure-relief valve (8) that vents to the atmosphere if the gas pressure downstream of the regulator is too high. After the gas-pressure regulator, there is a low-gas-pressure switch (9) with a manual reset that causes the combustion controller to secure the safety shutoff valves if the gas pressure is too low.

The safety shutoff valves (10 and 11) are typically slow-opening, fast-closing, hydraulic-electric valves. The safety shutoff valves should have an internal proof-of-closure switch. A capped leakage test valve (12) is used to determine if the gas is leaking past the valve seat of the first safety shutoff valve. Another capped leakage test valve (13) is used to determine if the gas is leaking past the valve seat of the second safety shutoff valve.

Codes may require that a manual shutoff valve (14) be positioned before the firing-rate valve (15) that controls the volume of gas to the burner. A high-gas-pressure switch (16) with a manual reset causes the combustion controller to secure the safety shutoff valves if the gas pressure is too high. Air is supplied by a forced draft fan (17). The air mixes with the gas in the burner register (18).

## HIGH-PRESSURE GAS SYSTEMS

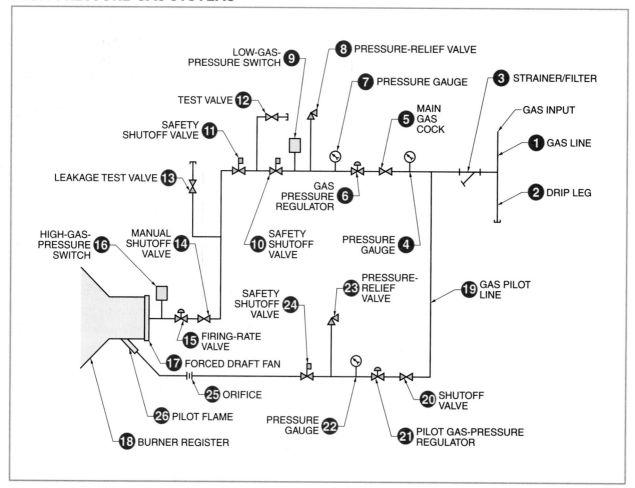

**Figure 2-5.** The basic controlling elements of a high-pressure gas system are the gas-pressure regulator, the safety shutoff valves, and the firing-rate valve.

*Draft moves the gases of combustion through the breeching and into the stack.*

A pilot flame ignites the air-gas mixture in the burner register. A separate system of valves and regulators controls the flow of gas to the pilot. A gas pilot line (19) supplies gas through a shutoff valve (20) to the pilot gas-pressure regulator (21). The shutoff valve is used to isolate the gas pilot line for repairs or testing.

A pressure gauge (22) shows the gas pressure on the downstream side of the gas pressure regulator. Some systems have a pressure-relief valve (23) that is vented to the atmosphere if the gas pressure downstream of the regulator is too high. The safety shutoff valve (24) is usually a solenoid valve controlled by the combustion controller. The orifice (25) is usually a valve cock used to control the size of the pilot flame (26).

## FUEL OIL SYSTEMS

The purpose of any fuel oil system is to supply fuel oil to the burner at the proper temperature and pressure. The system must pump, heat, and regulate the flow of fuel oil before it reaches the burner. Once the fuel oil reaches the burner, it is the function of the burner to properly atomize the fuel oil into the combustion chamber of the boiler. **See Figure 2-6.** Fuel oil is usually available as No. 6 or No. 2 fuel oil. No. 6 fuel oil is very thick and needs to be heated so that it flows. No. 2 fuel generally flows freely and does not need a heater. The fuel oil systems are very similar except for the presence of a heating system for No. 6 fuel oil.

Most steam plants that use No. 6 fuel oil have two tanks (1). The size of the tanks depends on the number of gallons that a plant burns during an average day. The fuel oil is heated by a heating coil located in the fuel oil tank and controlled by steam regulators (2). A thermometer (3) is located on the suction line to indicate the temperature of the fuel oil coming from the tank.

Stop valves (4) isolate a tank when it is not in use. The duplex strainers (5) permit one strainer to be cleaned while the other is in service. The suction gauge (6) shows how much vacuum is on the suction side of the pump. The suction valves (7) and discharge valves (8), located before and after the fuel oil pumps, allow a pump to be isolated from the system for maintenance.

The fuel oil pumps (9) increase the pressure to move the fuel oil through the discharge line (10). In the past, it was common practice to have both an electrically driven and a steam-driven fuel oil pump in the system in case of a power failure. The relief valves (11) protect the system from excessive fuel oil pressure and discharge back to the fuel oil return line (12). All fuel oil return lines return directly to the suction side of the fuel oil pump or back to the fuel oil tanks.

A pressure gauge (13) indicates the fuel oil discharge pressure. The fuel oil is further heated by steam fuel oil heaters (14). Inlet valves (15) and outlet valves (16) allow a heater to be isolated for maintenance. Thermometers (17) indicate the temperature of the fuel oil leaving the heater.

The fuel oil then goes to an electric heater (18) where it is brought up to the temperature at which it will be burned. Another thermometer (19) shows the temperature of the fuel oil after it leaves the electric heater. A strainer (20) is used to clean the fuel before it goes to the burner.

A pressure gauge (21) indicates the pressure of the fuel oil at the burner (22). The pressure is controlled by a back pressure valve (23). The temperature and pressure interlock switches (24) secure the burner if the fuel oil temperature or pressure exceeds the safe operating conditions. All the fuel oil is not burned. Some fuel oil returns through a return line. A thermometer shows the fuel oil return temperature as the fuel oil returns to the storage tank.

## FUEL OIL SYSTEMS

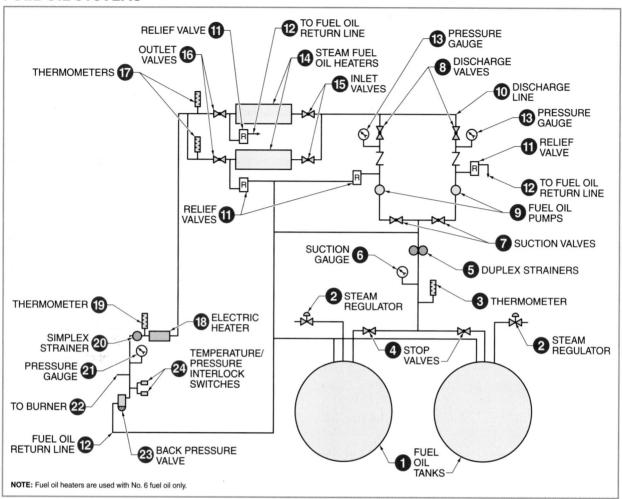

**Figure 2-6.** The fuel oil system must supply fuel oil to the burner at the proper temperature and pressure.

## GAS/FUEL OIL SYSTEMS

Combination gas/fuel oil burners are used in some boiler installations. These burners are actually two burners in one and are capable of burning gas or fuel oil. These combination burners have the advantage of being able to burn the cheapest or most available fuel. The gas part of a combination gas/fuel oil burner typically has the same components and configuration used in a high-pressure gas system. The gas pilot ignition system is common to both systems as well and is the same type used in the high-pressure gas system.

## COAL SYSTEMS

Modern coal boilers use pulverized coal that is ground into a fine powder and blown into the combustion chamber where it is burned in suspension. Special precautions must be taken when handling pulverized coal because it is explosive in the presence of air. Coal is pulverized for the same reason fuel oil is atomized (broken up). Pulverizing allows closer contact between the coal and the oxygen in the air, making it easier to achieve complete combustion.

In the basic pulverized coal system, the coal bunker (1) stores the coal before it flows onto the coal conveyor (2). **See Figure 2-7.** The coal drops into the coal scale (3) and is weighed before being dumped into the coal chute (4) that leads to the coal feeder (5). The coal feeder controls the flow of coal entering the pulverizer (6), which is driven by a motor drive (7). The pulverizer grinds the coal to a fine powder. Hot air (8) enters the pulverizer and mixes with the coal powder before passing to the exhauster (9). The exhauster discharges the mixture of coal and air to the burner (10).

## PULVERIZED COAL SYSTEMS

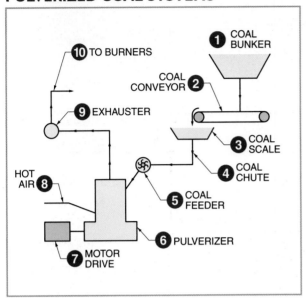

**Figure 2-7.** Pulverized coal is burned in suspension and requires special precautions when handled because of its explosive properties.

Older coal systems use stokers to feed the coal into the boiler. Stokers are classified by the method used to feed the coal into the boiler furnace. Coal stokers feed coal continuously or intermittently, depending on the type of stoker. The stoker must be able to regulate the amount of coal that is fed to the boiler furnace to maintain constant steam pressure in the boiler. An increase in steam demand requires more coal. A decrease in steam demand requires less coal. Many smaller coal boilers have been converted to gas or oil because of environmental requirements and because of the difficulty in handling coal and ash.

## SECTION 2.3 – CHECKPOINT

1. What is the purpose of the vaporstat in a natural gas system?
2. What is the purpose of gas-pressure regulator?
3. When does a natural-gas pressure-relief valve open?
4. When is a fuel oil heater used?
5. What is the purpose of duplex fuel oil strainers?
6. What is the purpose of a coal pulverizer?
7. Why have many smaller coal boilers been converted to gas or oil?

## SECTION 2.4
# DRAFT SYSTEMS

## NATURAL DRAFT

*Draft* is the flow of air or gases of combustion caused by a difference in pressure between two points. In order to burn a fuel in a furnace, sufficient quantities of air at the correct pressure must pass through the boiler. Without the proper draft, complete combustion of fuels cannot be achieved. Draft is classified as natural or mechanical. The type of draft system used will depend on the burner and boiler design. Draft systems contribute to the overall efficiency of a steam plant.

### TECH TIP

> Burners using natural draft are often termed "atmospheric burners." Burners using a forced draft fan are often termed "power burners."

Natural draft is produced by the natural action resulting from temperature differences between air and gases of combustion. Natural draft is produced without a fan using only a stack. The amount of available draft depends on the difference in temperature between the column of gas inside the stack and the column of air outside the stack. **See Figure 2-8.**

The flow of ambient (surrounding) air (1) into the boiler (3) is controlled by the inlet damper (2). The combustion process takes place in the boiler, and the flow of the gases of combustion is controlled by the outlet damper (4). The gases of combustion (5) flow into the stack (6), and then rise to the atmosphere (7).

Using a natural draft system has disadvantages. The rate of combustion (amount of fuel burned per hour) is limited because of the amount of draft available to pull air into the combustion chamber. The amount of draft can be increased by a taller stack, but this increases the capital cost of the boiler system. Natural draft also places limits on the degree to which the combustion process can be controlled.

## MECHANICAL DRAFT

Mechanical draft is produced using a mechanical device, such as a fan. Because of the limitations of natural draft, mechanical means were developed to supply the required airflow in larger boilers. When compared with natural draft, mechanical draft allows increased rates of combustion. Mechanical draft also offers greater control of the combustion process. The three types of mechanical draft are forced draft, induced draft, and combination forced and induced draft.

## NATURAL DRAFT

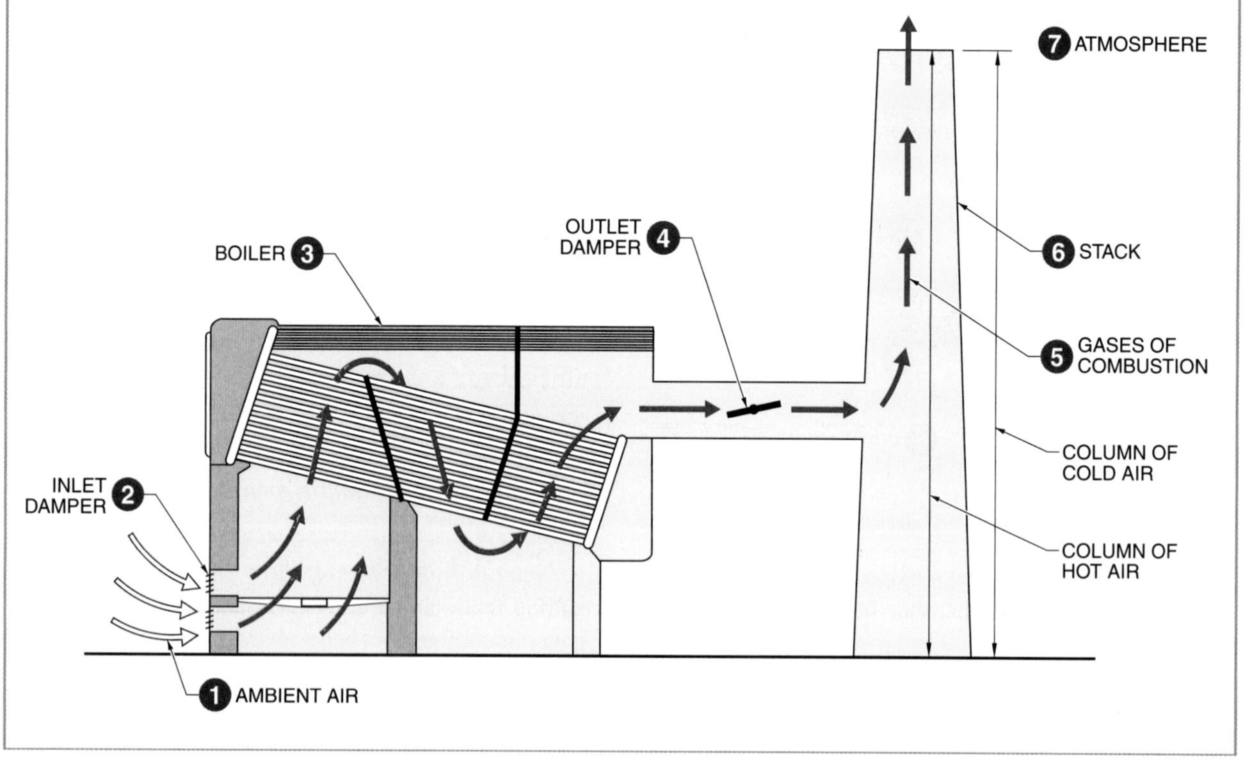

**Figure 2-8.** Natural draft does not require any fans to move the combustion air or gases of combustion.

### Forced Draft

*Forced draft* is mechanical draft produced by a fan supplying air to a furnace. **See Figure 2-9.** Forced draft systems produce a pressure in the furnace that is slightly above atmospheric pressure. Forced draft systems are often used with the small-size to medium-size package boilers found in industrial or commercial applications, such as hospitals, bakeries, or office towers. In this system, ambient air (1) is drawn into the forced draft fan (2), which blows air through the inlet dampers (3). Air passes through the dampers into the boiler (4), where combustion occurs. The gases of combustion (5) move through the passes of the boiler and leave the boiler through the breeching (6). The gases of combustion then pass into the stack (7). From the stack, the gases of combustion discharge out into the atmosphere (8).

### Induced Draft

*Induced draft* is draft produced by pulling air through the boiler furnace with a fan. **See Figure 2-10.** Induced draft systems produce a pressure in the furnace that is slightly below atmospheric pressure. In this system, ambient air (1) is drawn through the inlet damper (2) into the boiler (3) where combustion occurs. After the combustion process is completed, the gases of combustion flow through the outlet damper (4) into the induced draft fan (5). The induced draft fan is located between the boiler and the stack. The induced draft fan discharges the gases of combustion (6) into the stack (7). From the stack, the gases of combustion discharge out into the atmosphere (8).

### Combination Draft

*Combination draft (balanced draft)* is draft produced from one or more forced draft fans located before the boiler and one or more induced draft fans after the boiler. Combination draft systems produce a pressure in the furnace that can be controlled to be above or below atmospheric pressure. Combination draft systems are often used in large steam plants such as those in central heating plants, large industrial boilers, or utility boilers. **See Figure 2-11.**

## FORCED DRAFT

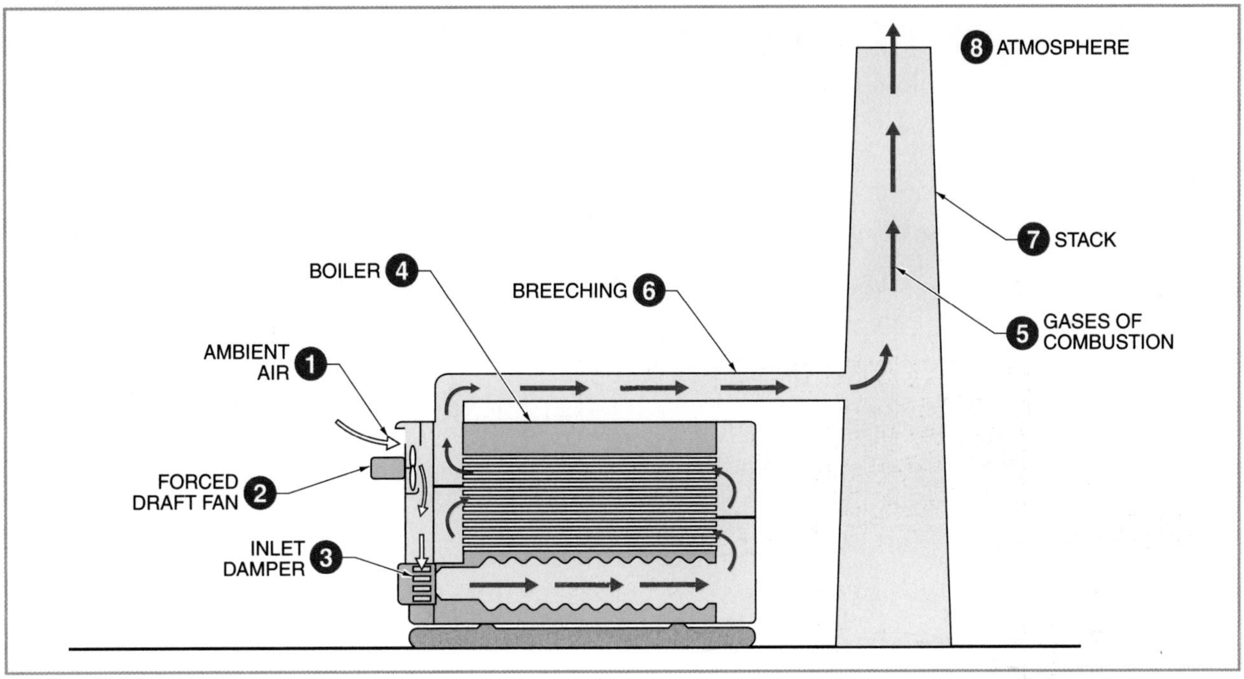

**Figure 2-9.** Forced draft is produced by a fan supplying air to the furnace.

## INDUCED DRAFT

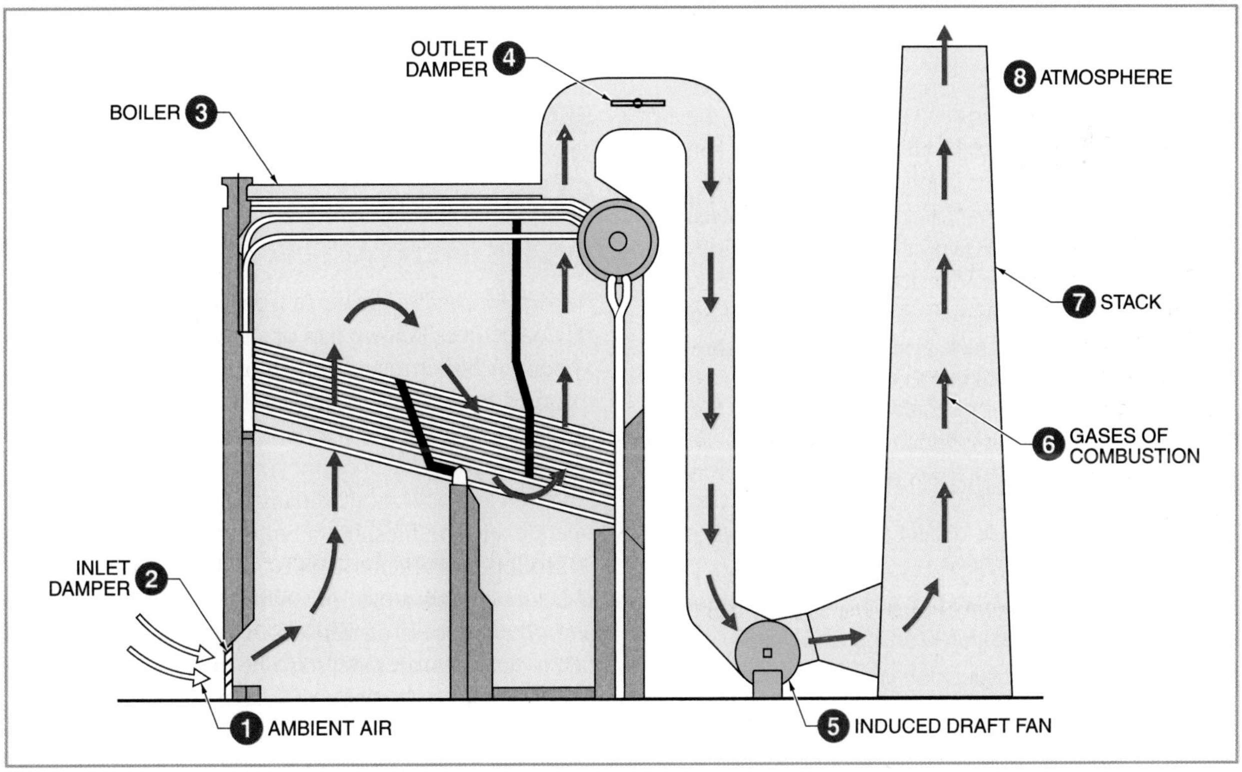

**Figure 2-10.** Induced draft produces a pressure below atmospheric pressure to create airflow in the furnace.

## COMBINATION DRAFT

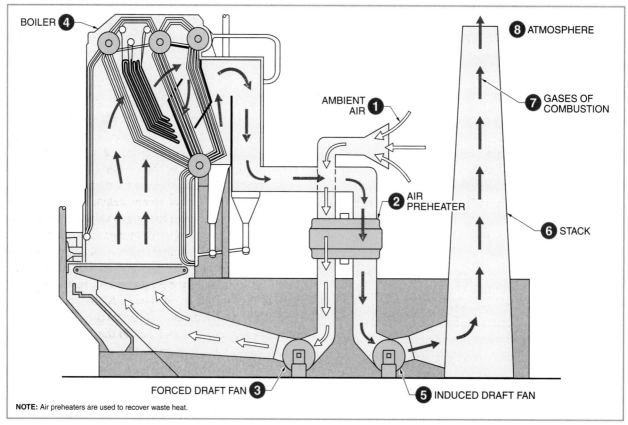

**Figure 2-11.** Combination forced and induced draft allows more effective control of combustion for maximum efficiency in large plants.

In this system, ambient air (1) is drawn through the air preheater (2) into the forced draft fan (3) and is discharged into the boiler (4) where combustion occurs. After the combustion process is completed, the gases of combustion flow through the air preheater into the induced draft fan (5). The induced draft fan is located between the boiler and the stack (6). The induced draft fan discharges the gases of combustion (7) into the stack. From the stack, the gases of combustion discharge out into the atmosphere (8).

---

### SECTION 2.4 – CHECKPOINT

1. What is the purpose of inlet and outlet dampers in a natural draft system?
2. Where is the fan located in a forced draft system?
3. Where is the fan located in an induced draft system?
4. Where are the fans located in a combination draft system?
5. How are the pressures in the furnaces of forced and induced draft systems different?

SECTION 2.5

## INSTRUMENTATION AND CONTROL SYSTEMS

### INSTRUMENTATION

*Instrumentation* is a group of measuring instruments and related devices that are part of a monitoring and control system. A boiler must be operated within design limits in order to produce steam safely and efficiently. The only way to know how a boiler is operating is to take measurements. An operator must understand the construction and operation of many instruments to properly operate and troubleshoot a boiler.

Boiler instruments consist of the devices used to measure temperature, pressure, level, flow, opacity, and other process conditions. After a measurement is taken, that measurement must be made available to the operator and to the control system. Many instruments have a local indicator where the measurement can be read directly by the operator. Other instruments use transmitters that send signals to controllers. In some cases, a transducer must be used to convert the signal from one form to another. For example, a transducer can be used to convert the movement of the internal parts of a pressure gauge into an electrical or digital signal that can be sent to a control system.

## CONTROL SYSTEMS

A *control system* is a system of measuring instruments and controllers that work together to control a process. Every boiler system has its own control system. In addition, many boilers have connections to larger control systems, such as a building automation system. A control system consists of the controllers that use measurements to determine optimum operating conditions. All control systems have differences because they control different systems, but they all share common concepts that must be understood by the boiler operator.

A control system consists of a control loop that executes a control strategy. A control strategy is the method a controller uses to manipulate the measured value into something that can be used to make changes to the process to keep it at the desired operating conditions. A control loop contains a primary element that measures a process condition, a control element that executes a control strategy, and a final element that controls the flow of material or energy to keep the process condition at the desired setpoint. There may be more than one element of any type in a control loop. Many different methods are used to communicate between the different elements.

For example, a typical control loop controls the steam pressure in the boiler by controlling the amount of natural gas and air for combustion going to the furnace. **See Figure 2-12.** The ratio of the fuel to the air must be controlled during load changes to optimize boiler safety and efficiency.

The control loop typically uses a steam pressure controller (1) as the control element. A pressure gauge (2) is a primary element that uses a transmitter to send a signal to the controller. The controller uses the signal from the steam pressure gauge to determine the amount of gas required to provide the amount of heat needed to generate enough steam to meet the demand.

When the amount of steam demand changes, the controller sends a signal to the gas valve (3) to change the amount of gas going to the burner. At the same time, the controller sends another signal to the damper (4) to change the amount of air going to the burner. The gas valve and damper are the final elements that control the flow of fuel and air to the burner. A gas flowmeter (5) and a combustion air flowmeter (6) are other primary elements that provide feedback to the controller to ensure the air-fuel ratio is correct.

### SECTION 2.5—CHECKPOINT

1. Why do some instruments use a transducer?
2. What are the three elements of a control loop?
3. List at least two examples of primary elements in a boiler control loop.

## INSTRUMENTATION AND CONTROL SYSTEMS

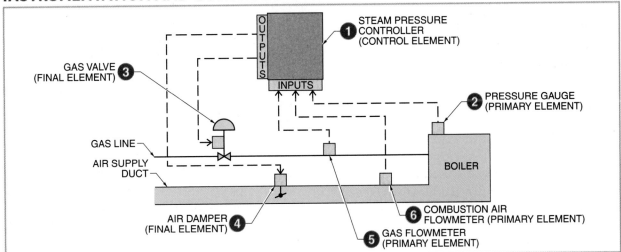

**Figure 2-12.** Instrumentation and control systems are a necessary part of operating a modern boiler.

## CASE STUDY—LOW-WATER CONDITION

### Situation

As an operating engineer in a shipyard, it was the engineer's responsibility to start, stop, operate, and log the boilers, air compressors, cooling towers, generators, etc. One afternoon, the engineer received a page from the plumbing foreman that a portable rental boiler was off-line and would not restart. The water supply line to the portable boiler was a ¾" hose connected to the city water supply.

This portable boiler was being used for a process called "float coating." A tank is filled with hot water, and a tank coating material floats on the water. When the water is drained from the tank, the coating material coats the tank as the water level drops. The portable boiler was being used to heat the water as it was pumped into the tank. A shell-and-tube heat exchanger was used to heat the water. The water temperature had to be maintained to make sure the coating material properly coated the inside of the tank.

### Problem

The plumbers performing this task thought it was taking too long to heat the water indirectly with the heat exchanger. They decided to bypass the heat exchanger and take the hot water directly from the boiler. This was done by connecting a 2" hose to the 2½" blowdown line at the bottom of the portable boiler. Each time they opened the valves to take water from the boiler, the water level quickly dropped. The water was being removed from the boiler faster than it was being replaced by the feedwater system. The boiler shut off on a low-water condition, and the plumbers closed the blowdown valve. They then waited for the boiler to refill and automatically restart. The plumbing foreman told the engineer that they used this method several times and each time the boiler restarted. When the boiler eventually failed to restart, they called the engineer.

### Solution

When the engineer arrived, the water level could not be seen in the gauge glass and the bottom blowdown valve was open. Apparently, the water level had been allowed to drop far enough that the manual-reset low-water fuel cutoff had tripped and needed to be reset in order for the boiler to restart. The plumbers had allowed the water level to fall even further than before because they needed a little more water to fill the tank.

According to the ASME Code, any time the boiler water level falls below the lowest visible level in the gauge glass, the boiler must be secured and inspected. The engineer secured the boiler and called the rental company to inform them of the situation. A second boiler was brought to the site, and the original boiler was taken back to their shop to be retubed. The shipyard was billed for the repairs. After losing two days while the boiler was being replaced, the plumbing foreman decided that using the heat exchanger was not too slow a method after all.

**Name** _____ **Date** _____

_____  **1.** If a boiler is generating 20,000 lb/hr of steam, it must be supplied with ___ lb/hr of feedwater.
  A. 10,000
  B. 20,000
  C. 30,000
  D. 40,000

_____  **2.** From the deaerator, oxygen and other noncondensable gases are separated from the feedwater and directed to the atmosphere ___.
  A. by using the closed feedwater heater
  B. through the economizer
  C. by using steam traps
  D. through the vent

_____  **3.** A(n) ___ system is a system of measuring instruments and controllers that work together to control a process.
  A. instrumentation
  B. feedwater
  C. control
  D. combustion

_____  **4.** One pound of water is needed to produce ___ of steam.
  A. 1 lb
  B. 3 lb
  C. 144 Btu
  D. 970.3 Btu

_____  **5.** In a low-pressure gas system, the ___ valve is an electric valve that cannot be opened until the gas pilot is lit.
  A. check
  B. pressure-regulating
  C. manual reset
  D. pressure-reducing

_____  **6.** ___ are used for loads that require dry steam, such as turbines.
  A. Superheaters
  B. Forced draft fans
  C. Reclaimed steam lines
  D. Automatic nonreturn valves

_____ **7.** In a feedwater system, the ___ valve is located closest to the shell of the boiler on the feedwater line so that the check valve can be repaired without draining (dumping) the boiler.
- A. blowdown
- B. manual reset
- C. stop
- D. automatic nonreturn

_____ **8.** A steaming boiler can have the water in the gauge glass go from half full to empty in a matter of ___ if water is not added to the boiler.
- A. minutes
- B. hours
- C. days
- D. weeks

_____ **9.** In a fuel oil system, the purpose of duplex strainers is to ___.
- A. isolate the fuel oil system in the event of fire
- B. remove water from the fuel oil
- C. prevent return flow to the tank
- D. permit one strainer to be cleaned while the other is in service

_____ **10.** Superheated steam has more ___ than saturated steam at the same pressure.
- A. heat energy
- B. volume
- C. pressure
- D. moisture

_____ **11.** The deaerator tank is located ___.
- A. above the feedwater pumps
- B. below the feedwater pumps
- C. after the closed feedwater heater
- D. before the centrifugal pump

_____ **12.** A gas pilot line on a high-pressure gas system supplies gas through a ___ valve to the pilot gas-pressure regulator.
- A. check
- B. shutoff
- C. butterfly
- D. throttling

_____ **13.** The purpose of any fuel oil system is to supply fuel oil to the burner at the proper ___.
- A. temperature and volume
- B. volume and pressure
- C. temperature and pressure
- D. viscosity and speed

_____ **14.** No. ___ fuel oil needs to be heated before it is pumped so that it can flow.
- A. 2
- B. 3
- C. 4
- D. 6

_____ **15.** Pulverizing coal allows ___.
    A. the coal to be stored easily
    B. for better burning on the grates
    C. for closer contact between the coal and oxygen for complete combustion
    D. for easy hand-firing

_____ **16.** Draft is ___ caused by a difference in pressure between two points.
    A. the balance of pressure
    B. the flow of air or gases of combustion
    C. a type of back pressure
    D. the stack height

_____ **17.** Natural draft is the result of temperature differences between ___.
    A. fuel oil and air
    B. burning coal and sulfur
    C. air and gases of combustion
    D. fuel oil and gases of combustion

_____ **18.** In a low-pressure gas system, the air mixes with the gas in the ___.
    A. injector
    B. gas pilot line
    C. burner register
    D. pulverizer

_____ **19.** A measuring instrument can use a ___ to send a signal to a controller.
    A. control valve
    B. final element
    C. transmitter
    D. primary element

_____ **20.** Pulverized coal is ground into a fine powder and blown into the combustion chamber where it is burned in ___.
    A. scales
    B. grates
    C. suspension
    D. large particles

_____ **21.** A control system uses a ___ to execute a control strategy.
    A. steam pressure gauge
    B. pressure-control station
    C. feedwater control valve
    D. control loop

_____ **22.** Water weighs approximately ___ lb/gal.
    A. 2.5
    B. 6.4
    C. 8.3
    D. 10.2

_____ **23.** Induced draft is produced by pulling air through the boiler furnace with a fan located in the ___.
   A. breeching
   B. inlet damper
   C. combustion chamber
   D. economizer

_____ **24.** A control loop typically uses a ___ as the control element.
   A. pressure transmitter
   B. modulating water valve
   C. temperature sensor
   D. steam pressure controller

_____ **25.** Codes may require that a(n) ___ be positioned before the firing-rate valve that controls the volume of gas to the burner.
   A. equalizer
   B. manual shutoff valve
   C. flue gas analyzer
   D. pressure-reducing governor

# Chapter 3 Objectives

### SECTION 3.I—FITTINGS FOR SAFETY

- Explain the purpose of safety valves.
- Describe how safety valves are connected.
- Describe the principles of safety valve design.
- Explain how safety valves operate and how they are tested.
- Describe safety valve adjustment and repair.
- List the information included on a safety valve data plate.

### SECTION 3.2—FITTINGS FOR OPERATION

- Describe steam pressure gauges, scales, and calibration.
- Explain the operation of steam separators.
- Describe stop valves, blowdown valves, and boiler vents.

## ARROW SYMBOLS

| AIR | GAS | WATER | STEAM | FUEL OIL | CONDENSATE | AIR TO ATMOSPHERE | GASES OF COMBUSTION |
|-----|-----|-------|-------|----------|------------|-------------------|---------------------|
| ⇨ | ⇨ | ➩ | ➡ | ⇨ | ➡ | ⇨ | ➡ |

**Digital Resources**
ATPeResources.com/QuickLinks
Access Code: **472658**

# STEAM BOILER FITTINGS

## INTRODUCTION

To operate a boiler safely and efficiently, certain fittings are required. Whether a boiler is a firetube or watertube type, the fittings are located in the same area of the boiler and serve the same purpose. Fittings must be constructed in accordance with the ASME Code. The ASME Code requires that all fittings used in boilers be constructed of materials that will withstand the temperatures and pressures to which boilers are subjected.

The ASME Code also specifies how these fittings must be attached to the boiler. This varies according to boiler temperature, pressure, and design. In addition, the ASME Code has suggested procedures for testing and operation of fittings found on boilers. These procedures will vary, depending on the temperatures and pressures at which the boilers operate.

---

## SECTION 3.1
## FITTINGS FOR SAFETY

### FITTINGS

A *fitting* is a component directly attached to a boiler that is required for the operation of the boiler. Boiler fittings include valves, gauges, and other components required for safe and efficient operation. The locations of the fittings necessary for operation are the same for firetube and watertube boilers. **See Figure 3-1.** There is no ornamental chrome on a steam boiler. Each fitting has a definite purpose. Each fitting is needed for safety, efficiency, or both.

The materials used for fittings depend on the temperatures and pressures involved. For example, cast iron is used for the construction of water columns up to 250 psi.

Malleable iron is used for water columns operating up to and including 350 psi. Steel is used for water columns operating above 350 psi.

Steam boilers are designed for a maximum allowable working pressure. The *maximum allowable working pressure (MAWP)* is the operating pressure of a boiler as determined by the design and construction of the boiler in conformance with the ASME Code. If this pressure is exceeded, a failure of the pressure vessel can occur.

---

**TECH TIP**

*Every pressure-containing part of a system must be able to withstand the pressure. Fittings must be marked with the ASME symbol stamp to verify that they are manufactured according to the ASME Code.*

---

## TYPICAL BOILER FITTINGS

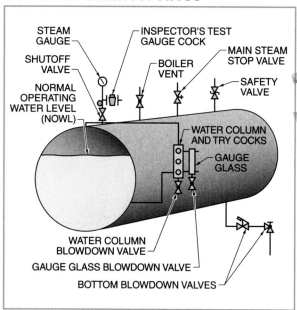

**Figure 3-1.** Fittings on boilers are located for visibility and accessibility.

## SAFETY VALVES

All fittings found on a boiler are necessary, but safety valves are one of the most important fittings. A *safety valve* is an automatic, full-open, pop-action valve that is opened by an overpressure in a boiler and used to relieve the overpressure before damage occurs. The purpose of the safety valve is to protect the steam boiler from exceeding its MAWP per the ASME Code. Safety valves can be set lower than the MAWP but must not allow the boiler to exceed it.

All steam boilers must be equipped with at least one safety valve. Steam boilers having more than 500 sq ft of heating surface require two or more safety valves. Smaller electric boilers require one safety valve. Electric boilers exceeding 500 kW of input require two safety valves.

### Safety Valve Connections

Boiler safety valves are placed at the highest part of the steam space of the boiler and on the superheater header outlet, if present. They must be connected to the boiler shell according to the ASME Code. One requirement is that the safety valve be connected as close as possible to the boiler shell or superheater header outlet without any other valves between them. The safety valve connection is a screw, flange, or welded connection, depending

upon the size, operating pressure, and temperature of the boiler. Low pressure boilers usually have screw connections. High pressure boilers have flange or welded connections.

When discharge piping is used in connection with a safety valve, the piping must be adequately braced and must not place a strain on the valve. Drains are placed at the lower sections of the discharge piping to remove any condensation that may form. **See Figure 3-2.** Discharge piping must be solidly connected to the building structure. Sufficient space must be provided between the safety valve discharge piping and the discharge piping above the drip pan to prevent strain on the safety valve. All drains and piping in the discharge system must be piped to a safe discharge area to prevent possible injury to personnel.

### Safety Valve Designs

Safety valves are designed to open fully at a predetermined pressure and remain open until there is a definite drop in pressure. *Popping pressure* is the predetermined pressure at which a safety valve opens and remains open until the pressure drops. *Safety valve blowdown,* also known as safety valve blowback, is a drop in pressure between popping pressure and reseating pressure as a safety valve relieves boiler overpressure. **See Figure 3-3.** Spring-loaded pop-type safety valves are of simple construction and are very reliable. Lever-type and deadweight-type safety valves cannot be used on steam boilers.

Several designs of safety valves are available for various pressures. Spring-loaded pop-type safety valves are used for pressures up to 600 psi. At higher temperatures and pressures, it becomes more critical that the safety valve seats more tightly. For saturated or superheated steam service, a shut-tight safety valve with a flat uniform seat is used. A high-pressure, high-capacity safety valve is used for saturated steam service on boilers with design pressures over 1500 psig. For steam pressures up to 3000 psig and steam temperatures up to 1000°F, a superjet safety valve is used.

*Safety valve capacity* is the amount of steam that can be relieved by a safety valve, measured in the number of pounds of steam per hour it is capable of discharging under a given pressure. The safety valve on a steam boiler must be capable of discharging all the steam the boiler can generate without the pressure increasing more than 6% above the set pressure. In no case can the pressure be greater than 6% above the MAWP for that boiler. The minimum safety valve relieving capacity for electric boilers is 3½ lb/hr/kW.

## SAFETY VALVE INSTALLATION

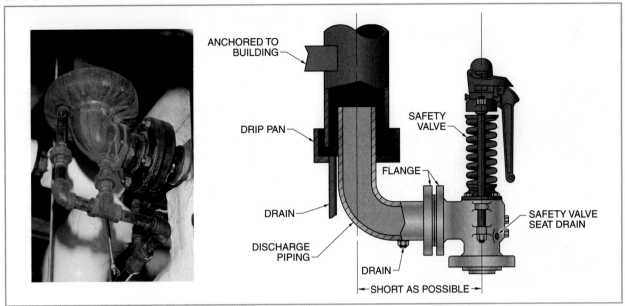

**Figure 3-2.** The piping layout for the safety valve must allow for proper drainage of condensate.

## SAFETY VALVE BLOWDOWN

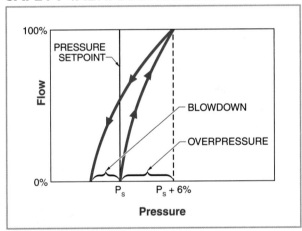

**Figure 3-3.** Safety valve blowdown is a drop in pressure between popping pressure and reseating pressure as a safety valve relieves boiler overpressure.

A safety valve must close tight without chattering (quick opening and closing). Once it has reseated, it must remain closed until its popping pressure is reached again. The blowdown is at least 2 psi and up to about 8 psi below the popping pressure but never more than 4% below the set pressure. For pressures between 200 psi and 300 psi, the blowdown cannot be less than 1% of the set pressure.

### Safety Valve Operation

The operation of a safety valve is relatively simple. A spring exerts a downward force to keep the valve closed. The steam pressure acts on the safety valve disc to exert an upward force against the valve. The safety valve opens when the total force exerted by the steam exceeds the force exerted by the spring. **See Figure 3-4.**

The total force from the steam pressure needed to overcome the spring force to open the safety valve is equal to the area of the safety valve disc multiplied by the steam pressure. To find the total force applied to a safety valve, apply the following procedure:

1. First find the area of the valve disc of the safety valve. Valve discs have a circular area.

$$A = \pi \times r^2$$

$$A = \pi \times \left(\frac{d}{2}\right)^2$$

$$A = \pi \times \frac{d^2}{4}$$

$$A = \pi \times \frac{1}{4} \times d^2, \text{ or } A = \frac{\pi}{4} \times d^2$$

where

$A$ = area of valve disc (in sq in.)
$r$ = radius of valve disc (in in.)
$d$ = diameter of valve disc (in in.)

## SAFETY VALVE OPERATION

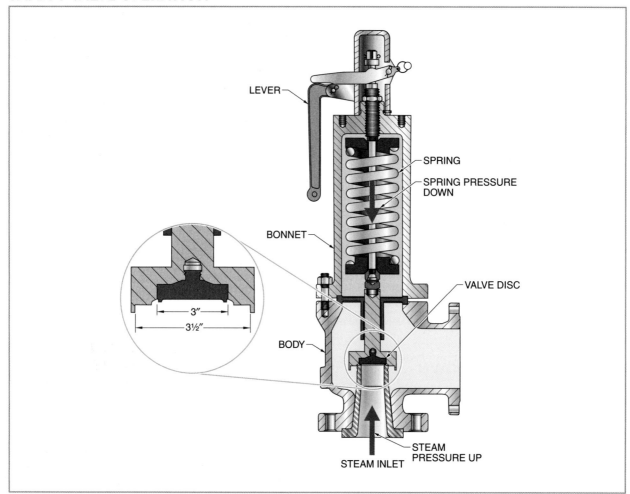

**Figure 3-4.** The safety valve opens when the total force exerted by the steam exceeds the force exerted by the spring.

For example, the area of a valve with a disc diameter of 3″ is 7.07 sq in.

$$A = \frac{\pi}{4} \times d^2$$

$$A = \frac{3.1416}{4} \times 3^2$$

$$A = 0.7854 \times 9$$

$$A = \textbf{7.07 sq in.}$$

2. Next, multiply the area by the pressure to find the total force.

$$TF = A \times P$$

where

$TF$ = total force applied to valve (in lb)
$A$ = area of valve disc (in sq in.)
$P$ = steam pressure (in psi)

For example, a boiler with an MAWP of 100 psi with a safety valve having a 3″ valve disc has a total force of 707 lb.

$$TF = A \times P$$
$$TF = 7.07 \times 100$$
$$TF = \textbf{707 lb}$$

At the MAWP, the maximum steam pressure has a total force of 707 lb applied against the valve disc. The safety valve must be designed with a spring that can be adjusted to apply an equal force of 707 lb to keep the valve closed until the pressure exceeds the safety valve setting.

Normally, a valve under pressure is opened slowly. This is not true with a safety valve. It pops open. This does not happen by accident, but by design. **See Figure 3-5.**

## POPPING ACTION

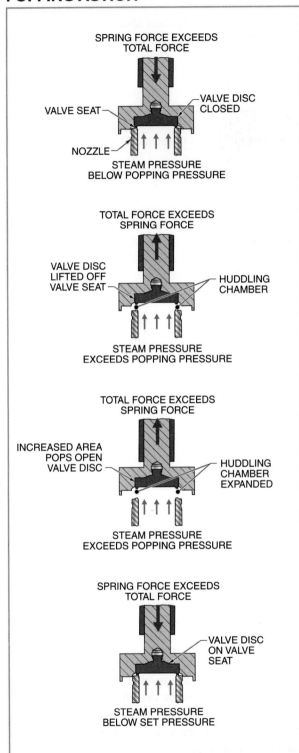

**Figure 3-5.** As soon as the steam pressure starts to overcome the force of the spring, the valve will slowly start to open. This allows the steam to enter the huddling chamber and force the valve to pop open.

The spring exerts a downward force equal to the upward force to keep the valve closed. As soon as the steam pressure starts to overcome the force of the spring, the valve slowly starts to open (feather). This allows the steam to enter a huddling chamber, which exposes a larger area for the steam pressure to act on. A *huddling chamber* is the part on a safety valve that increases the area of the safety valve disc, thus increasing the total force, causing the valve to open quickly, or pop. For example, the valve disc and huddling chamber have a diameter of 3½″ (3.5″). The total force is calculated as follows:

$$A = \frac{\pi}{4} \times d^2$$
$$A = \frac{3.1416}{4} \times 3.5^2$$
$$A = 0.7854 \times 12.25$$
$$A = \textbf{9.62 sq in.}$$
$$TF = A \times P$$
$$TF = 9.62 \times 100$$
$$TF = \textbf{962 lb}$$

The total force increases from 707 lb to 962 lb, an increase of 255 lb. This greater total force overcomes the spring force and causes the valve to pop open.

### Safety Valve Testing

Safety valves must be tested at regular intervals, either manually or by pressure. Testing ensures that the valves are functioning properly. The interval between tests will vary from plant to plant because of operating steam pressures, local codes, and plant routine. If it is not practical to test a safety valve while the boiler is operating, the safety valve may be bench tested. To bench test a safety valve, the valve is removed from the boiler and mounted on a test tank with a controlled pressure and volume of air. The popping pressure, blowdown, reseating pressure, and sealing capability are all tested.

Manufacturer recommendations should be followed for safety valve testing procedures. Typically, boilers 15 psi to about 400 psi should be manually tested once a month and pressure-tested once a year. Boilers above about 400 psi should be manually tested every six months (preferably when the boiler is going to be removed from service). They should also be pressure-tested once a year or completely overhauled by the factory. *Note:* The procedures provided by the manufacturer should always be followed. It is also good practice to have a replacement set of calibrated safety valves on hand whenever testing safety valves.

When testing a safety valve manually, the boiler needs to be operating at a pressure of at least 75% of the safety valve popping pressure. This prevents possible damage to the safety valve. Manual testing of the safety valve is accomplished by pulling the test lever by hand and releasing it to allow the safety valve to snap closed. **See Figure 3-6.** There should be no leakage or dripping when the valve is closed. If leakage occurs, the manual test should be repeated to clean the seating surface. Safety valves should only be tested to ensure proper operation. If a valve is tested too frequently, steam and fuel would be needlessly wasted.

## SAFETY VALVE TESTING

**Figure 3-6.** Manual testing of the safety valve is accomplished by pulling the test lever by hand and releasing it to allow the safety valve to snap closed.

A *pop test* is a safety valve test performed to determine if a safety valve opens at the correct pressure. The pop test must be performed with at least two qualified persons present. The pop test requires the boiler operator to slowly increase the pressure in the boiler until the setpoint of the safety valve is reached. As soon as the valve pops, the firing rate is reduced or the burner is shut off to allow the pressure to drop back to the normal operating range.

An *accumulation test* is a test used to establish the relief capacity of boiler safety valves. The accumulation test determines if the safety valve capacity is large enough to protect the boiler. This test should not be conducted without a boiler inspector present. Boilers equipped with superheaters should not be subjected to an accumulation test because the superheater tubes could be damaged from overheating due to the lack of steam flow through them.

The accumulation test is performed by shutting off all steam outlets from the boiler and increasing the firing rate to maximum. The safety valves should relieve all the steam without the pressure increasing more than 6% above the set pressure. During this test, the boiler operator must maintain a normal operating water level. The *normal operating water level (NOWL)* is the water level designated by the manufacturer as being the proper water level for safe boiler operation. It takes 1 lb of water to make 1 lb of steam. The steam lost during this test must be replaced with makeup feedwater. An adequate water supply must be available during this test.

### Safety Valve Adjustments and Repairs

The setting or adjusting of the popping pressure and the blowdown of a safety valve should only be done by a qualified person who is familiar with the construction, operation, and maintenance of safety valves. Repairs must be done by the manufacturer or a manufacturer-authorized representative. Local code requirements should always be checked to verify the proper setting, adjustment, and testing procedures for safety valves.

### Safety Valve Data Plates

A *data plate* is a plate that is attached to a piece of equipment that provides important information about that equipment. Each boiler safety valve has a data plate attached. **See Figure 3-7.** The information shown on the data plates may vary slightly from manufacturer to manufacturer. However, all data plates must provide the following information:

- manufacturer name or trademark
- manufacturer design or type number
- size of valve in in., seat diameter
- popping pressure setting in psig
- blowdown in lb/sq in.
- capacity in lb/hr
- lift of the valve in in.
- year built or code mark
- ASME symbol
- serial number

## SAFETY VALVE DATA PLATES

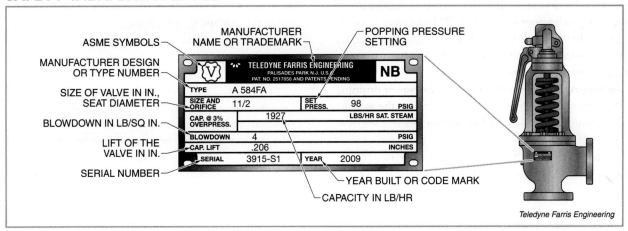

ASME SYMBOLS

MANUFACTURER NAME OR TRADEMARK

POPPING PRESSURE SETTING

MANUFACTURER DESIGN OR TYPE NUMBER

SIZE OF VALVE IN IN., SEAT DIAMETER

BLOWDOWN IN LB/SQ IN.

LIFT OF THE VALVE IN IN.

SERIAL NUMBER

| | | | |
|---|---|---|---|
| V | TELEDYNE FARRIS ENGINEERING<br>PALISADES PARK N.J. U.S.A.<br>PAT. NO. 2517858 AND PATENTS PENDING | NB | |
| TYPE | A 584FA | | |
| SIZE AND ORIFICE | 11/2 | SET PRESS. | 98 PSIG |
| CAP. @ 3% OVERPRESS. | 1927 | LBS/HR SAT. STEAM | |
| BLOWDOWN | 4 | | PSIG |
| CAP. LIFT | .206 | | INCHES |
| SERIAL | 3915-S1 | YEAR | 2009 |

YEAR BUILT OR CODE MARK

CAPACITY IN LB/HR

*Teledyne Farris Engineering*

**Figure 3-7.** The ASME Code requires certain data on all safety valve data plates.

### SECTION 3.1 – CHECKPOINT

1. What is a safety valve?
2. Where is a safety valve located on a boiler?
3. What is the purpose of a drain on safety valve piping?
4. What is safety valve popping pressure?
5. What is safety valve blowdown?
6. What is a huddling chamber?
7. What is a safety valve pop test?
8. What information is typically placed on a safety valve data plate?

### SECTION 3.2
# FITTINGS FOR OPERATION

## STEAM PRESSURE GAUGES

Fittings on the boiler include gauges for monitoring boiler operation and other required components. A constant steam flow with steady steam pressures and temperatures improves the boiler operation.

A *steam pressure gauge* is a boiler fitting that displays the amount of pressure inside a boiler, steam line, or other pressure vessel. A common design for a steam pressure gauge uses a Bourdon tube. A *Bourdon tube* is an oval metal tube inside a mechanical pressure gauge that is shaped like a question mark and has a tendency

to straighten when pressurized. **See Figure 3-8.** Other types of steam pressure gauges use a diaphragm. The displacement of the diaphragm corresponds to the pressure. The steam pressure gauge should be plainly visible to the operator.

The pipe that connects the pressure gauge comes from the highest part of the boiler steam drum. Connections to the boiler must not be less than ¼″ in diameter. If steel or wrought iron is used, the connections must not be less than ½″ in diameter. The steam pressure gauge must be protected by a siphon to prevent steam from damaging the gauge. Pipe siphons may be pigtail or U-tube. A pressure gauge should have a range of about 2 times the safety valve setting. It should never be less than 1½ times the safety valve setting. **See Figure 3-9.**

### Pressure Scales

Boiler plants use pressure gauges to indicate all the various pressures an operator must be aware of to ensure safe and efficient plant operation. **See Figure 3-10.** *Gauge pressure (psig)* is the pressure above atmospheric pressure. *Absolute pressure (psia)* is gauge pressure plus atmospheric pressure. Steam pressure is often expressed in pounds per square inch (psi), where it is implied that the measurement is actually in psig. Other pressure gauges are calibrated in pounds per square inch (psi), inches of water column (in. WC), or inches of mercury (in. Hg). The psi scale is used for higher positive pressures, in. WC for low positive pressures, and in. Hg for negative pressures (pressure below atmospheric pressure or vacuum).

## STEAM PRESSURE GAUGE OPERATION

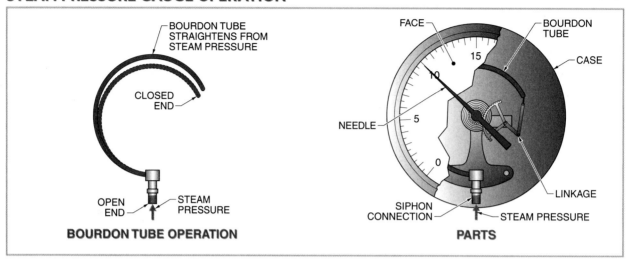

**Figure 3-8.** A Bourdon tube and adjustable linkage make the setting and correction of a steam pressure gauge possible.

## STEAM SIPHONS

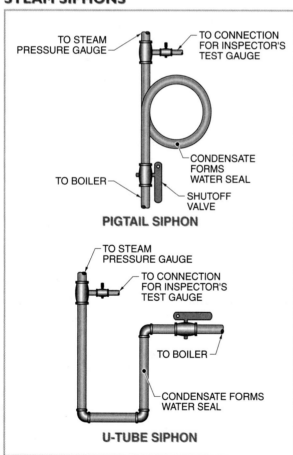

**Figure 3-9.** A steam siphon protects the Bourdon tube from the high temperature of the steam. An inspector's test gauge is connected when checking the pressure gauge for accuracy.

## PRESSURE SCALES

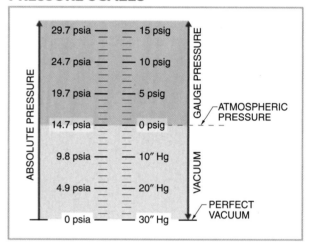

**Figure 3-10.** Absolute pressure and gauge pressure are two common pressure scales.

### Gauge Calibration

A connection for an inspector's test gauge is recommended by the ASME Code. A gate valve may be installed to facilitate changing the gauge in the event of a failure. However, this gate valve may not be allowed, depending on state and local codes. If a steam pressure gauge is not accurate to within 2% of the working pressure, it must be recalibrated or replaced. A steam pressure gauge can be recalibrated using an electronic test gauge or a deadweight tester. **See Figure 3-11.**

## ELECTRONIC TEST GAUGES

**Figure 3-11.** An electronic test gauge is used to recalibrate steam pressure gauges.

### Hydrostatic Pressure

When a steam pressure gauge is located below the steam drum, the reading must be corrected to compensate for the vertical height of the condensate above the pressure gauge. *Hydrostatic pressure* is the pressure caused by the weight of a column of water. The correction factor for hydrostatic pressure is 0.433 psi per vertical foot. **See Figure 3-12.** The hydrostatic pressure is calculated as follows:

$$P_h = V \times 0.433$$

where

$P_h$ = hydrostatic pressure (in psi)

$V$ = vertical feet (in ft)

For example, a pressure gauge mounted 20 ft below the water level shows a pressure of 100 psi. The hydrostatic pressure is calculated as follows:

$$P_h = V \times 0.433$$
$$P_h = 20 \times 0.433$$
$$P_h = \textbf{8.66 psi}$$

The actual steam pressure is 91.34 psi. The pressure at the lower gauge is the sum of the steam pressure and the hydrostatic pressure (91.34 + 8.66 = 100).

## HYDROSTATIC PRESSURE

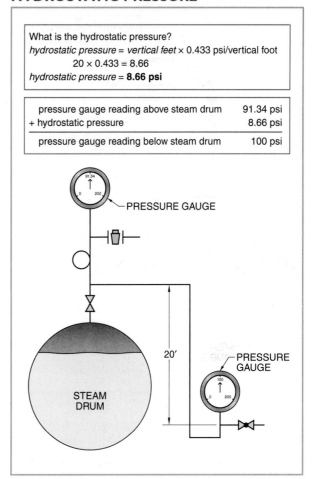

What is the hydrostatic pressure?
*hydrostatic pressure = vertical feet* × 0.433 psi/vertical foot
20 × 0.433 = 8.66
*hydrostatic pressure* = **8.66 psi**

| pressure gauge reading above steam drum | 91.34 psi |
|---|---|
| + hydrostatic pressure | 8.66 psi |
| pressure gauge reading below steam drum | 100 psi |

**Figure 3-12.** A pressure gauge reading made below the steam drum must be corrected for hydrostatic pressure (0.433 psi per vertical foot).

## STEAM SEPARATORS

A *steam separator,* or drum internal, is a device that is located in the steam drum of a boiler and used to increase the quality of steam. Separators work on the basic principle of changing the direction of the flow of steam, causing the heavier water droplets to separate from the steam. The purpose of the steam separator is to remove as much moisture from the steam as possible. Through this removal, a steam separator does the following:

- conserves the energy of the steam

- prevents water hammer or turbine damage due to the carryover of water

- protects valves, pistons, cylinders of reciprocating engines, and turbine blades from the erosive action of wet steam

When a steam boiler is operated at high loads or with a high water level, there is a tendency for water droplets to be carried in the steam. One method of controlling this carryover is to use a dry pipe separator. Another method is to use a cyclone separator.

### Dry Pipe Separators

A *dry pipe separator* is a closed-end pipe that is perforated at the top, has drain holes on the bottom, and removes moisture from steam. **See Figure 3-13.** The dry pipe separator is located in the upper part of a steam drum. The top half is drilled with many small holes and is connected at top center to the main steam outlet from the boiler. The drains on the bottom allow the trapped moisture to return to the boiler drum. The operation of the dry pipe separator is based on the change in direction of the steam flow.

## DRY PIPE SEPARATORS

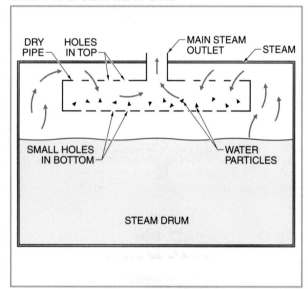

**Figure 3-13.** A dry pipe separator is a hollow pipe with holes in it that cause any water particles in the steam to drop out.

Steam enters the small holes on top. The steam has to change direction to leave through the steam outlet. This change in direction causes any entrained water droplets to separate and return to the boiler through the drains. The dry pipe separator is very effective for small boilers or boilers with light steam loads. However, as the capacity of a boiler increases, the efficiency of the dry pipe separator is reduced.

### Cyclone Separators

A *cyclone separator* is a cylindrical device that separates water droplets from steam using centrifugal force. **See Figure 3-14.** A cyclone separator consists of a number of cyclones set side by side along the length of the steam drum and a baffle arrangement that directs steam into the cyclones. Moisture is removed by centrifugal force as the steam is forced to rotate when it passes through the cyclones. For improved separation of steam and moisture, scrubbers are set up over the top of each cyclone. Cyclone separators are essential when a boiler has a superheater and when it is necessary to keep carryover to a minimum. Scrubbers remove any moisture and solids remaining in the steam before it leaves the steam drum.

## CYCLONE SEPARATORS

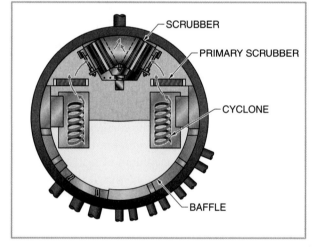

**Figure 3-14.** A cyclone separator sends steam through a circular drum in a spiral motion to remove moisture from the steam.

## STOP VALVES

A *stop valve* is a valve that is opened or closed by the operator and typically used to isolate the feedwater to the boiler or the steam discharge from the boiler. Typical types of stop valves include main steam stop valves and automatic nonreturn valves.

## Main Steam Stop Valves

A main steam stop valve is used to place a boiler in service or isolate it from the system for cleaning, inspection, or repairs. **See Figure 3-15.** The ASME Code states that high pressure boilers in battery (two or more boilers connected to a common header) and equipped with manhole openings must have two main steam stop valves with an ample free-blowing drain between them.

Additionally, these valves should be of the outside stem and yoke (os&y) type, which will show by stem position whether the valve is open or closed. The valve should be a gate valve because a gate valve offers no resistance to the flow of steam. It must always be either wide open or completely closed.

## MAIN STEAM STOP VALVES

**Figure 3-15.** The main steam stop valve should be outside stem and yoke (os&y) type because the stem indicates when the valve is open or closed.

## Automatic Nonreturn Valves

An automatic nonreturn valve may be used in place of one of the stop valves. An automatic nonreturn valve acts like a check valve to allow steam to flow out of the boiler only. However, the automatic nonreturn valve must be located as close to the shell of the boiler as possible. **See Figure 3-16.** The automatic nonreturn valve improves the safety and efficiency of the plant. It can cut a boiler in on-line automatically and protect the system in the event of a failure on the pressure side of any boiler on-line. If the pressure in a boiler were to drop below the header pressure, the nonreturn valve would close, taking the

boiler off-line. This prevents the steam from any boiler still operating at normal header pressure from flowing into the damaged boiler.

Both the main stop valve and the automatic nonreturn valve should be dismantled, inspected, and overhauled annually. During boiler inspection, these valves are closed, locked, and tagged while the drain between them is open.

## AUTOMATIC NONRETURN VALVES

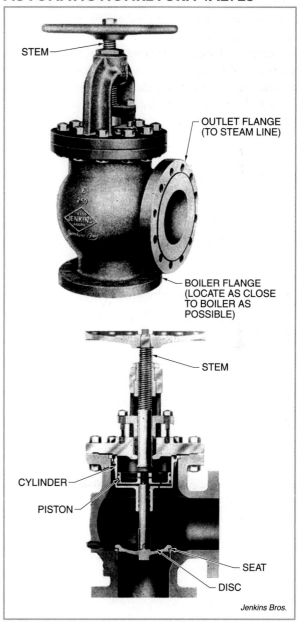

Jenkins Bros.

**Figure 3-16.** The automatic nonreturn valve should be located as close to the boiler shell as possible. It cuts the boiler in on-line and off-line automatically.

## Stop Valve Operation

In order to cut a boiler in on a line that is equipped with an automatic nonreturn valve, apply the following procedure:

1. After making all necessary safety checks and warming up the boiler slowly, open the bypass around the main steam stop valve to warm up the line and equalize the pressure on both sides of the valve.

2. When about 85% of the line pressure is on the incoming boiler, slowly open the main steam stop valve.

3. Open the free-blowing drain to remove any condensate trapped between the two valves.

4. Open the automatic nonreturn valve.

5. Slowly bring the pressure of the incoming boiler up to line pressure and let the automatic nonreturn valve cut the boiler in on-line.

6. Close the free-blowing drain. The automatic nonreturn valve will start to open when the pressure of the incoming boiler is slightly higher than the line pressure.

For two valves that are hand-operated, steps 1, 2, and 3 should be followed. Then, when the pressure of the incoming boiler is about 5 psi below header pressure, the main stop valve nearest the shell of the boiler should be cracked open. The header pressure, being a little higher than the incoming boiler pressure, will force any trapped condensate in the line back into the boiler. This helps to prevent carryover and helps maintain the water level in the boiler being brought on-line.

The water in the boiler being brought on-line is at a lower temperature and pressure than the steam in the header and will not boil. If the boiler that is being brought on-line were at a higher temperature and pressure than the steam in the header, the water would boil rapidly. This would make it difficult to maintain the water level in the boiler.

---

### TECH TIP

*Gate valves should always be used for the main steam stop valve. A gate valve opens and closes completely as a gate moves to allow or block flow.*

---

## BLOWDOWN VALVES

Boilers have bottom blowdown valves and may have surface blowdown valves and continuous blowdown valves. A *bottom blowdown valve* is a valve located

at the lowest part of the water side of a firetube boiler or the mud drum of a watertube boiler so that sludge and sediment can be removed from the bottom of the boiler. Boilers operating at 100 psi or more are required to have two bottom blowdown valves mounted on the same blowdown line.

When two valves are used, they may both be slow opening valves, which require five full turns of the handwheel to open or close fully, or one quick-opening valve and one slow-opening valve. If a quick-opening valve is used, it must be the valve closest to the shell of the boiler. **See Figure 3-17.** Higher pressures (up to 2500 psi) require valves to be seatless, high-pressure tandem blowdown valves or hard-head sealing valves. Bottom blowdown valves are designed for intermittent use, usually once per shift or once per day.

---

**BLOWDOWN VALVES**

**Figure 3-17.** Blowdown valves are used to clear out sludge and sediment from the bottom of a boiler.

The blowdown valves must close tightly. A leaking valve is indicated by a hot blowdown line and a drop in chemical concentrations. With a small leak, the boiler can still be operated if the water level can be easily maintained. However, the boiler should be removed from service for repairs as soon as possible. If maintaining a normal operating water level is difficult, the boiler should be secured immediately. Any leaks on the blowdown line are considered dangerous. The boiler must be removed from service and cooled slowly. Before repairs are made, the boiler inspector must be notified and proper recommendations followed. The repair must be made in accordance with the ASME Code.

## TECH TIP

*The amount of blowdown is often determined by historical use. The amount of bottom blowdown needed in modern boilers is often much less than in the past due to advances in water treatment technology. For this reason, the amount of bottom blowdown used is often excessive.*

## BOILER VENTS

A *boiler vent* is a valve connected to the top of a boiler that allows air to be removed from the boiler when filling and heating and allows air to be drawn in when the pressure drops during cool-down or when draining the boiler. **See Figure 3-18.** The boiler vent is connected to the highest part of the steam drum with piping and valves that meet the temperature and pressure requirements of the ASME Code.

When warming up a boiler, the boiler vent is in the open position until the boiler gauge registers about 10 psi to 25 psi. At this point, all air has been removed from the boiler drum, and the vent can be manually closed. After removing the boiler from service and allowing the steam pressure to drop to 10 psi to 25 psi, the vent should be opened manually. This prevents a vacuum from developing as the steam in the drum condenses.

## BOILER VENT VALVES

HIGHEST POINT

VENT VALVE

**Figure 3-18.** A boiler vent valve is used to remove air from a boiler as it is being filled with water.

## SECTION 3.2—CHECKPOINT

1. What is a Bourdon tube?
2. Where is a steam gauge typically located on a boiler steam drum?
3. What is the difference between absolute and gauge pressure?
4. What is the acceptable out-of-tolerance range of a steam gauge?
5. What is a dry pipe separator?
6. What is a cyclone separator?
7. What is the purpose of a stop valve?
8. Why are bottom blowdown valves used?
9. What is a boiler vent?

## CASE STUDY—WATER HAMMER IN A HEATING COIL

### Situation

A consulting engineer was called to a homeless shelter to determine why loud banging and clanging could be heard from the steam heating system. The shelter was located in the basement of a large warehouse and office complex. Low-pressure saturated steam was supplied by three large boilers to the entire facility. The noises were not present in the remainder of the facility.

The air handling unit was mounted in a tall ceiling space above the laundry facility for the shelter. It was a single-zone heating/cooling unit that supplied approximately 2500 cfm of conditioned air to the space. The heating coil was about 4′ × 8′ with four rows of tubes. The steam pipe into the coil was 4″, and the condensate pipe from the coil was ¾″ going to a float thermostatic steam trap.

### Problem

Originally, the thermostat from the occupied space controlled an electrically operated steam valve on the steam pipe. The steam valve had stopped working during the previous heating season, and the maintenance staff had decided to replace the 4″ valve in the steam line with a less expensive solenoid valve in the condensate return line in order to save money.

As steam entered the coil, heat was given up to the air passing over the coil. When the room temperature was satisfactory, the thermostat called for the solenoid valve to close. The consulting engineer determined that condensate would build up in the coil, condense the steam as it entered, and cause a vacuum to form. The vacuum pulled condensate to that area, resulting in water hammer.

### Solution

The solution to this problem was to remove the solenoid valve on the condensate return pipe and install a new valve and actuator on the steam line. It cost several hundred dollars to purchase and install the new valve, and it probably would have been much less expensive to rebuild the old valve and actuator, but they were thrown away after they were removed. An even more cost-effective solution would have been to rebuild and reuse the old valve and actuator after removal.

**Name** _____ **Date** _____

_____ **1.** Boiler fittings are necessary for ___.
    A. appearance
    B. efficiency
    C. aesthetic purposes
    D. controlling air flow

_____ **2.** To prevent failure from pressure over the MAWP, all steam boilers must be equipped with at least one ___.
    A. safety valve
    B. blowdown valve
    C. gauge glass
    D. feedwater heater

_____ **3.** The purpose of a safety valve is to ___.
    A. control the boiler operating range
    B. control high or low water
    C. prevent the boiler from exceeding its MAWP
    D. control high and low fire

_____ **4.** Safety valves are designed to ___.
    A. open slowly to prevent water hammer
    B. pop open
    C. open only by hand
    D. be opened by the pressure controller

_____ **5.** When testing a safety valve manually, there must be ___.
    A. at least 75% of the safety valve popping pressure on the boiler
    B. no boiler pressure
    C. an inspector present
    D. a pressure greater than the safety valve popping pressure on the boiler

_____ **6.** Any repairs to the safety valve must be done by the ___.
    A. chief engineer
    B. state inspector
    C. manufacturer-authorized representative
    D. maintenance department

_____ **7.** A(n) ___ test is used to test the relief capacity of a safety valve.
    A. accumulation
    B. try lever
    C. blowdown
    D. bench

_____ 8. ___ pressure is the pressure caused by the weight of a column of water.
    A. Pump
    B. Hydrostatic
    C. Steam
    D. Suction

_____ 9. A cyclone separator is a cylindrical device that separates water droplets from steam using ___.
    A. pressurized steam
    B. pressurized air
    C. centrifugal force
    D. automatic nonreturn valves

_____ 10. A steam pressure gauge must be protected by a siphon to prevent ___.
    A. steam from entering the mud drum
    B. boiler pressure from damaging the gauge
    C. air from entering the Bourdon tube
    D. steam from damaging the gauge

_____ 11. The correction factor for hydrostatic pressure is ___ psi per vertical foot.
    A. 0.433
    B. 8.3
    C. 14.7
    D. 34.5

_____ 12. The boiler vent is open when warming up a boiler until ___.
    A. the boiler is cut in on-line
    B. there is 10 psi to 25 psi of steam pressure on the boiler
    C. the superheater drain is closed
    D. the boiler begins to overheat

_____ 13. Separators work on the basic principle of ___.
    A. potential energy
    B. pressure drop separation
    C. trapping steam and letting water through
    D. changing the direction of the steam flow

_____ 14. The dry pipe separator is located in the ___.
    A. upper part of the steam drum
    B. superheater outlet header
    C. main steam line
    D. line before the turbine throttle

_____ 15. The ASME Code states that high pressure boilers in battery (two or more boilers connected to a common header) and equipped with manhole openings must have ___.
    A. one main steam stop valve
    B. two main steam stop valves
    C. no valve between the boiler and header
    D. one ball valve between the boiler and header

_____  **16.** A main steam stop valve should be a(n) ___ valve.
  A. os&y gate
  B. os&y globe
  C. quick-closing lever
  D. plug

_____  **17.** Boiler safety valves are placed at the highest part of the ___.
  A. water drum
  B. boiler vent
  C. steam siphon
  D. steam space

_____  **18.** During boiler inspection, the main steam stop valves are ___.
  A. left fully open
  B. closed, locked, and tagged
  C. left partially open
  D. replaced with globe valves

_____  **19.** A pressure gauge must be recalibrated if it is not accurate to within ___% of the working pressure.
  A. 0.5
  B. 1
  C. 2
  D. 5

_____  **20.** During normal usage, ___ are used to allow for the expansion and contraction of the main steam line.
  A. flanged fittings
  B. welded braces
  C. expansion bends
  D. rigid pipe hangers

_____  **21.** A main steam stop valve should be a gate valve because a gate valve ___.
  A. can be opened in ¼ turn
  B. can be used as a throttling valve
  C. offers no resistance to the flow of steam
  D. does not have to be fully open

_____  **22.** A safety valve must be capable of discharging all the steam without the pressure increasing more than ___% above the maximum MAWP.
  A. 1
  B. 6
  C. 14.7
  D. 25

_____  **23.** A safety valve opens when the ___ exerted by the steam exceeds that of the spring.
  A. total force
  B. temperature
  C. area gain
  D. huddling chamber force

_____ **24.** ___ fittings are used for water columns operating above 350 psi.
   A. Cast iron
   B. Malleable iron
   C. Steel
   D. Brass

_____ **25.** The ___ is the part on a safety valve that increases the area of the safety valve disc.
   A. bonnet
   B. huddling chamber
   C. body
   D. popping valve

# Chapter 4 Objectives

## SECTION 4.1—SUPERHEATERS AND DESUPERHEATERS

- Describe saturated and superheated steam and steam quality.
- Describe superheaters and their designs.
- Describe desuperheaters and pressure-reducing stations and their designs.

## SECTION 4.2—STEAM TURBINES

- Describe common types of turbines.
- Explain steam turbine operation.

## SECTION 4.3—STEAM TRAPS

- Describe common types of steam traps and their selection and maintenance.
- Explain testing of steam traps.

## SECTION 4.4—CONDENSATE RETURN

- Describe condensate return tanks and pumps.
- Explain condensate surge tanks.

## ARROW SYMBOLS

| AIR | GAS | WATER | STEAM | FUEL OIL | CONDENSATE | AIR TO ATMOSPHERE | GASES OF COMBUSTION |

**Digital Resources**
ATPeResources.com/QuickLinks
Access Code: **472658**

# CHAPTER 4

# STEAM SYSTEMS

## INTRODUCTION

Typical components required for the operation of a steam system include superheaters, desuperheaters, pressure reducers, turbines, steam traps, and condensate return components. Superheaters increase the heat content of steam for use in steam turbines. Desuperheaters reduce the temperature of steam to be used by auxiliary equipment. Pressure-reducing stations reduce high-pressure steam to the lower pressure required for a process. Turbines use steam to generate electricity. Steam traps increase the efficiency of plant operation. Condensate tanks with booster pumps collect the condensate for return to the boiler.

---

## SUPERHEATERS AND DESUPERHEATERS

### STEAM

*Saturated steam* is steam that is in equilibrium with water at the same temperature and pressure. Auxiliaries such as reciprocating pumps, saturated steam turbines, and heat exchangers are designed for saturated steam. Process steam is also saturated. *Superheated steam* is steam that has been heated above the saturation temperature. Superheated steam contains more heat energy than saturated steam because it has a higher sensible heat content.

### Saturated Steam

The boiling point of water depends on the pressure. When the pressure increases, the boiling point also increases. **See Figure 4-1.** For example, water boils at 212.0°F at 0 psi, 227.4°F at 5 psi, and 338°F at 100 psi. In all cases, the steam and the water are at the same temperature.

Saturated steam is in equilibrium with water at the same temperature and pressure. *Saturation pressure* is the pressure at which water and steam are at the same temperature. *Saturation temperature* is the temperature at which water and steam are at the same pressure. The saturation temperature increases as the pressure increases.

## PROPERTIES OF SATURATED STEAM

| Gauge Pressure* | Absolute Pressure† | Temperature‡ | Heat Content‖ | | | Specific Volume Steam $V_g$§ |
|---|---|---|---|---|---|---|
| | | | Sensible $h_f$ | Latent $h_{fg}$ | Total $h_g$ | |
| 27.96 | 1 | 101.7 | 69.5 | 1032.9 | 1102.4 | 333.0 |
| 25.91 | 2 | 126.1 | 93.9 | 1019.7 | 1113.6 | 173.5 |
| 23.87 | 3 | 141.5 | 109.3 | 1011.3 | 1120.6 | 118.6 |
| 21.83 | 4 | 153.0 | 120.8 | 1004.9 | 1125.7 | 90.52 |
| 19.79 | 5 | 162.3 | 130.1 | 999.7 | 1129.8 | 73.42 |
| 17.75 | 6 | 170.1 | 137.8 | 995.4 | 1133.2 | 61.89 |
| 15.70 | 7 | 176.9 | 144.6 | 991.5 | 1136.1 | 53.57 |
| 13.66 | 8 | 182.9 | 150.7 | 987.9 | 1138.6 | 47.26 |
| 11.62 | 9 | 188.3 | 156.2 | 984.7 | 1140.9 | 42.32 |
| 9.58 | 10 | 193.2 | 161.1 | 981.9 | 1143.0 | 38.37 |
| 7.54 | 11 | 197.8 | 165.7 | 979.2 | 1144.9 | 35.09 |
| 5.49 | 12 | 202.0 | 169.9 | 976.7 | 1146.6 | 32.35 |
| 3.45 | 13 | 205.9 | 173.9 | 974.3 | 1148.2 | 30.01 |
| 1.41 | 14 | 209.6 | 177.6 | 972.2 | 1149.8 | 28.00 |
| **Gauge Pressure**\*\* | | | | | | |
| 0 | 14.7 | 212.0 | 180.2 | 970.6 | 1150.8 | 26.80 |
| 1 | 15.7 | 215.4 | 183.6 | 968.4 | 1152.0 | 25.20 |
| 2 | 16.7 | 218.5 | 186.8 | 966.4 | 1153.2 | 23.80 |
| 3 | 17.7 | 221.5 | 189.8 | 964.5 | 1154.3 | 22.50 |
| 4 | 18.7 | 224.5 | 192.7 | 962.6 | 1155.3 | 21.40 |
| 5 | 19.7 | 227.4 | 195.5 | 960.8 | 1156.3 | 20.40 |

\* in Hg vac   ‡ in °F   § in cu ft/lb
† in psia   ‖ in Btu/lb   \*\* in psig

**Figure 4-1.** The boiling point of water depends on the pressure.

Initially, steam power plants that used reciprocating engines were designed to operate with saturated steam. Steam separators were introduced between the boiler and the engine to remove condensate from the steam. This improved the quality of the steam but did not remove all moisture. Drawbacks to using this wet steam included erosion of engine parts, loss of heat to metal causing condensation of steam, and erosion of lines and fittings caused by impingement of the steam. *Steam impingement* is the condition where steam strikes a metal surface, causing erosion of that surface. These drawbacks were almost eliminated by the use of superheated steam.

### Steam Quality

Steam may be either wet or dry. Boilers generate steam to be used in a load. Heat energy is extracted from the steam when it is put to work. When any heat is removed from the saturated steam, a portion of the steam condenses into a liquid state. If there are water droplets in the steam, it is wet steam. If there is only steam with no condensation or carryover, it is dry steam. *Steam quality* is the ratio of dry steam to the total amount of water evaporated. If steam has a quality of 98%, it is 98% vapor and contains 2% water.

A throttling calorimeter is used to directly measure the moisture content in steam. The heat content of steam does not change when steam expands without doing work. When dry steam is discharged to the atmosphere, it becomes superheated because the temperature is above the saturation temperature. When wet steam is discharged to the atmosphere, some work is done as the water evaporates to steam. This reduces the temperature of the steam. The reduced temperature at atmospheric pressure can be compared to the saturation temperature to determine the steam quality.

### Superheated Steam

Superheated steam has been heated above the saturation temperature without a change in pressure. Superheaters are used for loads that require dry steam. Superheated steam contains more heat energy than saturated steam because it has a higher heat content than saturated steam. Therefore, superheated steam can produce more work than saturated steam can produce at the same pressure. For example, saturated steam at 100 psi has a temperature of 338°F. If this steam at 100 psi had a temperature higher than 338°F, it would be superheated steam.

The degree of superheat is the difference in temperature between the saturated steam and the superheated state of the same steam. In an average steam turbine, there is a gain of about 1% in efficiency for every 35°F of superheat. For example, if saturated steam at 100 psi and 338°F were heated to 500°F, it would have a superheat of 162°F (500 − 338). To find the increase in efficiency, the gain is divided by 35. This equals about a 4.6% increase. In some designs, the increase can go as high as 15% for 200°F of superheat in larger turbines.

## Thermal Expansion

Steam pipes expand and contract as the temperature changes during normal use. Steam pipes can break loose from their mountings if they are not allowed to move to accommodate this expansion and contraction. Steam pipes hanging from overhead supports can be set on rollers to allow this movement. **See Figure 4-2.** A steel plate is typically placed between the rollers and the insulation to prevent damage to the insulation.

## THERMAL EXPANSION ROLLERS

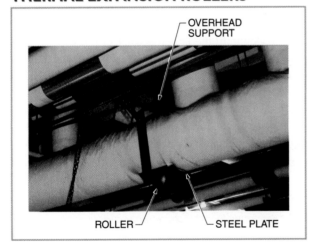

Figure 4-2. Thermal expansion rollers allow a steam line to move during expansion or contraction.

Steam lines attached to solid objects, such as a steam line coming out of a boiler, should have some type of expansion bend. This prevents excessive stress on the boiler and header. An expansion bend may be a simple bend or a set of elbows that allow the pipe to move slightly. This is common in the main steam line coming out of a boiler. **See Figure 4-3.** Longer runs of straight pipe may require an expansion loop in the shape of a U-bend.

## EXPANSION BENDS AND LOOPS

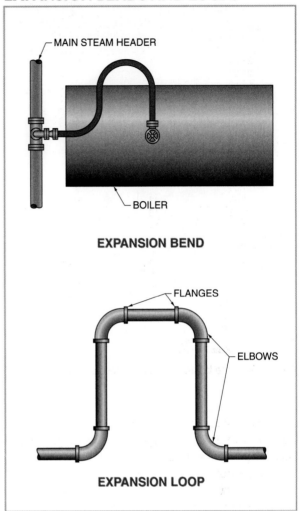

Figure 4-3. Expansion bends and loops allow a steam line to expand or contract without causing damage.

## TECH TIP

*Steel piping needs to expand and contract with changes in temperature in order to prevent damage. An engineering analysis must be done to determine the proper supports.*

## SUPERHEATERS

Watertube boilers can be fitted with superheaters to increase the heat content of steam. A *superheater* is a bank of tubes through which steam passes after leaving the boiler where additional heat is added to the steam. **See Figure 4-4.** This causes the steam temperature to rise significantly above its saturation temperature.

## SUPERHEATERS

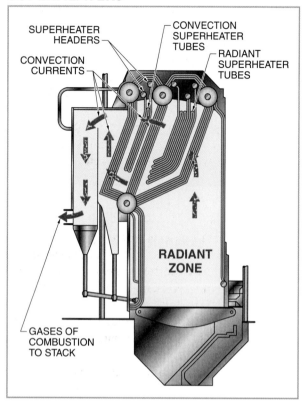

**Figure 4-4.** A superheater is a bank of tubes used to heat steam above its saturation temperature.

### Superheater Locations

Superheaters are located in the radiant zone (furnace area) or in the convection section of a boiler. Therefore, superheaters may be classified as radiant or convection. A *radiant superheater* is a superheater that is directly exposed to the radiant heat of the boiler. A *convection superheater* is a superheater that receives heat from convection currents in the gases of combustion. A consistent steam temperature during fluctuations in load can be maintained by using both radiant and convection superheaters in the same unit.

### Superheater Designs

Superheaters come in many different shapes. This variety is necessary because of the difference in boiler sizes and designs and the amount of space available for the superheater in the boiler. Superheaters can have either smooth or extended surfaces. Smooth surface superheaters consist of bare tubes. Extended surface superheaters have fins or grills mounted on each smooth tube to increase the heating surface.

A multiloop superheater is made up of many tubes bent back a number of times between the inlet and outlet headers. By placing the superheater headers outside the radiant heat zones, the tube connections are readily available for maintenance. This also minimizes the possibility of leaks. Detachable metal-to-metal ball joints connect the tubes to the header and are held in place by steel studs and clamps. This arrangement makes it relatively simple to remove any joint of a superheater tube for cleaning or inspection.

A safety valve is required at the superheater outlet header. The safety valve must be of sufficient capacity to relieve 25% of the steam that the boiler can produce. Superheater safety valves are set at a lower popping pressure than the safety valves on the steam drum to ensure this flow of steam. Superheaters should also be fitted with a drain on the outlet side.

### Superheater Operation

When operating a steam boiler equipped with a superheater, a flow of steam must be maintained through the superheater at all times. The flow of steam should never be less than 25% of boiler capacity. Failure to maintain steam flow will result in overheating, warping, or burning of superheater tubes.

The drain remains open when warming up a boiler and is left open until the boiler is placed on-line. The drain must also be opened as soon as the boiler is taken off-line. This ensures a flow of steam through the superheater at all times. This flow of steam keeps the superheater tubes from overheating during startup and shutdown. In addition, superheaters must be kept free of soot to maintain the transfer of heat. The manufacturer recommendations for startup and shutdown must be followed.

**TECH TIP**

*Superheaters increase the temperature of steam but not the pressure. This ensures dry steam is available.*

## DESUPERHEATERS

A *desuperheater* is an accessory used to remove heat from superheated steam to make it suitable for a process. In plants requiring both superheated and saturated steam, having one boiler for superheated steam and another for saturated steam is impractical. It is much easier to draw off some superheated steam, desuperheat it, and then send it to the auxiliaries or process lines. To desuperheat superheated steam, enough heat must be removed to bring the steam back to its saturation point.

## Line Desuperheaters

One type of desuperheater is a line desuperheater. **See Figure 4-5.** It is typically used in conjunction with a pressure-reducing station. A *line desuperheater* is a desuperheater that injects feedwater into a super-heated steam line to reduce the steam temperature. This desuperheater uses a chamber in the steam line that is fitted with one or more nozzles. The nozzles deliver a fine spray of feedwater into the superheated steam. The water absorbs heat from the superheated steam and turns to steam itself, thus reducing the temperature.

## LINE DESUPERHEATERS

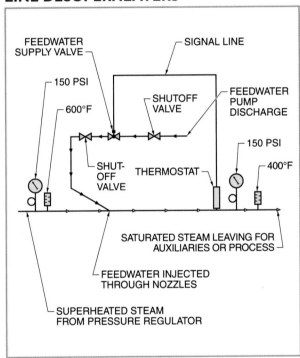

**Figure 4-5.** A line desuperheater injects feedwater into a superheated steam line to reduce the steam temperature.

The quantity of steam increases when the feedwater changes state after absorbing the superheat. A thermostat is placed in the line on the saturated steam side of the desuperheater. This thermostat sends a signal that controls the feedwater supply valve to the nozzles. If the temperature of the steam rises, the feedwater supply valve to the nozzles begins to open. If the temperature of the steam falls, the feedwater valve starts to close.

## Drum Desuperheaters

A drum desuperheater, which is used in marine boilers, is another type of desuperheater. A *drum desu-perheater* is a desuperheater that routes steam back through coils of submerged piping within the steam drum. **See Figure 4-6.** Some drum desuperheaters divert part of the superheated steam through a heat exchanger in the boiler mud drum. The superheated steam gives up some of its heat to the boiler water, which is at the saturation temperature. The steam leaves the coil at or near the temperature of the saturated steam.

## Pressure-Reducing Stations

Plants often require steam at different pressures for different purposes in a process. For example, high-pressure steam may be required for process operations and low-pressure steam for heating, cooking, and laundry operations. A pressure-reducing station reduces steam pressure as the steam passes through a valve from the high-pressure side to the low-pressure side of the valve. **See Figure 4-7.**

The pressure-sensing line for the valve positioning controller is located several feet down the line on the low-pressure side to eliminate pressure variations caused by the valve modulation. A safety valve must be installed on the low-pressure side to protect the lighter piping and equipment. Because of the small openings in the valve, there should be a strainer installed before the valve to remove foreign material that might plug these openings or damage the valve seat. The pressure-reducing valve and strainer require periodic maintenance to ensure proper operation.

### TECH TIP

*The flow of steam through a superheater should never be less than 25% of the capacity of the boiler. The steam flowing through the superheater protects the superheater tubes from damage caused by the heat of the flue gases.*

The low-pressure steam cannot dissipate all of its heat immediately to reach the saturation temperature corresponding with its pressure and becomes superheated. The steam eventually reaches its saturation temperature as it loses heat energy downstream. When directing steam to remote loads, the steam will not condense as readily as saturated steam. It also allows for the piping and equipment on the low-pressure side to be much lighter, reducing installation costs.

## DRUM DESUPERHEATERS

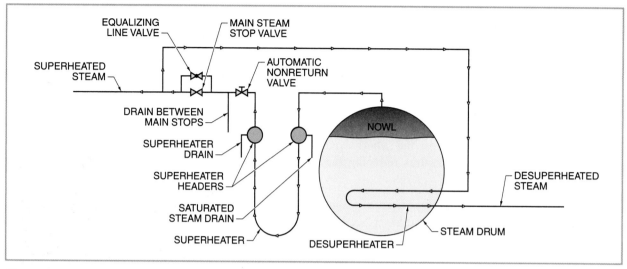

**Figure 4-6.** Drum desuperheaters are used primarily on marine boilers.

## PRESSURE-REDUCING STATIONS

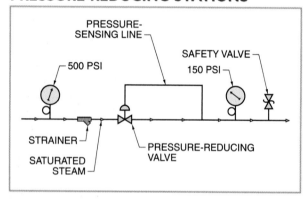

**Figure 4-7.** Plants often require steam at different pressures. Steam at the required pressure is supplied by using pressure-reducing stations.

### SECTION 4.1–CHECKPOINT

1. What is the difference between saturated steam and superheated steam?
2. What is steam quality?
3. What are common types of superheater designs?
4. What is the minimum flow rate of steam through a superheater?
5. Why does a line desuperheater use spray nozzles?
6. How does a drum desuperheater reduce superheat?
7. Why is a safety valve used with a pressure-reducing station?

### SECTION 4.2
# STEAM TURBINES

## STEAM TURBINE TYPES

A *steam turbine* is a rotary mechanical device used to drive rotating equipment, such as a generator, by extracting thermal energy from pressurized steam. The amount of heat energy in the steam determines the amount of work performed. The greater the heat energy in the steam, the greater the amount of work that can be performed by the steam turbine. Steam turbines have fewer moving parts, require less maintenance, and are more efficient than reciprocating steam engines. Common uses of steam turbines include generating electricity and driving auxiliaries such as pumps, fans, and compressors.

A steam turbine can consist of several stages, each having a set of revolving blades. The size of the blades increase with each stage. Steam turbines require dry superheated steam. Many types of loads need only moderately superheated steam.

### Noncondensing Steam Turbines

A *noncondensing steam turbine* is a steam turbine that exhausts at atmospheric pressure or above. Noncondensing steam turbines are normally used in steam plants that require large amounts of low-pressure process or heating steam. If the exhaust steam were not used and allowed to go to waste in the atmosphere, the steam turbine would have a very low thermal efficiency and would waste a considerable amount of energy. The exhaust steam from small, noncondensing steam turbines is sometimes used to heat feedwater in an open feedwater heater. Noncondensing steam turbines must use a large amount of the available exhaust steam to maintain maximum efficiency. In addition, noncondensing steam turbines do not have condensate returns.

## Condensing Steam Turbines

A *condensing steam turbine* is a steam turbine that allows condensate to be reclaimed for use in the system. The condensate is returned with some heat still in it. The heat is then reclaimed at an open feedwater heater. Condensing steam turbines are commonly used on medium to large units and operate with an exhaust pressure of 26 in. Hg to 29 in. Hg vacuum. The higher the vacuum is, the greater the efficiency is. This low exhaust pressure allows the steam to expand to a greater volume and release more heat to perform more work. Plant conditions determine the steam turbine best suited for an operation. Plants that are designed for maximum efficiency may use both condensing and noncondensing steam turbines.

### TECH TIP

*Steam from boilers is used to drive steam turbines. Steam turbines drive a shaft that turns within a generator to produce electricity. Step-up transformers are used to increase the voltage to the level required for transmission lines.*

## Impulse Steam Turbines

Impulse steam turbines use steam velocity as a force acting in a forward direction on a blade mounted on a wheel. **See Figure 4-8.** Impulse steam turbines may contain many sets of revolving blades that are divided into stages. To produce the force, a nozzle (1) is located before each stage. The steam velocity increases and pressure decreases as steam flows through the nozzle. The nozzle directs the steam to the turbine's revolving blades (2) on the first-stage wheel. This causes the shaft (3) to rotate. Steam is then routed through fixed blades (4), which redirect the flow of steam to another set of revolving blades (5) on the same wheel as the first stage. The steam then enters the second-stage nozzle (6), gaining velocity before striking the revolving blades (7) of the second-stage wheel. Steam exits through the exhaust opening (8).

## IMPULSE STEAM TURBINES

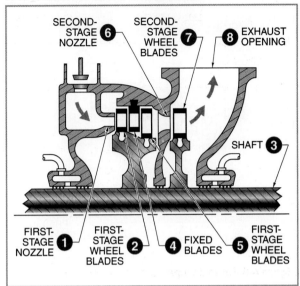

**Figure 4-8.** An impulse steam turbine uses steam velocity to force the turbine to rotate.

The pressure drop incurred in an impulse turbine is through the nozzles and not the blade area. **See Figure 4-9.** The fixed blades are used to redirect the flow of steam between each set of revolving blades without any change in pressure or velocity. Steam enters the first-stage nozzle (1) and drops slightly in pressure as it passes through, but the steam velocity increases. The steam strikes the revolving blades (2), imparting energy to the blades but losing velocity in the process. The steam then enters the fixed blades (3) and changes direction without losing pressure or velocity. Once again, the steam strikes a second set of revolving blades (4), with the corresponding drop in velocity but not in pressure.

The steam velocity returns to its initial value after passing through the blades before the next stage. To increase the velocity, the steam passes through the second-stage nozzle (5). The velocity-pressure relationship is then repeated as the steam passes through the second-stage revolving blades (6).

The pressure drops through the first-stage and second-stage nozzles before being exhausted after the second stage. The diameter of the wheels increases in size from the first stage to subsequent stages to allow for the drop in pressure and an increase in volume. Most steam turbines are designed to have a steam velocity that is twice as great as the blade velocity.

## STEAM FLOW THROUGH STAGES

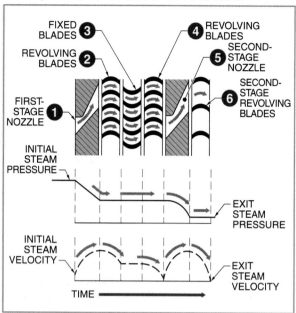

**Figure 4-9.** Steam velocity increases as steam flows through a nozzle.

The velocity of a turbine blade is directly proportional to the velocity of steam flowing past the blade. In other words, the turbine blades turn faster when the steam velocity is higher. In an extreme case, the steam energy could be extracted by one set of blades. However, if the steam pressure drop is very high, the steam velocity will also be very high. This causes the blades and shaft to rotate very fast. This can cause vibration that can damage the turbine. Therefore, the turbine uses velocity compounding to reduce the speed of the blades. *Velocity compounding* is a method in which the energy from steam is extracted using multiple stages in a turbine to reduce the steam velocity and pressure in a short time.

### Reaction Steam Turbines

Newton's law states that for every action there is an equal and opposite reaction. Reaction steam turbines operate on this principle for part of their action. Instead of nozzles, a reaction steam turbine uses fixed blades for its first stage. **See Figure 4-10.** Fixed blades are designed so that each blade pair acts as a nozzle. Steam expands between each blade pair. The expanding steam gains velocity in the same manner as the impulse nozzle.

This steam, with its high velocity, enters the blades of the moving element, imparting a direct impulse to it.

Steam is then directed into the nozzle-shaped blade passages. The steam expands in the nozzle-shaped passages and gains additional velocity. When leaving the steam turbine blades at a high velocity, a backward kick, or reaction, is applied to the blades. This is where the term "reaction steam turbine" comes from.

## REACTION STEAM TURBINES

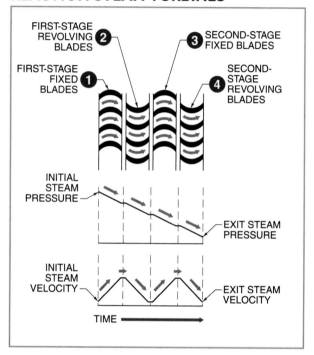

**Figure 4-10.** A reaction steam turbine allows steam to expand between blades in order to increase velocity.

The steam at its initial pressure enters the fixed blades (1) and increases in velocity while losing pressure. The steam strikes the revolving blades (2), giving up energy and losing velocity. As the steam leaves, it gives a reactive force to the revolving blades, and a loss of pressure occurs. The second stage begins as the steam enters the next row of fixed blades (3). Again, velocity increases and some pressure is lost in the fixed blades. Then, velocity and pressure decrease through the revolving blades (4).

Because of the difference in pressure between the entrance and exit sides of both the fixed and revolving blades, the reaction steam turbine is a full-admission steam turbine. Admission of steam takes place completely around the wheel. In impulse steam turbines, only certain nozzles are opened to produce the flow required.

Because certain characteristics of reaction steam turbine operation differ from impulse steam turbine operation, physical characteristics of the reaction steam turbine differ from the impulse steam turbine. Reaction steam turbines have a low mechanical clearance between the tips of the fixed blades and the shaft and a low mechanical clearance between the moving blades and the casing. They have a relatively large number of elements in order to allow a small pressure drop per row of blades. The large axial thrust, resulting from the differences in pressure between the stages, must be balanced by balancing pistons of various sizes, a dummy piston, and a double flow of steam.

### Impulse-Reaction Steam Turbines

Impulse steam turbines work primarily at a high speed resulting from large drops in pressure. They are smaller and weigh and cost less than reaction steam turbines of the same capacity. Reaction steam turbines are primarily low-pressure steam turbines because of losses around the blades at high pressures. Reaction steam turbines have small drops in pressure per stage, resulting in small amounts of steam expansion with low output of work per stage. Many stages must be used to obtain the full range of pressure drops throughout the steam turbine.

Impulse-reaction steam turbines combine impulse and reaction blading. Usually there are several stages of impulse blading in the high-pressure end of the steam turbine with large pressure drops through the nozzles. The reaction blading in the low-pressure end causes small pressure drops across each row of blades. In addition, impulse-reaction steam turbines are commonly run on high vacuum for even greater efficiency.

In some impulse-reaction steam turbines, nozzles are used. In others, the fixed-reaction blades act as a series of nozzles directing steam against the impulse blades. The same parts are required in an impulse-reaction steam turbine as those in the separate steam turbines. The governor controls steam admission to the first stage. The overspeed and low-oil pressure trip shuts off steam to the steam turbines if necessary.

### TECH TIP

*Superheated steam is usually used for driving steam turbines to prevent moisture droplets from entering the turbine. Moisture in the steam can erode the turbine rotor.*

*Small generators powered by steam turbines can be part of the green energy goals for a facility.*

## STEAM TURBINE OPERATION

In steam turbines, the feedwater flows into the steam drum (1). **See Figure 4-11.** This boiler water is heated by the gases of combustion, evaporates into steam, and leaves the boiler drum. For a boiler generating steam to drive a turbine, the steam is directed through a superheater (2). The steam is further heated in the superheater to a temperature above its corresponding saturation pressure and becomes superheated steam.

The superheated steam leaves the superheater through the main steam line (3). Some steam is sent out to the auxiliaries and the desuperheating and pressure-reducing station (4). The bulk of the steam goes to the steam turbine (5), which is connected to the generator (6). The generator produces electricity.

The steam turbine may have steam extraction lines (7). *Extraction steam* is steam that is removed from the turbine at a controlled pressure after it has passed through some of the turbine stages. Extraction steam is used for process operations in the plant. A second extraction line extracts steam at a lower pressure than the first extraction line.

After the steam has passed through the rest of the turbine stages, it enters the surface condenser (8). In a surface condenser, water passes through tubes, causing the steam to cool and condense. During this process, a vacuum is formed in the condenser. This vacuum on the exhaust side of the turbine reduces the back pressure on the turbine, thereby reducing the amount of steam needed to run the turbine. When the back pressure is reduced, the steam or water rate is reduced.

The condensate collects in the hot well (9). The condensate pump (10) sends the condensate back to the feedwater surge tank. The cycle is now ready to begin again. The air ejectors (11) on the condenser are used to remove air and other noncondensable gases from the condenser.

## STEAM TURBINE OPERATION

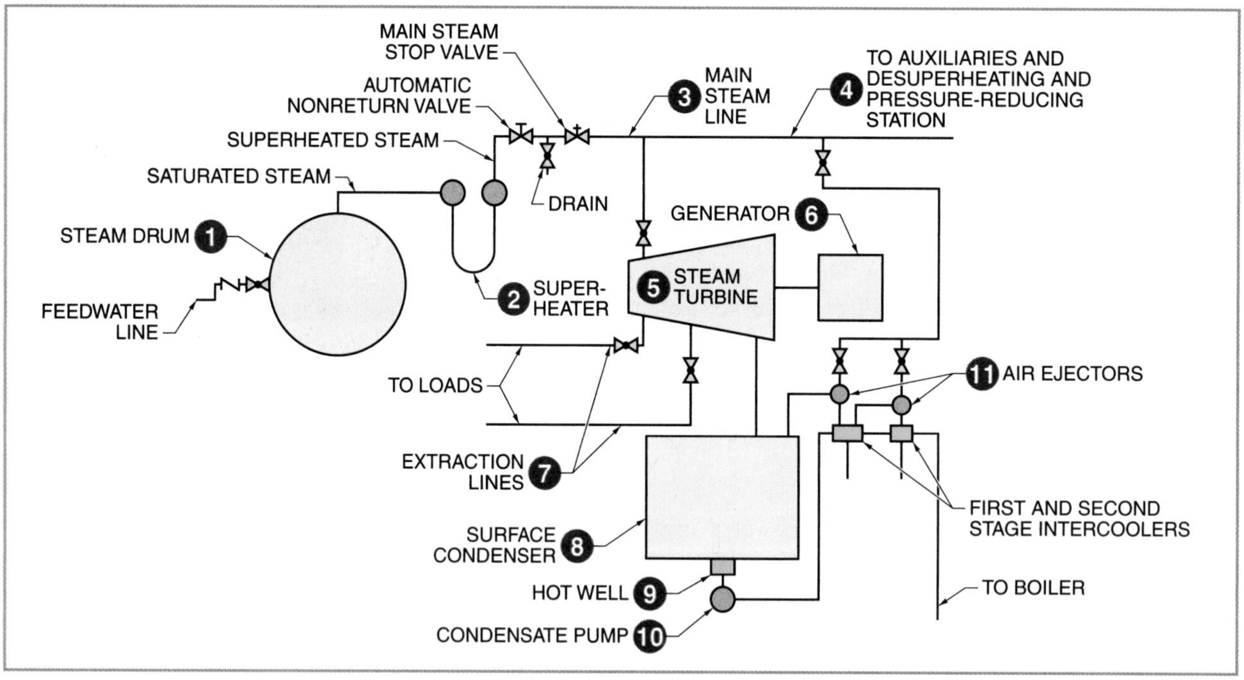

**Figure 4-11.** Steam provides power to a turbine and heat to a facility. Clean condensate is returned to the boiler to achieve maximum efficiency.

### SECTION 4.2–CHECKPOINT

1. What are common steam turbine types?

2. What is the difference between noncondensing and condensing steam turbines?

3. Where does the pressure drop occur in impulse steam turbines?

4. Where does steam increase in velocity in a reaction steam turbine?

5. What is extraction steam?

## SECTION 4.3
# STEAM TRAPS

## STEAM TRAP TYPES AND SELECTION

A *steam trap* is an accessory that removes air and noncondensable gases and condensate from steam lines and heat exchangers without a loss of steam. Steam traps increase the overall efficiency of a steam plant. If condensate is not removed from the steam lines and heat exchangers, it could lead to water hammer or cause the heat exchangers to become waterlogged, thus reducing the heat transfer rate. Water hammer is caused by mixing steam and water and could lead to a line rupture.

### Steam Trap Locations

Steam traps are located as necessary to achieve the most efficient condensate removal. **See Figure 4-12.** The main steam header in the boiler room must be equipped with a steam trap. A steam trap must also be installed at the base of a steam riser. Heat exchangers such as fuel oil heaters, hot water heaters, radiators, and steam jackets require steam traps.

### Steam Trap Types

Several types of steam traps are available. These operate in different ways, yet they all perform the same function. A steam trap usually has a steam strainer installed on the inlet line to prevent scale or other solid particles from entering the trap. The temperature drop across a steam trap varies considerably for different types of traps. This temperature drop is called subcooling. The amount of subcooling can vary from about 2°F to 100°F. Some traps continuously modulate the condensate flow and have very little subcooling. Other traps back up condensate in the steam lines and have more subcooling.

## STEAM TRAP LOCATIONS

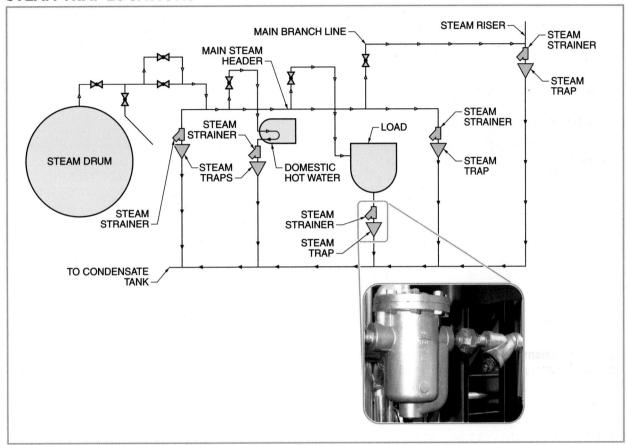

**Figure 4-12.** Steam traps are located in a steam system wherever steam releases its heat and condenses. Steam strainers are installed before steam traps.

**Thermostatic Steam Traps.** A *thermostatic steam trap* is a steam trap that contains a temperature-operated element, such as a corrugated bellows. A thermostatic steam trap is also known as a balanced-pressure thermostatic steam trap. **See Figure 4-13.** The bellows is typically filled with an alcohol and water solution that boils at a few degrees below the saturation temperature. The amount of alcohol in the solution can be increased to increase the sensitivity of the trap. Most thermostatic steam traps operate with 20°F to 40°F subcooling. Traps with less subcooling back up more condensate. Some designs may have a built-in strainer.

When steam heats the bellows, the solution boils and the pressure increase causes the bellows to expand. This pushes the valve head up to close the trap. If air or condensate is in contact with the element, it cools and contracts and opens the valve to allow the air or condensate to leave the trap. This type of steam trap is good for venting air during startup and operation.

**Float Thermostatic Steam Traps.** A *float thermostatic steam trap* is a steam trap that contains a thermostatic bellows or another thermostatic element and also contains a steel ball float connected to a discharge valve by a linkage. **See Figure 4-14.** When condensate enters the body of the trap, the float rises and the valve opens to allow the condensate to flow into the condensate return line. If there is no condensate in the body, the valve closes. The trap modulates the discharge to match the condensate flow at saturated steam temperature.

The thermostatic element opens only if there is a temperature drop around the element caused by air or noncondensable gases in the body of the trap. When steam enters the body of the trap, the thermostatic element expands and closes. In this type of trap, the thermostatic element opens and closes only to discharge air and noncondensable gases. The float opens and closes the valve to discharge condensate. These traps are generally used in plants with pressures up to 125 psi.

## THERMOSTATIC STEAM TRAPS

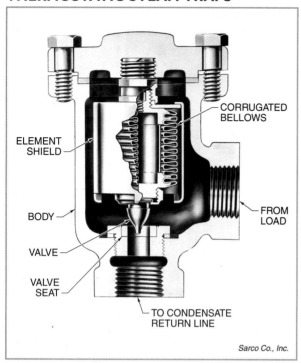

Figure 4-13. A thermostatic steam trap contains a temperature-operated device, such as a corrugated bellows, that controls a small discharge valve.

## FLOAT THERMOSTATIC STEAM TRAPS

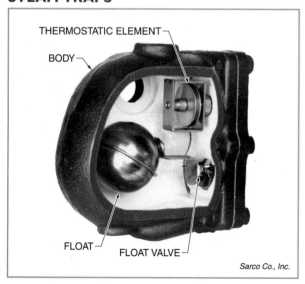

Figure 4-14. A float thermostatic steam trap contains a thermostatic bellows or another thermostatic element and also contains a steel ball float connected to a discharge valve by a linkage.

**Inverted Bucket Steam Traps.** An *inverted bucket steam trap* is a steam trap that contains an inverted bucket connected to a discharge valve. When condensate fills the trap, the bucket loses buoyancy and sinks to open the discharge valve. **See Figure 4-15.** Steam and condensate enter the trap from the bottom of the inverted bucket. If only steam enters the trap, the bucket will become buoyant and the valve will close. This type of trap has no condensate backup.

## INVERTED BUCKET STEAM TRAPS

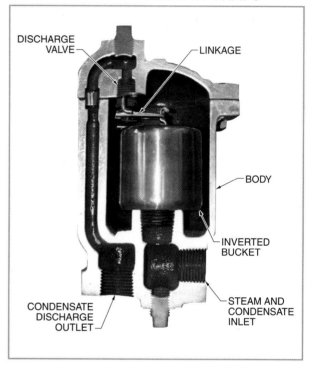

Figure 4-15. An inverted bucket steam trap contains an inverted bucket connected to a discharge valve.

When condensate enters the trap, the bucket sinks, pulls the linkage, and the valve opens. A small hole at the top of the bucket removes air that becomes trapped in the bucket. Air and condensate leave the body of the trap when the valve opens.

**Bimetallic Steam Traps.** A *bimetallic steam trap* is a steam trap in which a temperature-sensitive bimetallic element controls a small discharge valve. **See Figure 4-16.** The two parts of the element expand and contract differently when exposed to temperature changes, thus opening and closing the valve. The opening and closing of the valve depends on the temperature differential within the chamber.

## BIMETALLIC STEAM TRAPS

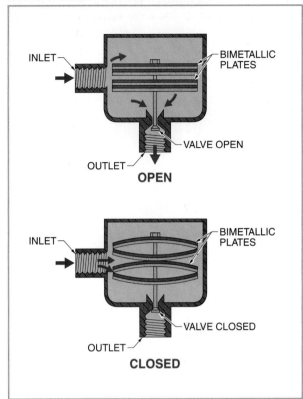

**Figure 4-16.** A bimetallic steam trap opens because of a drop in temperature across the bimetallic element as condensate or air enters the control chamber.

Bimetallic steam traps normally operate with 50°F to 100°F subcooling. Because of this, a bimetallic trap can back up a large amount of condensate. These traps are generally used in industrial or commercial applications where hot water is needed for a process, such as heating.

**Thermodynamic Steam Traps.** A *thermodynamic steam trap* is a steam trap that has a single movable disc that rises to allow the discharge of air and cool condensate. **See Figure 4-17.** A thermodynamic steam trap is also known as an impulse floating disc or a kinetic energy floating disc steam trap. Pressurized air or condensate enters in the inlet port under the center of the disc and lifts the disc from its seat. The discharge of condensate continues until the flashing condensate approaches steam temperature. The high-velocity steam under the disc recompresses, building up pressure in the control chamber above the disc. This pressure forces the disc down to the seat, ensuring a tight closure without steam loss.

Steam pressure in the control chamber acting on the total disc area from above holds the disc closed against the inlet pressure from the smaller inlet seat below. As condensate collects in the trap, it reduces heat transfer to the control chamber. The pressure in the control chamber decreases as the steam in the control chamber condenses. The disc is then lifted by the inlet pressure, and the condensate is discharged to repeat the cycle. Thermodynamic steam traps open and close quickly with no condensate backup.

### Moisture Separators

A *moisture separator* is a device placed in a wet steam line used to remove entrained water droplets. Steam separators have barriers placed in the path of the wet steam flow that cause the steam to change direction. However, the water droplets are not able to change direction, strike the barrier, and fall out of the steam flow. A variable orifice moisture separator contains manually adjustable compartments that reduce the pressure at which wet steam enters and condensate discharges. **See Figure 4-18.** The manual orifice is adjusted to maintain a constant water level in the sight glass. Some variable orifice moisture separators work automatically because they use a wax that expands and contracts.

### Selection and Maintenance of Steam Traps

Steam traps are selected for a specific function in a steam system. When selecting a steam trap, size and application are considered for best efficiency. A reliable manufacturer should be contacted for specific application requirements.

Steam traps are often neglected in a plant. A regular maintenance schedule should be established for checking each steam trap for proper operation. A steam trap that is stuck closed will lead to a waterlogged line and could cause water hammer. A steam trap that is stuck open will cause steam to blow through. If steam is allowed to enter the return line, it can result in a loss of plant efficiency. All steam traps must be in proper working order for maximum plant efficiency.

### TESTING STEAM TRAPS

Steam traps should be tested as a part of overall boiler maintenance tasks and also at the first sign of underheating heating units or heat exchangers, a temperature increase in condensate returning to the condensate return tank, or a pressure buildup in the condensate return tank. Steam trap testing is best performed by trained personnel using the proper testing equipment.

## THERMODYNAMIC STEAM TRAPS

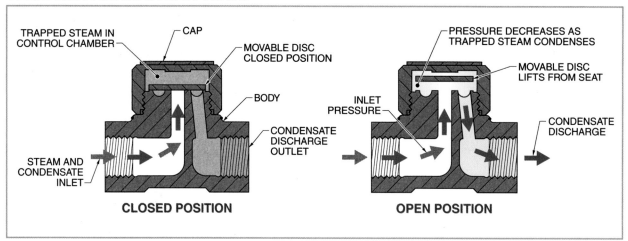

**Figure 4-17.** A thermodynamic steam trap has a single movable disc that rises to allow the discharge of air and cool condensate.

## MOISTURE SEPARATORS

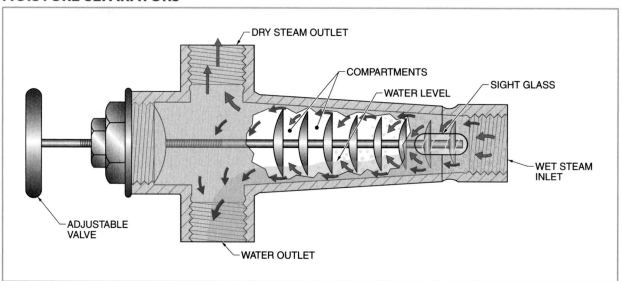

**Figure 4-18.** In a moisture separator, wet steam and condensate enter the body of the separator and pass through a series of compartments that reduce the pressure.

Steam traps can be tested by sight, sound, temperature, or conductivity. The sight method consists of a visual observation of the discharge of the steam trap or an inspection of a sight glass installed in the piping before or after the steam trap. With the sound method, a simple mechanic's stethoscope can be used or more sophisticated ultrasonic devices with diagnostic capabilities can be employed. Testing methods using temperature include temperature-indicating crayons, contact thermometers, infrared thermometers, and thermal imagers. The latest method of testing steam traps uses a sensor fitted inside a sensing chamber to detect the presence of steam or condensate by measuring its conductivity.

Accuracy in the testing of steam traps can be affected by several factors. Live steam before and flash steam after the steam trap can make it difficult to determine proper operation. Another factor is the design of the steam trap. Some steam traps maintain a flow of condensate continuously, and some discharge condensate intermittently. The steam trap design also affects the normal inlet and outlet temperature differential. Baseline temperature measurements previously established during normal steam trap operation should be referred to before steam trap testing.

# SECTION 4.4
# CONDENSATE RETURN

## CONDENSATE RETURN TANKS AND PUMPS

In many steam boiler plants, clean condensate is pumped back to the condensate tank or a surge tank. The condensate is usually a considerable distance from the boiler room and cannot flow back by gravity. Accessories used with condensate return systems include condensate tanks, pumps, and surge tanks.

A *condensate return tank* is an accessory that collects condensate returned from the point of use. Condensate tanks and pumps are installed at the lowest points in each building where steam is being used. **See Figure 4-19.** A gauge glass may be attached to the condensate tank to indicate level. The condensate tank collects the condensate from the heating or process condensate return lines. Condensate pumps are used to pump condensate from the tank and discharge it to a surge tank or feedwater heater.

---

**TECH TIP**

*Reusing hot condensate increases efficiency. It recovers heat that would be wasted if the condensate were dumped.*

---

A pressure buildup in this tank is prevented by an atmospheric vent line. Although the condensate tank is vented, it is possible to have a pressure buildup caused by leaking steam traps. When this occurs, it is necessary to trace back, locate, and repair or replace any faulty steam traps.

## CONDENSATE TANKS AND PUMPS

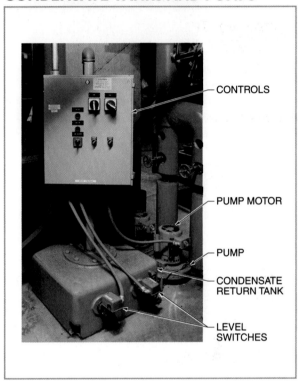

CONTROLS

PUMP MOTOR

PUMP

CONDENSATE RETURN TANK

LEVEL SWITCHES

**Figure 4-19.** Condensate pumps are used to return condensate from various parts of the plant to a condensate return tank and then to a surge tank or a feedwater heater.

## CONDENSATE SURGE TANKS

In larger steam boiler plants, there are many sources from which condensate returns to the boiler room. In these plants, a surge tank is necessary to provide the extra capacity required to handle changing loads and peak flows of condensate. Normally, the surge tank is fitted with raw makeup water fittings and controls along with transfer pumps and controls. A feedwater regulator, also referred to as a pump controller, is used to operate the pump. The surge tank is vented to the atmosphere and is equipped with a manhole for cleaning and repairs. **See Figure 4-20.**

## SURGE TANKS

GAUGE GLASS

LOW-WATER ALARM — LEVEL CONTROLS

**Figure 4-20.** A surge tank provides the extra capacity required to handle changing loads and peak flows of condensate in larger steam boiler plants.

The surge tank discharges to transfer pumps located below the tank. Transfer pumps are used to transfer the mixture of condensate and raw makeup water to the deaerator or directly to the boiler feedwater pumps.

The raw makeup water system generally has a float chamber located at the minimum water level. The float operates a makeup water valve to maintain the minimum water level in the tank. If the water level drops below the minimum water level, a low-water alarm is activated. If the water rises above a maximum acceptable water level, a high-water alarm is activated.

---

### SECTION 4.4—CHECKPOINT

1. What is a condensate return tank?
2. What is a condensate surge tank?

## CASE STUDY—FINDING AN UNDERGROUND LEAK

### Situation

An HVAC technician was assigned to do routine safety checks on several boilers located in various employer-owned buildings around the area. One particular building was a slab-on-grade structure. This meant that there was no crawl space because the concrete floor was poured directly on the ground. A hot water boiler was used for building heating. The boiler was a closed system where all the water returned to the boiler. Makeup water would normally only be used for small losses from leaks and blowdowns.

The piping for this type of system is generally buried in the ground about 12″ inside the perimeter of the slab before the slab is poured. The supply and return piping is stubbed through the floor for connections to the heating radiators located around the inside perimeter for each room.

### Problem

During routine inspection, it was observed that the city water makeup pipe was dripping water. Further observation indicated that the boiler was running continuously. It became obvious that there was a lot of makeup water being used for this small hot water heating system. It was the fall of the year, and the temperatures were cooling off to nearly freezing at night and warming up a few degrees above that during the day, with winter temperatures expected to be much colder. Based on these conditions, the boiler was expected to be operating but not overloaded.

### Solution

Using a pocket thermometer, the technician proceeded to take the ground temperature around the perimeter of the building. A warmer ground temperature was observed along one area of the perimeter. The technician went inside the building to take the concrete floor temperature adjacent to the perimeter area where the ground was warmest. It was difficult to measure the floor temperature with a thermometer, but the technician was able to determine that the floor was in fact warmest near the area where the ground was warmest outside the building.

The technician assumed that a hot water line was leaking under the slab. A work order was made out for the maintenance plumbers to break up the concrete floor and dig up the leaking pipe for repair. The maintenance department used an infrared camera to get a better picture of where the temperature was the warmest so they could minimize the digging they needed to do.

Name _____ Date _____

_____ 1. A line desuperheater is used in conjunction with a ___.
   A. steam trap
   B. draft system
   C. condensate tank
   D. pressure-reducing station

_____ 2. Some plants are equipped with a ___ instead of just a condensate return tank.
   A. ball float separator
   B. condensate surge tank
   C. desuperheater
   D. pressurizing tank

_____ 3. If steam has a quality of ___%, it is 98% vapor and contains 2% water.
   A. 2
   B. 49
   C. 96
   D. 98

_____ 4. Saturated steam is ___.
   A. at a higher temperature than its saturation temperature
   B. at a lower temperature than its saturation temperature
   C. in equilibrium with water at the same temperature and pressure
   D. at the superheater outlet

_____ 5. The two parts of a bimetallic element expand and contract differently when exposed to changes in ___.
   A. temperature
   B. pressure
   C. condensate saturation
   D. steam saturation

_____ 6. A throttling calorimeter is used to directly measure the ___.
   A. temperature of steam
   B. moisture content in steam
   C. rate of heat transfer
   D. flue gas temperature

_____ **7.** The flow of steam maintained through a superheater should never be less than ___%
of boiler capacity.
A. 5
B. 15
C. 25
D. 100

_____ **8.** When water is boiling, the water in the boiler and steam above it are at the same ___.
A. temperature and volume
B. volume and pressure
C. temperature and pressure
D. volume and density

_____ **9.** Water hammer is caused by ___ and can lead to a line rupture.
A. mixing steam and water
B. removing condensate from steam lines
C. starting and stopping a feedwater pump
D. bringing heat up to the saturation point

_____ **10.** A steam trap usually has a ___ installed on the inlet line.
A. thermostat
B. brewing kettle
C. steam strainer
D. pressure gauge

_____ **11.** Steam turbines usually require ___ steam.
A. extraction
B. desuperheated
C. saturated
D. dry superheated

_____ **12.** The air ejectors on the condenser of a steam turbine are used to remove air and other
noncondensable gases from the ___.
A. tubes
B. condenser
C. steam and water drum
D. feedwater surge tank

_____ **13.** Superheater tubes are protected from warping or burning out by ___.
A. circulation of gases of combustion
B. circulation of steam and water
C. a continuous flow of steam
D. circulation of hot air

_____ **14.** When ___ fills an inverted bucket steam trap, the bucket loses buoyancy and sinks to open the discharge valve.
    A. air
    B. steam
    C. sludge
    D. condensate

_____ **15.** Failure to maintain steam flow will result in ___ superheater tubes.
    A. corrosion of
    B. scale buildup in
    C. condensate buildup in
    D. overheating of

_____ **16.** Watertube boilers can be fitted with ___ to increase the heat content of steam.
    A. turbines
    B. superheaters
    C. deaerators
    D. steam traps

_____ **17.** The bellows of a thermostatic steam trap is usually filled with a(n) ___ and water solution to increase the sensitivity of the trap.
    A. salt
    B. ammonia
    C. alcohol
    D. carbon dioxide

_____ **18.** Although the condensate tank is vented, it is possible to have a pressure buildup caused by ___.
    A. leaking steam traps
    B. water hammer
    C. reduced heat transfer
    D. failed safety valves

_____ **19.** A desuperheater is an accessory used to ___ to make it suitable for a process.
    A. vaporize condensate
    B. condense steam
    C. remove heat from superheated steam
    D. add heat to superheated steam

_____ **20.** A steam trap can be tested with a(n) ___ to determine if it is working properly.
    A. pressure gauge
    B. ultrasonic device
    C. ball float
    D. thermodynamic float

_____ **21.** In large steam boiler plants, a(n) ___ is necessary to provide the extra capacity required to handle changing loads and peak flows of condensate.
     A. flash tank
     B. overflow trap
     C. surge tank
     D. open feedwater heater

_____ **22.** A(n) ___ steam trap is a steam trap that has a single movable disc that rises to allow the discharge of air and cool condensate.
     A. thermodynamic
     B. thermostatic
     C. inverted bucket
     D. variable orifice

_____ **23.** Some ___ moisture separators automatically open and close because they use a wax that expands and contracts with temperature changes.
     A. impulse
     B. inverted bucket
     C. thermodynamic
     D. variable orifice

_____ **24.** The raw makeup water system generally has a float chamber located at the ___ water level in a surge tank.
     A. minimum
     B. normal operating
     C. maximum
     D. average

_____ **25.** Transfer pumps are used to transfer the mixture of ___ from a surge tank to the deaerator or directly to the boiler feedwater pumps.
     A. air and gases of combustion
     B. condensate and raw makeup water
     C. steam and salt
     D. raw makeup water and soot

# Chapter 5 Objectives

### SECTION 5.1 — WATER LEVEL CONTROL

- Describe low-water fuel cutoffs.
- Explain the operation and testing of low-water fuel cutoffs.
- Describe water level measurement and water columns.
- Explain shrink and swell.
- Explain the operation of feedwater regulators.

### SECTION 5.2 — FEEDWATER HEATERS

- Describe the purpose of using feedwater heaters.
- Explain deaerators.
- Explain the difference between open and closed feedwater heaters.
- Describe economizers.
- Describe high-pressure receivers.

### SECTION 5.3 — FEEDWATER PUMPS AND LINES

- Explain centrifugal and turbine feedwater pumps.
- Describe reciprocating feedwater pumps.
- Explain variable-speed drives.
- Describe feedwater lines and valves.

## ARROW SYMBOLS

| AIR | GAS | WATER | STEAM | FUEL OIL | CONDENSATE | AIR TO ATMOSPHERE | GASES OF COMBUSTION |
|-----|-----|-------|-------|----------|------------|-------------------|---------------------|
| ⇨ | ⇨ | ⇨ | ⮕ | ⇨ | ⮕ | ⇨ | ⮕ |

**Digital Resources**
ATPeResources.com/QuickLinks
Access Code: **472658**

# CHAPTER 5

# FEEDWATER SYSTEMS

## INTRODUCTION

A feedwater system includes feedwater accessories required for the safe and efficient operation of a boiler. Maintaining the correct level of water in a boiler is critical for safe operation. Feedwater accessories are used to control the supply of water to the boiler. Feedwater regulators maintain a constant water level in a boiler. Accessories also heat and store feedwater. Pumps are used to deliver water to boilers at the proper pressure.

## SECTION 5.1
## WATER LEVEL CONTROL

### LOW-WATER FUEL CUTOFFS

An understanding of the ASME Code requirements is important for the safe operation of equipment in a plant. A low water level can cause boiler explosions. A high water level can cause low efficiencies as well as moisture carryover, which can cause damage to a steam turbine. Following the requirements of the ASME Code is key for those specifying, installing, maintaining, and operating high pressure boiler controls, such as the various types of water level controls installed on a boiler steam drum.

To maintain proper operation and safety of a steam boiler, it is necessary to control the water level in the boiler at the normal operating water level (NOWL). The *normal operating water level (NOWL)* is the water level designated by the manufacturer as being the proper water level for safe boiler operation. By maintaining a consistent water level, fuel consumption and thermal shock to boiler metal can be reduced.

A *low-water fuel cutoff* is a boiler control that secures the burner in the event of a low-water condition. The primary function of a low-water fuel cutoff is to deenergize the burner limit circuit and shut down the burner if the water level in the boiler drops below the safe operating level. **See Figure 5-1.**

## LOW-WATER FUEL CUTOFFS

**Figure 5-1.** A low-water fuel cutoff secures the burner in the event of a low-water condition.

An automatic makeup water feeder normally starts and stops the feedwater flow into the boiler. However, ASME CSD-1, *Controls and Safety Devices for Automatically Fired Boilers,* requires all steam boilers to have at least one low-water fuel cutoff even though the boiler may be equipped with an automatic makeup water feeder. An automatic makeup water feeder can malfunction or there could be an interruption in the water supply. A loss of water can lead to the tubes burning out and/or a boiler waterside explosion.

The low-water fuel cutoff may be independently mounted on the boiler or it may be integrated into the water column on the boiler. The low-water fuel cutoff is located slightly below the NOWL. The top line connects to the highest part of the steam side of the boiler. The bottom line connects to the water side well below the NOWL. A water column blowdown line and valve are used to keep the float chamber free of sludge and sediment.

Other functions may be included on a low-water fuel cutoff. For example, an alarm may sound at the same time the burner is shut down or an ON/OFF control can start and stop the boiler feedwater pump to maintain water at a safe operating level in the boiler. A high-water alarm may sound when the water level in the boiler exceeds a predetermined level. The type of low-water fuel cutoff required is specified by the boiler manufacturer. In general, the larger the boiler, the more sophisticated the low-water fuel cutoff controls.

Some states and jurisdictions require that boilers be equipped with an auxiliary low-water fuel cutoff. **See Figure 5-2.** An auxiliary low-water fuel cutoff is designed and wired in series with the primary low-water fuel cutoff. The auxiliary low-water fuel cutoff is installed slightly below the primary low-water fuel cutoff with separate piping. This allows the auxiliary low-water fuel cutoff to act as a backup if the primary low-water fuel cutoff fails to shut the burner off in a low-water condition.

## AUXILIARY LOW-WATER FUEL CUTOFFS

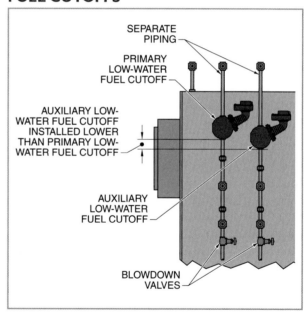

**Figure 5-2.** An auxiliary low-water fuel cutoff provides additional safety.

If only one low-water fuel cutoff is used, it must have a manual reset. If more than one low-water fuel cutoff is used, the lowest low-water fuel cutoff must have a manual reset. Auxiliary low-water fuel cutoffs should be tested at the same time as the primary low-water fuel cutoff.

## Low-Water Fuel Cutoff Operation

Low-water fuel cutoff designs vary, but all low-water fuel cutoffs have the same purpose. The burner continues to fire while the water level in the boiler is at the NOWL. If the water level drops to an unsafe level, the float or sensor in the low-water fuel cutoff opens the electric circuit to shut the burner off. The burner shuts off to prevent damage to the boiler from overheating. The water is still visible in the gauge glass when the burner shuts off.

A float-type low-water fuel cutoff consists of a bowl mounted at the minimum water level in the boiler. The float is connected to a lever that actuates a switch when the float drops below the minimum water level. The switch deactivates the control circuit that allows the fuel to flow.

### TECH TIP

*A low-water fuel cutoff must shut down a boiler in the event of low water. Boiler overheating due to low water is a common cause of damage. Water must not be added to the boiler during low water conditions. It can cause a boiler explosion when feedwater touches the hot surface and flashes to steam.*

A probe-type low-water fuel cutoff performs the same function, except the probe uses the conductivity of the water to complete the burner control circuit. The metal probes vary in length and are suspended in a sleeve or chamber. Current flows through the boiler water in the chamber to complete a circuit. When the water level in the boiler drops below the end of the probe, the circuit is opened and the burner control relay is deenergized. In a high-water condition, a different circuit is energized. In a probe-type low-water fuel cutoff, there are no moving parts within the probe chamber.

A displacer-type low-water fuel cutoff consists of a cylindrical object that is immersed in the water and connected to a spring or torsion device that twists in response to the weight of the cylinder. **See Figure 5-3.** The weight changes according to the amount of the cylinder immersed in the water. The displacer is usually mounted in a housing called a cage that eliminates fluid turbulence. A rod running down the center of the torsion tube transfers the amount of rotation at the sealed end of the torque tube to the various output devices. An output device can be a pneumatic transmitter, a pneumatic controller, or a 4 mA to 20 mA transmitter.

## DISPLACER-TYPE LOW-WATER FUEL CUTOFFS

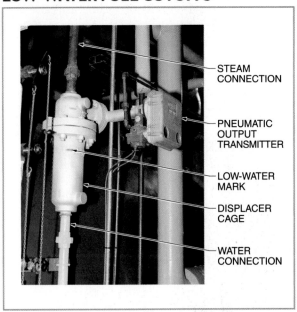

STEAM CONNECTION

PNEUMATIC OUTPUT TRANSMITTER

LOW-WATER MARK

DISPLACER CAGE

WATER CONNECTION

**Figure 5-3.** A displacer is an alternate type of low-water fuel cutoff.

## Low-Water Fuel Cutoff Testing

Low-water fuel cutoff controls are tested by blowing down the control or by performing an evaporation test. The simplest method is to blow down the controls daily or more often. This removes most accumulated sediment and sludge. However, this is not representative of normal operation where the water level in a boiler drops slowly. Therefore, an evaporation test needs to be performed at least every 30 days.

## WATER LEVEL MEASUREMENT

An important step in controlling the water level in a boiler is to measure the water level. A water column is used to minimize movement of the water surface to make it easier to measure. There are several types of instruments used to measure the water level. Alarms are often included in water level measurement systems.

## Water Columns

A *water column* is a boiler fitting that reduces the turbulence of boiler water to provide an accurate water level in the gauge glass. **See Figure 5-4.** The water column also serves as a fitting to which the following may be attached: gauge glass, try cocks, high- and low-water alarms, and gauge glass and water column blowdown lines. Some water columns have pump controls and a low-water fuel cutoff attached. Water columns are made of cast iron, malleable iron, or steel. Cast iron can be used for pressures up to 250 psi, malleable iron for pressures up to and including 350 psi, and steel for pressures higher than 350 psi.

## WATER COLUMNS

**Figure 5-4.** A water column reduces turbulence in the boiler water.

**Water Column Location.** The water column is connected to the highest part of the steam space and to the water side of the boiler. The connecting pipe size must be a minimum of 1″ and must meet the requirements of the ASME Code for both pressure and temperature. Provisions must be made for cleaning and inspection by installing cross tees in pipe connections.

The location of the water column differs slightly in firetube and watertube boilers. **See Figure 5-5.** On firetube boilers, it is located so that the lowest visible part of the gauge glass is 2″ to 3″ above the top of the tubes. On watertube boilers, the water column is located so that the lowest visible part of the gauge glass is 2″ above the lowest permissible water level as specified by the manufacturer. The position of the steam and water connection for a water column is specified in the ASME Code.

## WATER COLUMN LOCATION

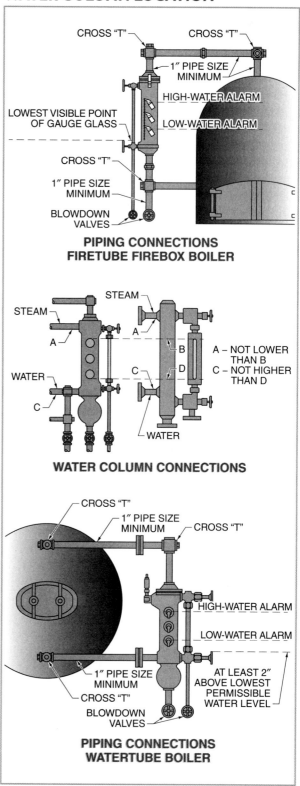

**Figure 5-5.** A water column is attached to the steam space and to the water below the NOWL.

**Valves.** Valves are permitted between the water column and boiler if they are as specified in the ASME Code. The valves must be either outside stem and yoke (os&y) valves or stopcocks with levers permanently fastened. Both types of valves are to be sealed or locked in an open position. Some state and local codes do not allow any valves installed between the water column and the boiler. The appropriate codes should always be checked for valve installation requirements. Lines cannot connect to the water column that would cause a flow of steam or water because this would result in a false water-level reading.

### Level Indicators

Section I of the ASME Code has requirements to ensure water level measurements are accurate. For boilers with an MAWP up to 400 psi, the minimum ASME Code requirement for gauge glasses is to use at least one gauge glass that is in service at all times. For boilers with an MAWP above 400 psi, one of the following must be used:

- two gauge glasses that are in continuous service and visible to the operator
- two remote water level indication systems independent of each other that are continuously displayed, along with a single gauge glass, which may be isolated from continuous operation but must be maintained in an operable condition
- one independent remote water level indication system and one gauge glass, both of which must be in continuous operation at all times

According to the ASME Code, if the boiler operator in a control room cannot see the gauge glass level on the boiler there must be two remote water level indication systems independent of each other that are continuously displayed for the operator. A common type of level measurement device is a gauge glass. Other devices are available.

**Gauge Glasses.** A *gauge glass* is a water level indicator that consists of a glass column that indicates water level. **See Figure 5-6.** A gauge glass is normally attached to a water column by a screw or flanged fitting. Most are equipped with quick-closing stop valves that can be closed in the event of a glass failure. Quick-closing stop valves are frequently chain-operated so that they can be closed from a location where the boiler operator is out of danger. Some gauge glasses have ball check valves incorporated into the body of quick-closing stop valves that close automatically in the event of a gauge glass failure.

## GAUGE GLASS TYPES

**TUBULAR GAUGE GLASS**

**FLAT GAUGE GLASS**

**Figure 5-6.** Two common designs of gauge glasses are tubular and flat glass.

**Try Cocks.** On older boilers, try cocks are a secondary water level indicator. *Try cocks* are valves located on the water column of a boiler that are used to determine the boiler water level if the gauge glass is not functioning. **See Figure 5-7.** Try cocks are used while a broken gauge glass is being replaced or if there is doubt as to the true water level in the boiler. Try cocks are somewhat effective up to 250 psi. Above that point, it is difficult to distinguish between the water and the flash steam that blows out of the try cock. As specified by the ASME Code, if there are two gauge glasses mounted on the boiler, try cocks are not required.

## TRY COCKS

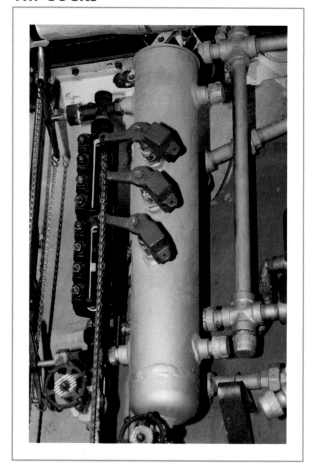

**Figure 5-7.** Try cocks are an alternate method of determining water level and are used on older boilers.

**Alarms.** Alarms may be mounted on a water column to alert the boiler operator of a high-water or low-water condition. Common types of alarms are whistles, bells, and horns. An alarm may be activated by either float or probe actuators.

With a float actuator, the float is connected to a set of switches that make or break the switches, depending on the water level. The actuator may include switches for high water, pump controls, and low water. An alarm sounds in the event of a high-water or low-water condition. A probe actuator performs the same functions, but

with varying-length probes instead of a float. The probes use electrical conductance through the boiler water to make or break switches.

### Water Column and Gauge Glass Blowdown

*Blowdown* is the process of opening valves to blow water or steam through a fitting or from a boiler in order to remove any sludge, sediment, or other undesirable particles. **See Figure 5-8.** The water column and gauge glass should be blown down once every shift or whenever the operator is in doubt as to the true water level. The water column and gauge glass are both equipped with valves used to remove any sludge or sediment from the gauge glass lines and piping connections. A buildup of sludge or sediment can cause a false water level reading.

## BLOWDOWN VALVES

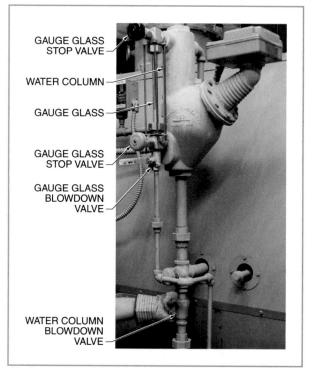

GAUGE GLASS STOP VALVE

WATER COLUMN

GAUGE GLASS

GAUGE GLASS STOP VALVE

GAUGE GLASS BLOWDOWN VALVE

WATER COLUMN BLOWDOWN VALVE

**Figure 5-8.** Gauge glasses and water columns use blowdown valves to remove sludge and sediment.

The water column should be blown down first and then the gauge glass. If the gauge glass is blown down first, the sediment from the water column can create a blockage in the gauge glass stop valves. On boilers equipped with flat gauge glasses, the manufacturer recommendations for blowing down must be followed. Some flat glasses have mica between the glass and steam or water to prevent the steam or water from coming into direct contact with the glass. The mica prevents etching and discoloration of the glass. The mica in the gauge glass can be damaged when steam is blown across it.

## Swell and Shrink

It takes time for the boiler controls to change the rate of steam generation when there are changes in steam demand. When the steam demand increases, steam is removed faster than it is being generated. This results in a temporary drop in steam pressure and causes swell.

*Swell* is a process where the water level in a boiler momentarily rises with an increase in steam demand. When the steam load suddenly increases, more steam leaves the boiler. This causes a sudden decrease in pressure. With the lower pressure, more water turns to steam, making the steam bubbles larger and taking up more space in the boiler. This causes the water level in the boiler to momentarily rise from the loss in steam pressure. A false signal is sent to the feedwater regulator to decrease feedwater flow to the boiler. In this instance, the feedwater regulator should be directing more feedwater to the boiler as more steam is leaving the steam and water drum. Before recovering from the false signal, the boiler water level can fall below the NOWL.

When the steam demand decreases, steam is not removed as fast as it is generated. This results in a temporary increase in pressure and causes shrink. *Shrink* is the process where the water level in a boiler momentarily drops with a decrease in steam demand. This causes a sudden increase in steam pressure. With the higher pressure, less water turns to steam, making the steam bubbles smaller and taking up less space in the boiler. Less steam leaves the boiler and the water level drops, resulting in a false signal to a feedwater regulator to increase feedwater flow to the boiler. Before recovering from the false signal, the boiler water level can rise above the NOWL, which can cause priming and/or carryover.

## FEEDWATER REGULATORS

A *feedwater regulator* is a device that maintains the water level in a boiler by controlling the amount of feedwater pumped to the boiler. The feedwater regulator modulates the amount of feedwater in order to maintain the water level at or close to the NOWL.

## One-Element Feedwater Regulators

A *one-element feedwater regulator* is a feedwater regulator that regulates the amount of feedwater by measuring only the actual water level in the boiler. It does not respond well to shrink and swell. A one-element feedwater regulator is typically used on smaller package boilers.

Basic types of one-element feedwater regulators are float, thermoexpansion, and thermohydraulic feedwater regulators. Float regulators are found on package fire-tube or watertube boilers where feedwater pumps run either intermittently or continuously. Thermoexpansion and thermohydraulic feedwater regulators are found on larger watertube boilers where feedwater pumps run continuously.

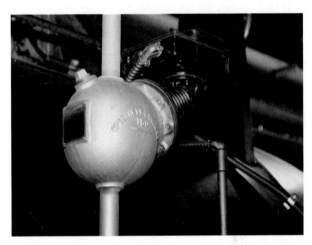

*A float feedwater regulator is used to control the water level in a boiler.*

**Float Feedwater Regulators.** A *float feedwater regulator* is a feedwater regulator that contains a steel or copper float ball connected to a switch by a linkage. The switch is connected to the feedwater pump. A float feedwater regulator consists of a float chamber, a float, and a mercury switch or a microswitch. **See Figure 5-9.** The float chamber is connected to the steam and water side of the boiler. It is located at the NOWL. A blowdown line and valve remove sludge and sediment from the float chamber. Low-water fuel cutoffs are often integrated with this type of regulator.

When the boiler water level drops, the float in the float chamber drops and mechanically actuates the switch. The switch is connected electrically to a feedwater pump starter relay that energizes the pump motor. The pump will continue to run until the float rises and actuates the switch in the opposite direction.

## FLOAT FEEDWATER REGULATORS

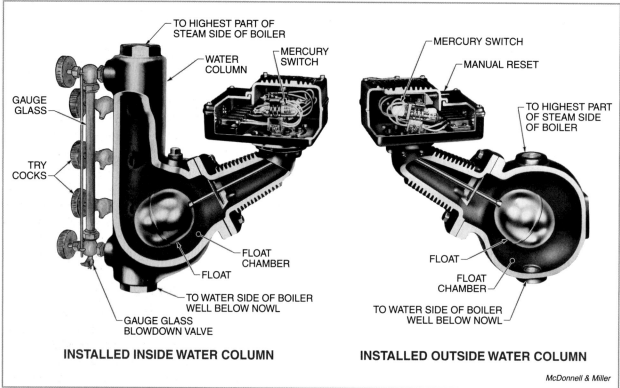

INSTALLED INSIDE WATER COLUMN

INSTALLED OUTSIDE WATER COLUMN

*McDonnell & Miller*

**Figure 5-9.** Float feedwater regulators are installed inside or outside the water column, depending on their design.

With float controls, it is essential that the float chambers be blown down once every shift in order to prevent any sludge or sediment from accumulating. Float feedwater regulators should be opened once a year for cleaning and inspection.

**Thermoexpansion Feedwater Regulators.** A *thermoexpansion feedwater regulator* is a feedwater regulator with a thermostat that expands and contracts with exposure to steam and moves a linkage that modulates a regulator valve. **See Figure 5-10.** The thermostat is located at the NOWL and is connected to the steam and water side of the boiler. If the water level drops, the steam space within the thermostat increases. This increases the temperature, causing the thermostat to expand.

The increase in the length of the thermostat moves the mechanical linkage, which opens the regulator valve. As the regulator valve on the main feedwater line opens, water enters the boiler and increases the water level. The opposite result occurs if the boiler water level is high. The thermostat contracts and the linkage moves the regulator valve toward the closed position.

## THERMOEXPANSION FEEDWATER REGULATORS

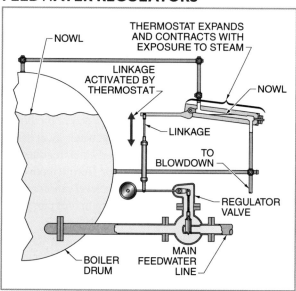

**Figure 5-10.** A thermoexpansion feedwater regulator has a thermostat that expands and contracts with exposure to steam and moves a linkage.

**Thermohydraulic Feedwater Regulators.** A *thermohydraulic feedwater regulator* is a feedwater regulator that has a regulating valve, a bellows, a generator, and stop valves and varies feedwater flow in direct response to changes in the boiler water level by using changes in temperature to create changes in hydraulic pressure. The change in pressure operates the feedwater regulating valve. **See Figure 5-11.** A blowdown valve is used to keep the inner generator tube free of sludge and sediment.

## THERMOHYDRAULIC FEEDWATER REGULATORS

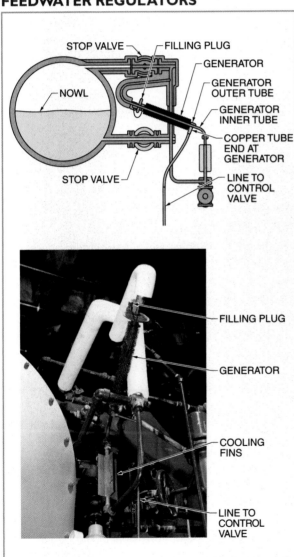

STOP VALVE — FILLING PLUG
— GENERATOR
— GENERATOR OUTER TUBE
— NOWL
— GENERATOR INNER TUBE
— COPPER TUBE END AT GENERATOR
STOP VALVE —
— LINE TO CONTROL VALVE

— FILLING PLUG

— GENERATOR

— COOLING FINS

— LINE TO CONTROL VALVE

**Figure 5-11.** A thermohydraulic feedwater regulator works by creating changes in hydraulic pressure that are then used to operate a feedwater valve.

The control element is the generator, which is a tube within a tube. The generator is located at the NOWL and the inner tube is connected to the steam and water side of the boiler. Stop valves are used to isolate steam and water to the generator. The outer tube of the generator is connected to the bellows with copper tubing and is filled with distilled water. When the level of the boiler water drops, the inner tube of the generator has a larger steam space. The heat from the steam in that space is given up to the distilled water in the outer tube. The expansion of the distilled water causes the bellows to expand.

When the water level in the boiler rises, the reverse action takes place. The connection to the steam and water side of the boiler should be blown down once a month to remove any sludge or sediment. After blowing down the regulator, approximately one hour should be allowed for the regulator to stabilize before being returned to service.

### Two-Element Feedwater Regulators

A *two-element feedwater regulator* is a feedwater regulator that regulates the amount of feedwater by measuring the steam flow from the boiler in addition to the water level. It controls shrink and swell. A two-element feedwater regulator is typically used on larger package boilers with loads that do not fluctuate dramatically but need to be closely followed.

There are several types of sensors used to measure steam flow. A common type of sensor consists of an orifice plate in the steam line and a differential pressure meter. A diaphragm assembly in the differential pressure meter moves in response to a pressure differential. The pressure differential can be used to calculate the steam flow. Other types of steam-flow sensors include venture meters and coriolis meters.

### Three-Element Feedwater Regulators

A *three-element feedwater regulator* is a feedwater regulator that regulates the amount of feedwater by measuring the steam flow from the boiler and the feedwater flow into the boiler in addition to the water level. **See Figure 5-12.** A three-element feedwater regulator is used for closer control of the water level in large watertube boilers where shrink and swell are a problem. Three-element feedwater regulators are effective when there are sudden fluctuations in the steam load and in boilers equipped with economizers where there is a considerable drop in feedwater pressure. A three-element feedwater regulator is typically used with power boilers that generate electricity.

## THREE-ELEMENT FEEDWATER REGULATORS

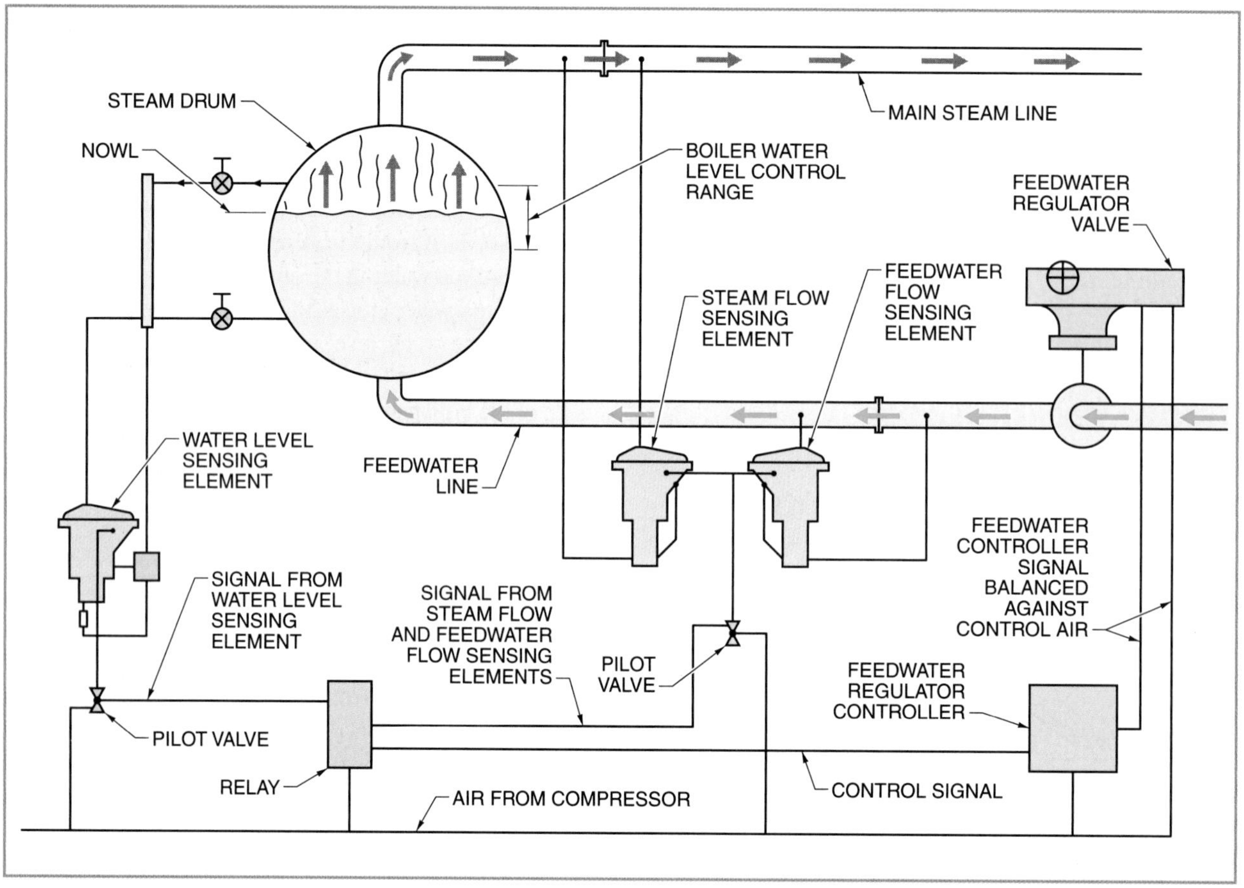

**Figure 5-12.** A three-element feedwater regulator measures the water level in the boiler, steam flow out of the boiler, and feedwater flow into the boiler.

---

### SECTION 5.1–CHECKPOINT

1. What is the purpose of a low-water fuel cutoff?

2. What are the differences between float-type, probe-type, and displacer-type low-water fuel cutoffs?

3. What are the two common methods of testing low-water fuel cutoffs?

4. What is the purpose of a water column?

5. What are the ASME requirements for the number of gauge glasses or other instruments used to measure boiler water level?

6. What is swell?

7. What is shrink?

8. What are the different types of one-element feedwater regulators?

9. What are two-element and three-element feedwater regulators?

---

### SECTION 5.2
# FEEDWATER HEATERS

### FEEDWATER HEATERS

A *feedwater heater* is a device used to heat feedwater before it enters the boiler. A feedwater heater is used to improve energy efficiency by recovering waste heat that would have been lost. A feedwater heater increases boiler plant efficiency when exhaust steam or waste heat is used to heat the feedwater. A rule of thumb is that for every 10°F rise in feedwater temperature, there is a 1% fuel savings.

Feedwater heaters can also improve feedwater quality by removing the dissolved oxygen and carbon dioxide that cause pitting and corrosion in the boiler drum. These gases are much less soluble in hot water than in cold water. As the feedwater heater heats the water, the dissolved gases come out of solution and are vented out of the feedwater heater. The cleaned water is sent to the boiler. The use of a feedwater heater also reduces thermal shock on the boiler metal and reduces boiler metal failure by preheating the feedwater to a temperature closer to that of the boiler water. In addition, a feedwater heater acts as a reservoir for available feedwater.

The raw water entering the feedwater heater contains varying degrees of scale-forming salts. At approximately 150°F, these salts begin to settle. When a feedwater heater is used, some scale caused by the salts will settle in the feedwater heater rather than the boiler. This benefits plant operation since scale is more easily removed from the feedwater heater than from the boiler.

The location of feedwater heaters in a system will vary, depending on the type used. Feedwater heaters can be located on the suction side or discharge side of the boiler feedwater pump or in the breeching. They may be heated by live steam, exhaust steam, or gases of combustion. The basic types of feedwater heaters are deaerators, open feedwater heaters, and closed feedwater heaters. Feedwater can also be heated using economizers.

The correct water level must be maintained in a feedwater heater. A low water level can cause a loss of suction pressure to the feedwater pump. This results in the feedwater pump running dry. A high water level can allow water to enter the steam or condensate lines and cause water hammer. The internal overflow in the feedwater heater controls this condition with a float that opens only when water enters its chamber. The water raises the float, which opens a dump valve that discharges to the sewer.

## Deaerators

A *deaerator* is a feedwater heater that operates under pressure and is used to separate oxygen and other gases from the steam before releasing the gases to the atmosphere through a vent. **See Figure 5-13.** Because a deaerator operates under pressure, the saturated temperature corresponds to the operating pressure. For example, a deareator operating at 5 psi has a saturated temperature of 227°F. Deaerators typically use steam to heat water.

There are two reasons deaerators are used. First, a deaerator is used to hold and preheat boiler feedwater before it is pumped into the boiler. This increases the sensible heat within the boiler water. Second, a deaerator is used to remove dissolved gases from the feedwater. Dissolved oxygen and carbon dioxide are corrosive and attack the boiler metal, steam piping, and condensate return lines. Deaerators can reduce dissolved oxygen levels to as low as 5 ppb, greatly reducing the amount of chemical treatment required. Three basic designs of deaerators are spray, tray, and column.

## DEAERATORS

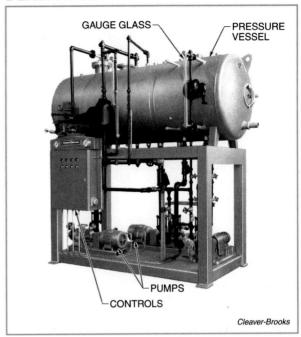

**Figure 5-13.** Deaerators separate air and other gases from feedwater.

## Open Feedwater Heaters

An *open feedwater heater* is a feedwater heater in which steam and feedwater mix with each other at atmospheric pressure to raise the temperature of the water. The external connections consist of a float-controlled makeup control valve on the city water line, a condensate return line, and a low-pressure steam line. **See Figure 5-14.**

The makeup valve is used to maintain a constant water level in the heater. The valve is controlled by a float mechanism that is actuated when insufficient condensate is returned to the heater and the water level starts to drop. Open feedwater heaters typically operate at about 180°F to 200°F. Open feedwater heaters are not as effective as deaerators at removing dissolved gases.

Internally, the steel vessel contains a feedwater distributing line, a set of steel or cast iron trays, and a filter bed. Steam enters the open feedwater heater. At the same time, feedwater and condensate flow onto the trays, where the water passes through small openings before dropping to the filter bed below. In this manner, the water is broken into small droplets and comes into close contact with the steam, thereby increasing the temperature of the feedwater.

## OPEN FEEDWATER HEATERS

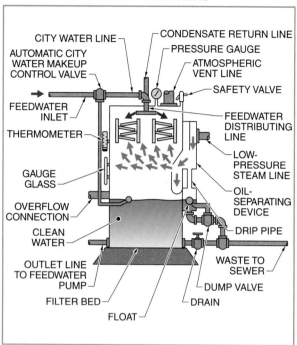

**Figure 5-14.** An open feedwater heater mixes live steam with feedwater and vents steam and noncondensable gases.

An open feedwater heater is located above the feedwater pump and is connected to the suction side of the pump. The location of the heater above the pump gives the suction side of the feedwater pump a positive head (pressure) caused by gravity. With a higher positive head on the suction side of the feedwater pump, it is possible to have a higher feedwater temperature. An increase in pressure will increase the boiling point of the water.

The correct water temperature must be maintained in a feedwater heater. Water that is too hot will cause the feedwater pump to become steambound, resulting in the water flashing into steam within the pump casing and causing cavitation in the feedwater pump. If this occurs, the pump will not discharge feedwater to the boiler, resulting in a dangerously low water level in the boiler. Feedwater temperatures that are too low will not remove the oxygen from the water, which causes pitting of the boiler drum. It can also increase the thermal shock to the boiler.

The heater is also equipped with a safety valve, pressure gauge, and thermometer. All are necessary for efficient and safe operation. The correct water temperature and level must be maintained when operating an open feedwater heating system.

### Closed Feedwater Heaters

A *closed feedwater heater* is a shell-and-tube feedwater heater used to capture heat from continuous blowdown water from a boiler or flash steam from a process. **See Figure 5-15.** A closed feedwater heater is basically a steel shell containing a large number of small tubes secured into two tube sheets. Closed feedwater heaters are designed for vertical or horizontal operations of single or multiple passes. The manufacture and installation of a closed feedwater heater must allow for expansion and contraction of the shell and tubes.

In this type of heater, the steam enters the shell, and feedwater passes through the tubes. The heat in the steam is transferred to the water to raise the water temperature. As the latent heat is removed from the steam, it condenses and falls to the bottom of the shell where it is removed by a steam trap.

This heater must be equipped with a safety valve on the shell and a relief valve on the feedwater side to protect the shell and heads from excessive pressure. A bypass piping arrangement is normally used when taking the heater out of service for cleaning or repairs. The closed feedwater heater is located on the discharge side of the feedwater pump and therefore applies a higher pressure on the feedwater than the open feedwater heater.

This higher pressure allows a higher water temperature without the water flashing into steam. The temperature of the water leaving the feedwater heater is determined by the rate at which feedwater passes through the heater and the steam pressure in the shell of the heater.

The purpose of the closed feedwater heater is to reduce thermal shock on the boiler metal. Correct feedwater temperature will reduce boiler metal failure. In addition, plant efficiency is increased with the use of a closed feedwater heater when steam that is bled or extracted from a steam turbine is used.

### Economizers

An *economizer* is a feedwater heater that heats feedwater by passing it through a finned-tube heat exchanger placed in the path of the gases of combustion. Economizers are used to reclaim heat from the hot gases of combustion. Heat that would normally be lost up the stack is reclaimed directly from the gases. Economizers are typically located in the breeching between the outlet damper and the stack, where gases of combustion make their last pass in the boiler. **See Figure 5-16.** The increase in boiler efficiency can be substantial when using an economizer.

## CLOSED FEEDWATER HEATERS

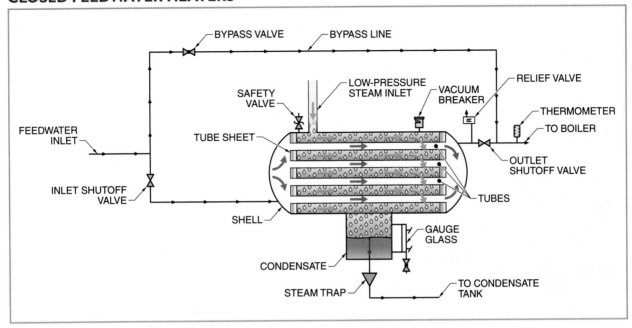

**Figure 5-15.** A closed feedwater heater keeps steam and feedwater separate.

## ECONOMIZERS

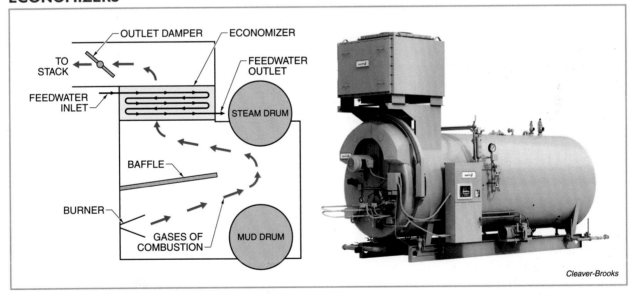

*Cleaver-Brooks*

**Figure 5-16.** In an economizer, feedwater passes through tubes placed in the path of the gases of combustion.

One economizer design is a two-stage condensing economizer. In the first stage, the economizer works like a standard economizer and recovers heat by preheating boiler feedwater with the heat from the gases of combustion. In the second stage, additional heat is recovered by preheating the incoming makeup water before it is fed to the deaerator. A boiler equipped with a condensing economizer and a high-efficiency burner and controlled by a fully integrated control system can achieve over 90% fuel-to-steam efficiency and reduce greenhouse gas emissions by up to 10%.

The temperature of the gases of combustion leaving a standard economizer must be maintained. A condensing economizer can decrease the stack gas temperature below the condensing temperature of water, causing condensate to form in the economizer. Some acidic components of the gases of combustion dissolve in the condensate. For this reason, a baffle is used to divert the condensate to the drain. Alternatively, the economizer coils must be made of a material that will not corrode.

---

**TECH TIP**

*Many high-efficiency boilers operate in condensing mode. The acidic condensate must be drained and not reused.*

---

Sufficient draft is needed to overcome the resistance caused by the economizer tubes. Induced draft fans are usually used in conjunction with forced draft fans to create this draft. Also, the economizer surface must be kept free of soot. Soot retards heat transfer because it insulates the surface of the economizer.

### High-Pressure Receivers

In some process applications, it is possible to recover high-pressure condensate at temperatures over 300°F (about 55 psi). This calls for a high-pressure receiver designed to withstand these pressures in accordance with ASME Code for unfired pressure vessels. High-pressure receivers are normally equipped with boiler feed pumps with high-temperature seals and are the primary unit used to feed the boiler. If makeup water is needed, a deaerator is used to remove dissolved gases before feeding it to the high-pressure receiver.

---

SECTION 5.2 – CHECKPOINT

1. What is a feedwater heater?
2. Why are feedwater heaters used?
3. What is a deaerator?
4. Why are deaerators used?
5. What is an open feedwater heater?
6. What is a closed feedwater heater?
7. What is an economizer?
8. How can an economizer cause condensation in the stack?
9. How does a high-pressure receiver provide hot feedwater?

---

## FEEDWATER PUMPS

A *feedwater pump* is a pump that takes treated water from a tank and delivers it to a boiler at the proper pressure. The pump must deliver the quantity of water necessary to maintain the NOWL in the boiler. Feedwater pumps are normally driven by electricity. In large plants, safety and flexibility may require a backup steam-driven feedwater pump.

The ASME Code states that boilers having over 500 sq ft of boiler heating surface must have at least two means of supplying feedwater to the boiler. Boilers fired with solid fuel (coal) may have one steam-driven feedwater pump. Boilers having large furnace settings (brickwork) that continue to radiate heat and generate steam after the fuel has been secured may also have at least one steam-driven feedwater pump. Feedwater pumps may be the centrifugal, turbine, or reciprocating type. Each of these types has its advantages and disadvantages.

### Centrifugal Feedwater Pumps

Centrifugal feedwater pumps are the most widely used feedwater pump today. A *centrifugal feedwater pump* is a pump that uses the centrifugal force of a rotating element to pressurize water so it can be added to a boiler. **See Figure 5-17.** Centrifugal pumps can be driven by electric motors. These pumps have few moving parts.

**Centrifugal Pump Designs.** Many different types of centrifugal pumps are used on boilers. Generally, the two main types used are the single-stage type and the multiple-stage type.

The basic parts of a centrifugal pump are the casing, impeller, impeller shaft, shaft bearings, and packing glands or mechanical seals. The casing is the outer part of the pump that directs the flow of water through the pump. The impeller is the rotating element that imparts centrifugal force to the liquid. The impeller shaft supports the impeller and transmits rotational force to the impeller from the drive. Shaft bearings support the impeller and shaft in a fixed axial and radial position. The packing glands or mechanical seals prevent leakage of liquid between the casing and shaft.

## CENTRIFUGAL FEEDWATER PUMPS

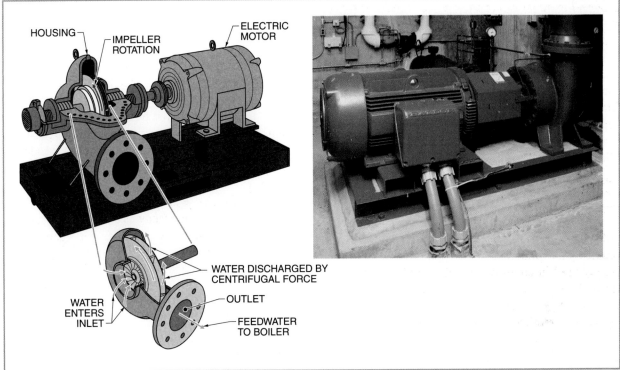

**Figure 5-17.** A centrifugal feedwater pump uses the centrifugal force of a rotating element to pressurize water.

**Centrifugal Pump Operation.** Water enters the rotating impeller at the center and builds up centrifugal force as it moves toward the outer edge of the impeller. Once the water leaves the impeller, the casing guides it to the discharge line. The discharge pressure depends on the centrifugal force that was built up by the impeller.

In a multiple-stage centrifugal pump, the discharge from the first impeller enters the suction of the second impeller and increases the pressure even more. As many as four impellers may be on a single shaft when higher pressures are needed. Each stage of a multiple-stage pump has a single impeller.

Centrifugal pumps will fail to supply sufficient water to a steam boiler if the water temperature is too high, if not enough water is supplied to the pump, or if the pump has a mechanical failure. Mechanical failure can be caused by the impeller shearing from the shaft, the bearings wearing excessively, the coupling that connects the pump and drive shearing off, or the mechanical seal badly wearing or breaking. Repairs to centrifugal pumps must be made by skilled technicians with proper training in pump repair.

### Turbine Feedwater Pumps

A *turbine feedwater pump* is a rotary positive-displacement pump that uses a flat impeller with small, flat, perpendicular fins machined into the impeller rim to discharge feedwater into the boiler. **See Figure 5-18.** Turbine pumps must be equipped with a relief valve on the discharge line to protect the pump from excessive pressure.

### Comparing Centrifugal and Turbine Pumps

A centrifugal feedwater pump should not be confused with a turbine feedwater pump. Although they may look somewhat alike, the basic difference between the centrifugal feedwater pump and the turbine feedwater pump is their design for discharging water. A centrifugal pump uses the shape of the casing to direct the discharge. A turbine pump uses a diffusing ring and casing. The axial clearance between the impeller and casing is smaller in the turbine pump than in the centrifugal pump.

Turbine pumps are positive-displacement pumps. They cannot be started with the discharge valve closed. Centrifugal pumps, however, can be started with the discharge valve closed. Also, turbine pumps have a higher operating efficiency than centrifugal pumps. However, both turbine and centrifugal pumps maintain high efficiencies with minimal maintenance.

### Reciprocating Feedwater Pumps

A *reciprocating feedwater pump* is a positive-displacement pump that uses steam to apply pressure to a large piston in one cylinder connected to a small piston in another cylinder that pumps the water. A specific amount of water is discharged when the piston moves through each stroke. The quantity of water delivered depends on the size of the piston and the length of its stroke.

The discharge valve must be open whenever the pump is running to prevent excessive pressure. A safety relief valve should be installed between the pump and the discharge stop valve to limit the pressure on the discharge side of the pump. Some boiler rooms use these pumps for reserve or back-up units in the event of an electrical power failure. However, backup generators are now commonly used for critical loads like feedwater pumps.

### Variable-Speed Drives

Variable-speed drives have been around for a long time as a way to control the speed of electric motors. A *variable-speed drive (VSD)* is a motor controller used to vary the frequency of the electrical signal supplied to an AC motor in order to control its rotational speed. **See Figure 5-19.**

## VARIABLE-SPEED DRIVES (VSDs)

**Figure 5-19.** A VSD can be used to modulate the speed of a motor that controls a blower or pump.

## TURBINE FEEDWATER PUMPS

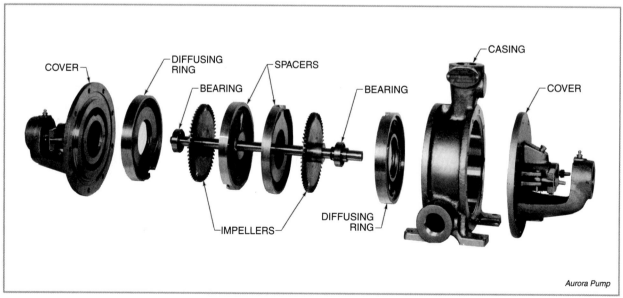

*Aurora Pump*

**Figure 5-18.** Turbine feedwater pumps are positive-displacement pumps and require an open discharge valve when starting.

A VSD on a feedwater pump can reduce energy use because it can precisely control the motor speed. The pump discharge pressure will more efficiently match the boiler operating pressure. The horsepower required is reduced by a factor in proportion to the cube of the speed reduction. In addition, the speed of a pump can be changed in order to deliver the correct amount of water to better match demand.

A conventional motor operates at a constant speed at 60 Hz, as supplied from the electrical utility. Some motors are designed to operate at variable speeds. A VSD converts the 60 Hz supplied from the utility to the frequency that a motor needs to maintain a specific pressure or flow. This is accomplished in the VSD through a rectifier that converts the supplied 60 Hz AC voltage to DC voltage. The DC is then fed to an inverter that converts it back to AC voltage at a different frequency. The new AC voltage and frequency are then supplied to the motor. For example, a constant speed motor may operate at 1800 rpm at 60 Hz but operates at 900 rpm when a frequency of 30 Hz is supplied to it.

---

**TECH TIP**

*In a VSD, the ratio between the voltage applied to a motor and the frequency of the voltage is normally kept constant. However, when operated very slowly at 15 Hz or slower, the voltage to the motor must be boosted to compensate for the power losses of low-speed AC motors.*

---

However, a pump cannot operate at all frequencies. Most pump manufacturers do not recommend operating a feedwater pump below approximately 40 Hz because the pump must be able to overcome the back pressure caused by the boiler's operating pressure. Also, the pump must maintain a minimum flow rate. If the VSD decreases the speed of the motor to the point where the minimum flow rate is not maintained, the pump may be damaged. Typical applications for VSDs in boiler feedwater pumps and deaerator systems are as follows:

• significantly varying boiler operating pressures such as for night setbacks, plant slowdowns, or significant process-load variances

• operating the boiler at a pressure more than 30 psig below the safety-valve setting

• continuously running centrifugal pump applications where the motor horsepower is 15 HP and greater

It is important to select the appropriate feedwater pump for the conditions the pump will encounter during its operation cycle. This will not only reduce the overall cost of a project, but will also significantly extend the life of the mechanical equipment, and it will reduce or eliminate pump maintenance and substantial replacement costs.

## FEEDWATER LINES AND VALVES

A *feedwater line* is the pipeline that carries the feedwater from a feedwater pump to a boiler. This line must have a check valve and a stop valve located near the shell of the boiler. **See Figure 5-20.**

**FEEDWATER LINE VALVES**

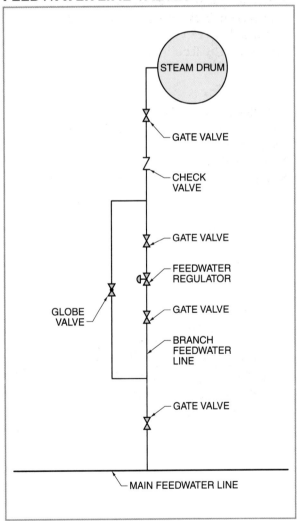

**Figure 5-20.** Many different types of valves are used in a feedwater line.

The stop valve is the valve closest to the boiler. A stop valve is required for isolation in case the check valve fails and requires repair or replacement. Without a stop valve, the boiler would have to be taken off-line to repair or replace the check valve.

Larger boilers use a gate valve as a stop valve. Gate valves are used for isolation only. When two or more boilers are fed from a common source, a gate valve is installed on each branch line between the check valve and the main line.

A globe valve can be used as a stop valve with small, older boilers where the feedwater flow into the boiler is manually controlled. A globe valve is used to meter the flow of feedwater into the boiler. A globe valve must be installed so that the pressure from the feedwater pump is under the seat of the valve. The direction of flow through the valve is marked on the outside of the valve.

A globe valve placed in parallel to the feedwater regulator can be used to regulate the flow of feedwater if the feedwater regulator fails or malfunctions. All valves and piping used on the feedwater lines must conform to the codes for the pressures and temperatures at which they will be operated.

## SECTION 5.3—CHECKPOINT

1. What is the ASME requirement for the number of feedwater pumps?
2. What situations can cause a feedwater pump to fail to supply sufficient water to a boiler?
3. What is a turbine feedwater pump?
4. What is a reciprocating feedwater pump?
5. What is a variable-speed drive?
6. How can a variable-speed drive reduce energy use?
7. What is a feedwater line?
8. What types of valves are commonly used on a feedwater line?

## CASE STUDY—FEEDWATER PUMP NOT PUMPING WATER

### Situation

The boiler operator in a shipyard was responsible for starting, stopping, operating, and logging the shipyard's boilers, air compressors, cooling towers, and generators. This particular shipyard was divided by a river. As the only individual in the maintenance department who had a vehicle, the engineer had to drive between the two areas of the shipyard to deliver tools, parts, and supplies to the two maintenance shops, and was thus kept very busy.

Six stationary boilers monitored at the guard shack were used to heat the buildings as well as for ongoing processes. Local codes required that the monitored boilers be checked and logged every 2 hr. At any given time, there were several portable boilers being used, two owned by the company and the others rented. These boilers were used to supply steam for various tasks in the shipbuilding process or to heat ships when their boilers were being worked on. These non-monitored boilers were required by local codes to be checked and logged every 20 min.

### Problem

One of the company-owned, non-monitored boilers developed a problem and automatically shut off. Every time the boiler was checked, it had shut off because of a low-water condition. At shutdown inspection, the boiler would have no measured pressure. In addition, the water level would be at about half the sight glass, which was normal. The engineer checked the boiler often, but most of the time it was shut off. Each time the boiler was restarted, the engineer would stay as long as possible to observe the boiler.

### Solution

Because the boiler operator was not able to observe the conditions at the time the boiler would shut off, the maintenance plumber was assigned to observe the boiler as time allowed. The plumber thought that the feedwater pump was not able to pump against the head pressure when the boiler was up to pressure. The plumber replaced the pump, but the problem persisted.

One day the engineer was able to observe the boiler shutting down in a low-water condition. The feedwater pump was running, but the water level was low. The engineer discovered the piping from the pump to the boiler was hot, which was not normal. The check valve between the pump and the boiler was leaking. This condition allowed hot water from the boiler to be pushed back through the check valve and into the pump. This had not been observed before because it took time for the boiler to build pressure on startup and for the hot water to fill the feedwater line.

Sometimes the feedwater pump would work properly as the water level dropped because of a demand for steam. This happened when the pump had operated recently and the hot water had not backed up to the pump. Sometimes the feedwater pump would be turned off long enough for the piping and pump to fill with hot water. In this case, the hot water flashed to steam at the inlet of the pump impeller. This caused the pump to become steambound and not pump water.

The check valve was replaced, the low-water cutoff was reset, and the boiler restarted with no further problems.

## CASE STUDY—FEEDWATER REGULATOR PROBLEM

### Situation

The boiler operator in a nursing home and retirement center was responsible for starting, stopping, operating, maintaining, and logging the boilers, air compressors, air handlers, and HVAC equipment throughout the buildings. In addition, the engineer was responsible for the maintenance and repair of the laundry and kitchen equipment.

The boilers were equipped with thermohydraulic feedwater regulators. A thermohydraulic feedwater regulator consists of a regulating valve, a bellows, a generator, and stop valves. The control element is the generator, which is a tube within a tube. The generator is located at the NOWL, with the inner tube connected to the steam side of the boiler. The outer tube of the generator is connected to a bellows by copper tubing and is filled with distilled water. The bellows is attached to the stem of the feedwater regulating valve. When the level of the boiler water drops, the inner tube of the generator has a large steam space. The heat from the steam in that space is given up to the distilled water in the outer tube. The expansion of the distilled water causes the bellows to expand, opening the feedwater regulating valve. When the water level in the boiler rises, the reverse action takes place.

These particular feedwater regulating valves had a handle used to open the valve. The feedwater line to the boiler was about six feet from the boiler room floor. The valve was mounted with the control head down to make it convenient to open the valve by hand. This also put the cover protecting the bellows upside down, allowing dust and dirt to fall between the bellows and the protective cover.

### Problem

The feedwater system operated normally for several years. Over time, the bellows developed a hole and needed to be replaced. After replacing the bellows, the engineer proceeded to fill the closed system (outer tube, copper tubing, and bellows) with distilled water, according to the manufacturer's recommendations. Upon completion, the feedwater regulator was put back into service. However, the regulator did not control the water level in the boiler.

### Solution

The manufacturer's recommendations said that if there was any air in the closed system, the air would compress and the regulator would not control the water level. The engineer tried filling the bellows by submerging it in distilled water and attaching the copper line while still submerged, but that did not work. After several attempts of filling and putting the regulator back into service, it was decided to mount the feedwater regulating valve in the correct position. The closed system was filled with distilled water, the air was evacuated, and the feedwater regulator was put in service. The feedwater regulator performed as designed after that.

**Name** _____ **Date** _____

_____  1. A makeup water control valve is used to ___.
  A. add condensate to the boiler if the water level is high
  B. maintain a constant water level in the feedwater heater
  C. start and stop the feedwater pump on boiler demand
  D. drain the boiler when needed

_____  2. The ___ is the water level designated as being the proper water level for safe boiler operation.
  A. normal operating water level (NOWL)
  B. safe water level (SWL)
  C. design operating water level (DOWL)
  D. low-water fuel cutoff (LWCO)

_____  3. The purpose of a feedwater regulator is to ___.
  A. shut down the boiler in the event of low water
  B. prevent a furnace shutdown
  C. control water level in the feedwater heater
  D. maintain a constant water level in the boiler

_____  4. Feedwater regulators control the feedwater pump to maintain the water level ___.
  A. on the steam line
  B. on the water line
  C. on the blowdown line
  D. at the NOWL

_____  5. ASME CSD-1 requires all steam boilers to have at least ___ low-water fuel cutoff(s) even though the boiler may be equipped with an automatic makeup water feeder.
  A. one
  B. two
  C. three
  D. four

_____  6. ___ are usually used to create the draft needed to overcome the resistance caused by economizer tubes.
  A. Natural draft conditions
  B. Air-flow meters
  C. Induced draft fans
  D. Flow regulators

_____ **7.** An economizer uses the heat energy from the gases of combustion leaving the boiler to heat the ___.
A. air
B. steam
C. fuel oil
D. feedwater

_____ **8.** Turbine feedwater pumps are ___.
A. negative-displacement pumps
B. positive-displacement pumps
C. pumps that can only be steam driven
D. straight mechanical-drive pumps

_____ **9.** A turbine feedwater pump must be equipped with a(n) ___ on the discharge line to protect the pump from excessive pressure.
A. bypass line
B. overspeed trip
C. pump governor
D. relief valve

_____ **10.** Water that is too hot in an open feedwater heater can cause ___.
A. oxygen pitting in the boiler
B. thermal shock to the boiler
C. the feedwater pump to become steambound
D. the feedwater pump to become waterbound

_____ **11.** Water temperatures that are too low in an open feedwater heater will cause ___.
A. oxygen pitting in the boiler
B. cavitation in the feedwater pump
C. the feedwater pump to become steambound
D. the feedwater pump to become waterbound

_____ **12.** The ___ of a centrifugal feedwater pump is the rotating element that imparts centrifugal force to the liquid.
A. casing
B. shaft bearing
C. impeller
D. shaft

_____ **13.** Plant efficiency is increased when a(n) ___ uses steam that is bled or extracted from a steam turbine.
A. economizer
B. closed feedwater heater
C. flue-gas heat recovery unit
D. blowdown heat recovery system

_____ **14.** If only one low-water fuel cutoff is used, it must have a(n) ___.
A. manual reset
B. automatic restart
C. high-level alarm
D. auxiliary backup

_____ **15.** One way to increase the efficiency of a plant is to ___.
A. burn as much fuel as needed
B. use a boiler with the highest capacity needed
C. recover as much heat as possible from the fuel
D. increase the temperature of gases leaving the boiler

_____ **16.** A ___ is a boiler fitting that reduces the turbulence of boiler water to make it easier to measure the water level.
A. water column
B. low-water fuel cutoff
C. try cock
D. flow-control valve

_____ **17.** The piping that connects a water column to the boiler must be a minimum of ___″ and must meet the ASME Code requirements for pressure and temperature.
A. ¼
B. ½
C. 1
D. 2

_____ **18.** ___ is a process where the water level in a boiler momentarily rises with an increase in steam demand.
A. Turbulence
B. Foaming
C. Shrink
D. Swell

_____ **19.** ___ feedwater regulators are found on package firetube or watertube boilers where feedwater pumps run either intermittently or continuously.
A. Float
B. Thermostatic
C. Impulse
D. Thermodynamic

_____ **20.** A three-element feedwater regulator measures the steam flow from the boiler, the ___ into the boiler, and the water level.
A. fuel flow
B. feedwater flow
C. steam pressure
D. orifice flow

_____ **21.** After blowing down a thermohydraulic feedwater regulator, approximately 60 min should be allowed for the regulator to ___ before being returned to service.
    A. stabilize
    B. cycle on and off
    C. cool off
    D. warm up

_____ **22.** For boilers with MAWP over 400 psi, at least ___ method(s) to measure water level must be used at all times.
    A. one
    B. two
    C. three
    D. four

_____ **23.** A ___ pump uses the shape of its casing to direct the discharge.
    A. turbine
    B. reciprocating
    C. centrifugal
    D. positive-displacement

_____ **24.** Feedwater heaters can improve water quality because ___ in warm feedwater.
    A. oxygen dissolves
    B. oxygen comes out of solution
    C. carbon dioxide becomes less reactive
    D. nitrogen comes out of solution

_____ **25.** A(n) ___ feedwater heater allows steam and feedwater to mix with each other at atmospheric pressure to raise the temperature of the water.
    A. horizontal
    B. vertical
    C. closed
    D. open

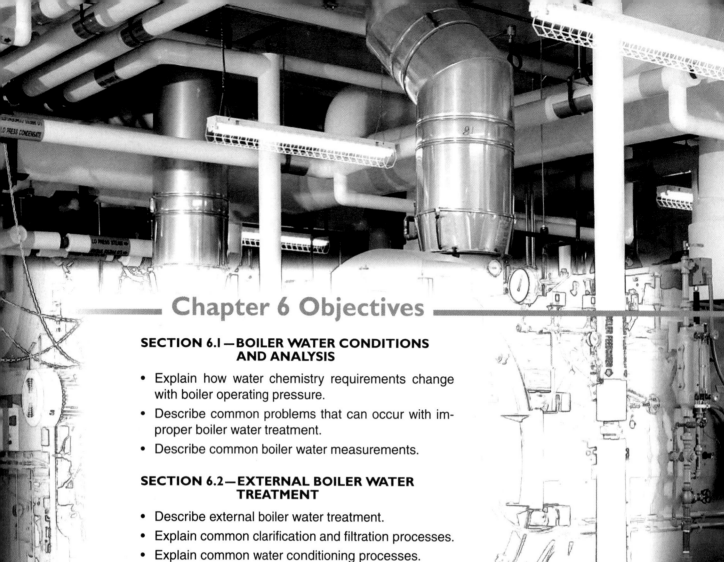

# Chapter 6 Objectives

### SECTION 6.1—BOILER WATER CONDITIONS AND ANALYSIS

- Explain how water chemistry requirements change with boiler operating pressure.
- Describe common problems that can occur with improper boiler water treatment.
- Describe common boiler water measurements.

### SECTION 6.2—EXTERNAL BOILER WATER TREATMENT

- Describe external boiler water treatment.
- Explain common clarification and filtration processes.
- Explain common water conditioning processes.
- Explain deaeration.

### SECTION 6.3—INTERNAL BOILER WATER TREATMENT

- Describe internal boiler water treatment.
- Describe common chemicals used in internal boiler water treatment.
- Explain boiler blowdown.

## ARROW SYMBOLS

| AIR | GAS | WATER | STEAM | FUEL OIL | CONDENSATE | AIR TO ATMOSPHERE | GASES OF COMBUSTION |
|-----|-----|-------|-------|----------|------------|-------------------|---------------------|
| ⇨ | ⇨ | ⇨ | ➡ | ⇨ | ➡ | ⇨ | ➡ |

**Digital Resources**
ATPeResources.com/QuickLinks
Access Code: **472658**

# CHAPTER 6

# WATER TREATMENT

## INTRODUCTION

Water is seldom pure. Rainwater combines with gases and particles as it falls to the ground. Groundwater accumulates dissolved solids such as carbonates of magnesium and calcium as it passes through the ground. Standing water accumulates solid and gaseous industrial pollutants on its surface. Drinking water has dissolved solids that must be removed before the water is used in a boiler. Water must also be tested for common contaminants.

Water used in a boiler must be treated to maintain optimum efficiency, reduce maintenance costs, increase reliability, and extend the useful life of the equipment. Water can be treated before use in the boiler (external treatment) or treated within the boiler (internal treatment). Boiler water is typically treated with a combination of external and internal treatments.

---

## SECTION 6.1
## BOILER WATER CONDITIONS AND ANALYSIS

### BOILER WATER CONDITIONS

Proper water treatment in a steam boiler can minimize impurity deposits, corrosion rates, and carryover of impurities with the steam. Deposits on the heating surface increase the tube wall temperature, potentially leading to tube failure. Corrosion causes damage to parts of the boiler in contact with the water, decreasing the expected life of the boiler. Impurities carried over with the steam can cause damage downstream of the boiler.

### Effect of Boiler Pressure

At higher boiler operating temperatures and pressures, boiler water chemistry has a more significant impact on boiler pressure vessel components. Therefore, tighter control of boiler water is needed. **See Figure 6-1.** Boiler water analysis and treatment become more complicated in high pressure industrial boilers, such as those used in electrical power production. In severe cases, poor control of boiler water chemistry can cause damage to a boiler. Boiler damage can lead to boiler explosions.

### Problems with Improper Treatment

Steam boilers must be protected from boiler water conditions that can occur as a direct result of using improperly treated supply water. Boiler inspectors look for conditions resulting from improperly treated boiler water during annual boiler inspections. Untreated or improperly treated supply water can cause scale, corrosion, caustic embrittlement, priming, carryover, and foaming.

**Scale.** *Scale* is an accumulation of compounds such as calcium carbonate and magnesium carbonate on the water side of the heating surfaces of a boiler. **See Figure 6-2.** Scale formation on tubes and boiler drums is one of the causes of boiler failure. The boiler metal overheats because scale acts as an insulating material between the combustion side of the boiler and the water. Overheating of the boiler metal can cause leaks, cracks, bags, blistering, distortion of tubes and tube sheets, and boiler explosion. Tubes free of scale absorb more heat from gases of combustion, thus increasing boiler efficiency. To prevent scale, calcium and magnesium should be removed from the water.

**Corrosion.** *Corrosion* is the deterioration of the boiler metal caused by a chemical reaction with oxygen and carbon dioxide in the water. **See Figure 6-3.** The metal is dissolved or eaten away by this reaction, causing pitting and channeling of the metal. Corrosion weakens the boiler structurally by thinning the boiler plates and tubes. If left unchecked, corrosion of boiler metal can lead to a boiler explosion. To prevent corrosion, oxygen and carbon dioxide should be removed from the water.

**Caustic Embrittlement.** *Caustic embrittlement* is a type of corrosion in which boiler metal becomes brittle because alkaline materials accumulate in cracks and crevices. Alkalinity of a solution is affected by the amount of carbonates, bicarbonates, and hydroxides present in the solution. Maintaining proper alkalinity of boiler water at all times can prevent caustic embrittlement.

## SCALE

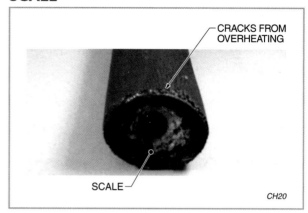

**Figure 6-2.** Scale acts like an insulator. Overheating can damage the heating surfaces.

## CORROSION

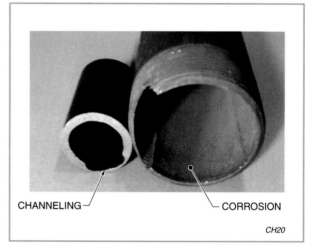

**Figure 6-3.** Corrosion damages metal surfaces.

## PRESSURE EFFECT ON RECOMMENDED WATER CHEMISTRY

| Operating Pressure* | Total Dissolved Solids† | Total Alkalinity† | Silica‡ | Total Suspended Solids† |
|---|---|---|---|---|
| 0 to 300 | 3500 | 700 | 150 | 15 |
| 301 to 450 | 3000 | 600 | 90 | 10 |
| 451 to 600 | 2500 | 500 | 40 | 8 |
| 601 to 750 | 1000 | 200 | 30 | 3 |
| 751 to 900 | 750 | 150 | 20 | 2 |
| 901 to 1000 | 625 | 125 | 8 | 1 |

* in psi
† in ppm
‡ in ppm $SiO_2$

**Figure 6-1.** Tighter control of boiler water chemistry is needed as the boiler pressure increases.

Caustic embrittlement can cause metal to crack along the seams and at the ends of tubes in the boiler. If such cracks occur, the boiler must be taken out of service for repairs. Caustic embrittlement has become relatively rare because modern boilers are now welded rather than riveted. Riveted joints in older boilers are more susceptible to caustic embrittlement because of the number of protruding locations where caustic embrittlement can occur.

**Priming, Carryover, and Foaming.** Dissolved solids and surface impurities such as oil must be minimized to prevent priming, carryover, and foaming. *Priming* is the act of large slugs of water and impurities being carried out of the boiler into the steam lines. **See Figure 6-4.** *Carryover* is the act of small water particles and impurities being carried out of the boiler into the steam lines. Priming is a very dangerous condition because it can lead to water hammer. Water hammer in the steam lines can result in steam header or line rupture. Priming and carryover are the result of high alkalinity, dissolved solids, and sludge in the boiler water.

## PRIMING AND CARRYOVER

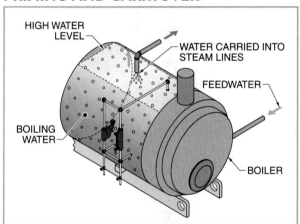

**Figure 6-4.** Priming and carryover can occur when a high water level causes water particles to be carried into steam lines.

*Foaming* is the rapid fluctuation of the water level that occurs when steam bubbles are trapped below a film of impurities on the surface of the boiler water. **See Figure 6-5.** Steam bubbles have difficulty breaking through the film. The water level in the boiler becomes very erratic, causing the sight glass to appear full one moment and empty the next. Maintaining the proper alkalinity of the boiler water prevents priming, carryover, and foaming.

## FOAMING

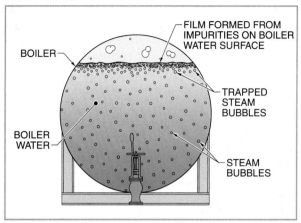

**Figure 6-5.** Foaming occurs as steam bubbles are trapped below a film of impurities on the surface of the boiler water.

## BOILER WATER ANALYSIS

Boiler water analysis is required to determine the condition of boiler water and the necessary external and internal boiler water treatment. **See Figure 6-6.** Boiler water is analyzed for alkalinity, phosphates, sulfite, and total dissolved solids.

## BOILER WATER TESTING

**Figure 6-6.** The specific boiler water treatment required is determined by regular testing of the boiler water.

Boiler water treatment requires periodic testing of the water and adjustment of the chemical treatment according to the test results. Certain minimum levels of chemicals are required in the boiler water to prevent scale formation, pitting, and corrosion of the boiler metal. Chemical concentrations required in the boiler will vary with the temperatures and pressures in the boiler. The manufacturer recommendations will indicate specific feedwater treatment and residual chemical levels appropriate for the boiler water.

Most facilities use an outside firm for advice on the proper boiler water treatment according to the conditions of their specific plant. There are a number of reputable companies prepared to equip plants with suitable test kits and supplies for the proper treatment indicated by the tests. Even though outside companies help with water treatment, plant personnel should have a basic understanding of boiler water analysis and treatment to ensure that water treatment procedures are understood and followed.

### Remote Boiler Water Monitoring

There are many boiler control systems available that remotely monitor boiler water conditions. Most of these systems integrate the functions of a monitor, data logger, and alarm notification. Many of these systems are customizable to the customer's specific needs. System status reports and data files can be accessed along with alarm notifications by phone, text message, or email. As the boiler water conditions change, the values continuously update to indicate the maximum, minimum, and average values. **See Figure 6-7.**

### pH and Alkalinity

The pH of boiler water and condensate can be measured with a pH meter. The *pH* of a substance is a measurement representing whether the substance is acidic, neutral, or alkaline. pH is measured on a scale from 0 to 14. A pH of 7 is neutral. An *alkaline* is any water with a pH greater than 7. An *acid* is any water with a pH less than 7. To test pH, a sample of the boiler water is cooled and drawn into a clean sample jar. A test probe is calibrated, dipped into the sample jar, and the pH is read on the meter. **See Figure 6-8.**

The term "alkalinity" can be confusing to a boiler operator because the term has two meanings when referring to boiler water. The first meaning of alkalinity is the condition where water has a pH above 7. The second meaning of alkalinity is the presence of materials dissolved in water that make the water alkaline.

## REMOTE BOILER WATER MONITORING

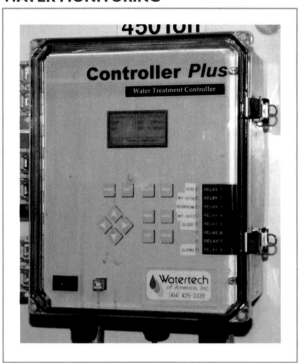

**Figure 6-7.** The results of boiler water testing can be shown at a remote location.

The dissolved materials that are of primary relevance to boiler operation are carbonates, bicarbonates, and hydroxides. Their presence is undesirable because they contribute to chemical reactions that form carbon dioxide. The carbon dioxide dissolves in condensate and forms carbonic acid, which causes corrosion in the condensate return lines. **See Figure 6-9.** Acidic water removes metal, leading to the weakening of the tubes and tube sheets, and causing them to leak. High alkalinity can lead to caustic embrittlement. Caustic embrittlement can cause metal to crack along the seams and at the ends of tubes in a boiler.

Alkalinity is normally measured by a titration test of the boiler water sample and is recorded as P, M, or T alkalinity. A *titration test* is a water treatment test in which a reagent is added to a sample to determine the concentration of a specific dissolved substance. A *reagent* is a chemical used in a chemical test to indicate the presence of a specific substance. P alkalinity is titrated with a phenolphthalein reagent. M alkalinity is titrated with a methyl orange reagent, and T alkalinity is titrated with a total alkalinity reagent. T alkalinity is sometimes called O alkalinity. Once the values for P, M, or T alkalinity are known, a treatment can be determined.

## MEASURING ALKALINITY—pH METERS

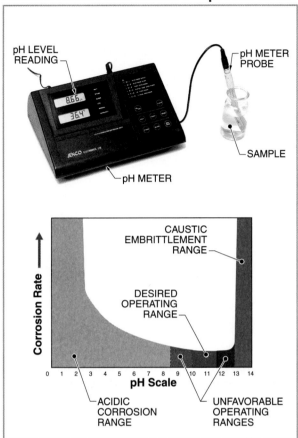

**Figure 6-8.** The pH value should be kept within the range recommended by the boiler manufacturer and the water treatment company.

## CARBONIC ACID DAMAGE

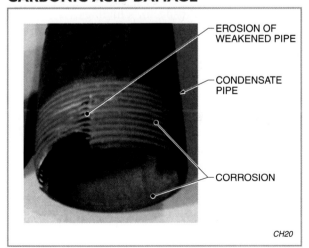

**Figure 6-9.** Carbon dioxide dissolves in condensate and causes corrosion in condensate return lines.

Boiler water pH requirements vary, but the pH value of boiler water should typically be between 8.5 and 11.5 to prevent scale formation. This range is based on a slightly alkaline water.

### TECH TIP

*High boiler water alkalinity may lead to foaming, carry-over, and scale formation. In addition, highly alkaline liquids can etch the inside of sight glasses. Etched glass is hard to see through and can cause erroneous readings.*

### Hardness

*Hardness* is a measurement of scale-forming minerals dissolved in water. *Hard water* is water that contains large quantities of dissolved scale-forming minerals, usually calcium and magnesium. Hardness is commonly checked with an electronic hardness tester. **See Figure 6-10.** A sample is sent through the instrument where a reagent is automatically added to the sample. The reagent causes the sample to change color. The color change is measured and the hardness is indicated on the instrument panel.

Another method of determining water hardness is by a manual titration test. A sample is collected and poured into a titration casserole and a reagent is added to the sample. A color change indicates hardness.

### Total Dissolved Solids

*Total dissolved solids* is a measurement of the concentration of dissolved impurities in boiler water. The total dissolved solids present in boiler water can be measured with a conductivity meter. A *conductivity meter* is an instrument that measures the electrical conductivity of a water sample to determine total dissolved solids present. **See Figure 6-11.** *Conductivity* is the ability of a material to allow the flow of electricity. The amount of solids in water affects the ability of water to conduct electricity. The more dissolved solids in the water, the higher the conductivity. Total dissolved solids are sometimes represented with the acronym "TDS."

### Additional Boiler Water Analysis

There are additional water analysis tests that can be performed. The choice of tests to perform depends on the raw water supply, the condition of the steam system, and the pretreatment equipment used. Water analysis often includes tests for residual chemicals in the boiler water. For example, a test can be performed to measure the concentration of sodium sulfite in the boiler water.

## MEASURING HARDNESS

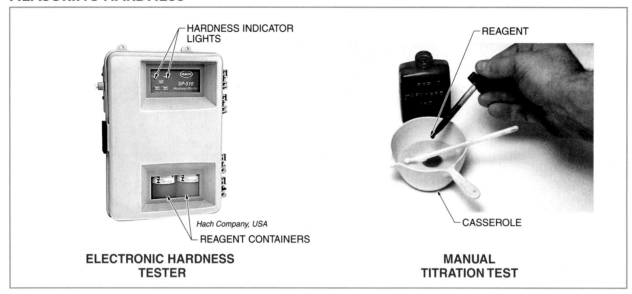

**ELECTRONIC HARDNESS TESTER**

**MANUAL TITRATION TEST**

**Figure 6-10.** Hardness testing indicates the amount of dissolved minerals and is a necessary part of boiler water testing.

## MEASURING TOTAL DISSOLVED SOLIDS—CONDUCTIVITY METERS

**Figure 6-11.** A conductivity meter can be used to measure the total dissolved solids present in boiler water.

### Water Treatment Logs

Boiler water is treated using external and internal treatment according to analysis data and plant water treatment procedures. Boiler water tests and water treatment activities are recorded in a water treatment log similar to a boiler room log. Boiler water samples are tested at the plant or sent to a water treatment company lab for analysis. The most efficient boiler water treatment program is based on accurate daily testing by the boiler operator and regular analysis by a lab. **See Figure 6-12.**

### Water Treatment Companies

Owners and operators of boilers depend on the technical service and expertise of their water treatment company. The service that the water treatment company offers must go further than merely analyzing the boiler water and selling water treatment chemicals. **See Figure 6-13.** The water treatment company should manage the entire water treatment program. The representative responsible for the customer's account must ensure that all parts of the water treatment program are in place. A company representative should be present at all boiler inspections and keep a record of the boiler inspector's findings.

The water treatment company must be carefully chosen to ensure successful protection of the boiler and steam system. At a minimum, steam systems require monthly water analysis. Larger steam systems should be serviced by the water treatment company representative at biweekly or weekly intervals, depending on the usage and makeup of the water used to supply the system. The representative should schedule site visits to analyze the water and make recommendations to ensure correct treatment.

### TECH TIP

*There is no single chemical solution that will address all boiler system water treatment challenges. Water treatment companies can develop a unique plan.*

## WATER TREATMENT LOGS

**WATER TREATMENT LOG**
**Heating and Chilling Plant**
**January**
**Boiler #1**

ALKALINITY TESTS · TOTAL DISSOLVED SOLIDS · SODIUM SULFITE PRESENT · CONTINUOUS BLOWDOWN SETTING AND DURATION · HARDNESS · WATER TREATMENT CHEMICALS

| Date | Boiler Water | | | | | Feedwater | Condensate | | | Products | | | | Blowdown |
|------|----|----|----|------|------------|------|------|-----|------|-----|------|-----|------|----------|
| | P | M | OH | TDS | $Na_2SO_3$ | TDS | pH | TDS | Hard | 938 | 8570 | 960 | 9980 | |
| Max | | | 400 | 3500 | 60 | | 10.8 | | | | | | | |
| Min | | | 200 | 2500 | 30 | | 10.0 | | | | | | | |
| 1 | 320 | 384 | 256 | 2700 | 45 | 45 | 10.8 | 15 | 0 | 44 | 4 | 4 | 12 | 3 / 1 hr |
| 2 | 376 | 440 | 312 | 2900 | 60 | 39 | 10.8 | 16 | 0 | 32 | 8 | 0 | 0 | 3 / 2 hr |
| 3 | 348 | 400 | 296 | 3000 | 55 | 36 | 10.9 | 16 | 0 | 32 | 6 | 0 | 8 | 3 |
| 4 | 324 | 380 | 268 | 2700 | 40 | 39 | 10.5 | 14 | 0 | 32 | 6 | 4 | 4 | 3 |
| 5 | 340 | 392 | 288 | 2900 | 45 | 34 | 10.7 | 14 | 0 | 32 | 6 | 4 | 8 | 3 |
| 6 | 272 | 328 | 216 | 2300 | 45 | 36 | 10.6 | 13 | 0 | 32 | 8 | 4 | 8 | – |
| 7 | 290 | 364 | 228 | 2600 | 35 | 34 | 10.8 | 13 | 0 | 36 | 4 | 4 | 8 | – |
| 8 | 311 | 396 | 336 | 2700 | 50 | 39 | 10.5 | 13 | 0 | 40 | 8 | 0 | 4 | 3 |

**Figure 6-12.** Water condition and treatment data from a water treatment log are used in developing and maintaining a water treatment program.

## WATER TREATMENT CHEMICALS

**Figure 6-13.** The water treatment program required is based on an initial analysis and treatment schedule to ensure protection of the boiler and steam system.

The water treatment service must take into consideration the system design, size, equipment specified, and the ability of the plant personnel to monitor and maintain the system. A representative should monitor the in-house water treatment analysis program performed by the plant personnel. Training sessions should be implemented to familiarize plant personnel with the water treatment system, practices, and procedures.

In addition, the water treatment company should establish the correct procedures and equipment used for adding chemicals to the boiler steam system. The representative should establish a daily water analysis schedule for plant personnel with a feedback system to ensure protection of the boiler and steam/condensate system. The representative should also regularly check the makeup water usage, boiler feedwater usage, water softener usage, and steam condensate return percentage and compare the actual consumption of chemicals against the expected consumption to determine if the usage is reasonable.

---

## SECTION 6.1–CHECKPOINT

1. How do the requirements for boiler water chemistry change with operating pressure?
2. What is the difference between scale, corrosion, and caustic embrittlement?
3. What is the difference between priming, carryover, and foaming?
4. What are common measurements taken while testing boiler water?
5. What is the role of a water treatment company?

---

## SECTION 6.2
# EXTERNAL BOILER WATER TREATMENT

## EXTERNAL BOILER WATER TREATMENT

*External boiler water treatment* is the treatment of boiler water before it enters a boiler. External treatment is commonly used to remove scale-forming salts, oxygen, and noncondensable gases. This allows for easier treatment of the boiler water internally.

For each plant operation, there is an optimum method of treatment. The type and amount of pretreatment used is determined by overall cost. Smaller boilers may not use any pretreatment equipment because of the low cost of low-quality water. A medium-sized package boiler may simply use a water softener and a deaerator. Larger boilers may use several pretreatment systems because it is more cost-effective to pretreat boiler water than it is to treat it internally. This is most important when a large amount of water is being used. External boiler water treatment often includes clarification and filtration to remove suspended solids, various forms of water conditioning to remove hardness, and deaeration to remove oxygen and other dissolved gases.

## CLARIFICATION AND FILTRATION

Water often contains suspended solids. Water clarification and filtration are used to remove these suspended solids. Common types of water clarification and filtration processes are coagulation, filtration, and reverse osmosis.

### Coagulation

*Coagulation* is a process in which the chemicals added to raw water cause suspended solids to adhere to each other, making them larger and heavier and causing them to settle out of the solution. Coagulation reduces the concentration of suspended solids and silt and reduces turbidity of the raw makeup water. Chemicals used for coagulation include aluminum sulfate, sodium aluminate, polymers, and ferrous sulfate. After the solids settle out, they can be removed by filtration.

### Filtration

*Filtration* is the process in which makeup water passes through a filter to remove sediment, particulates, and suspended solids. **See Figure 6-14.** A *filter* is a device that contains a porous substance through which fluid can pass but particulate matter cannot. Different filters are designed for removing specific materials. Filtration is also necessary for any external water treatment process to work properly. Removing these remaining solids is critical, particularly for the prevention of fouling or contamination during water softening.

---

**WATER FILTERS**

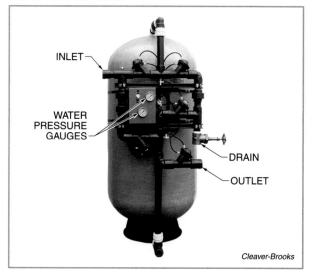

*Cleaver-Brooks*

**Figure 6-14.** Water filters remove chemicals, dissolved organic material, and sediment before water treatment chemicals are added.

## Reverse Osmosis

*Reverse osmosis (RO)* is a filtration process in which the water to be treated is pressurized and applied against the surface of a semipermeable membrane in order to demineralize it and remove its impurities. **See Figure 6-15.** RO filtration removes more material from water than standard filtration. The use of RO can reduce the amount of conventional chemical treatment needed for boiler feedwater. Room temperature water is required to maintain the ability of water to pass through the membranes at a controlled rate.

## WATER CONDITIONING

Water must be conditioned before it can be used in a boiler. A *water softener* is a boiler water pretreatment device used to reduce hardness. Common types of water softeners include ion-exchange softeners and lime-soda softeners. Other water-conditioning devices include dealkalizers, demineralizers, and condensate polishers.

## Ion-Exchange Softeners

An *ion-exchange softener* is a device that uses an ion-exchange resin, typically a zeolite, to exchange a sodium ion for an ion that causes hardness. A *zeolite* is one of a group of minerals including silicates of aluminum, sodium, and calcium that are used in ion-exchange softeners. Zeolites absorb all of the ions of minerals that cause water to be hard at ambient temperature. The device must be regenerated when the sodium in the softener is depleted and the zeolite molecules have hardness ions attached to them.

At the start of the process, each zeolite molecule in the ion-exchange softener has sodium ions loosely attached to it. During normal operation of the ion-exchange softener, water enters at the inlet (1) and passes to the raw water diffusers (2). **See Figure 6-16.** The water is distributed evenly over the resin bed (3) and then moves through the layers until it passes through the strainers (4). Zeolites soften the water by exchanging calcium and magnesium ions in the water for sodium ions from the zeolite. The softened water leaves through the outlet line (5).

To regenerate the softener, it must be removed from service and the resin bed backflushed with a salt solution (brine). The backwash is started by moving the multiport valve (6) to the flush position. This allows water to pass up through the bottom, flushing the resin bed. After flushing, the multiport valve is moved to the brine position. The brine inlet (7) introduces brine into the softener using the brine ejector (8) through the brine distributors (9). The brine slowly moves through the resin, reacting with the calcium and magnesium ions, and replacing the ions on the resin with sodium ions. This restores the resin to its original condition.

The final step in the regeneration of the softener is the rinsing out of excess brine and other impurities. The multiport valve is moved to the rinse cycle position and a slow flow of water is passed through the softener. When the brine is flushed out, the softener can be put back in service. Ion-exchange softeners are also made so that regeneration can be fully automatic. The ion-exchange method of external treatment is preferred in many boiler rooms today because of its compactness and simplicity of operation.

## REVERSE OSMOSIS (RO)

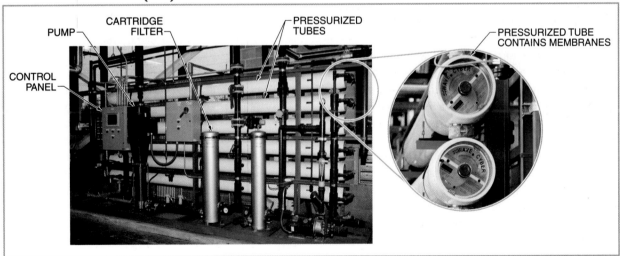

**Figure 6-15.** The RO process uses a pressurized system to force pure water through a semipermeable membrane.

## ION-EXCHANGE WATER SOFTENERS

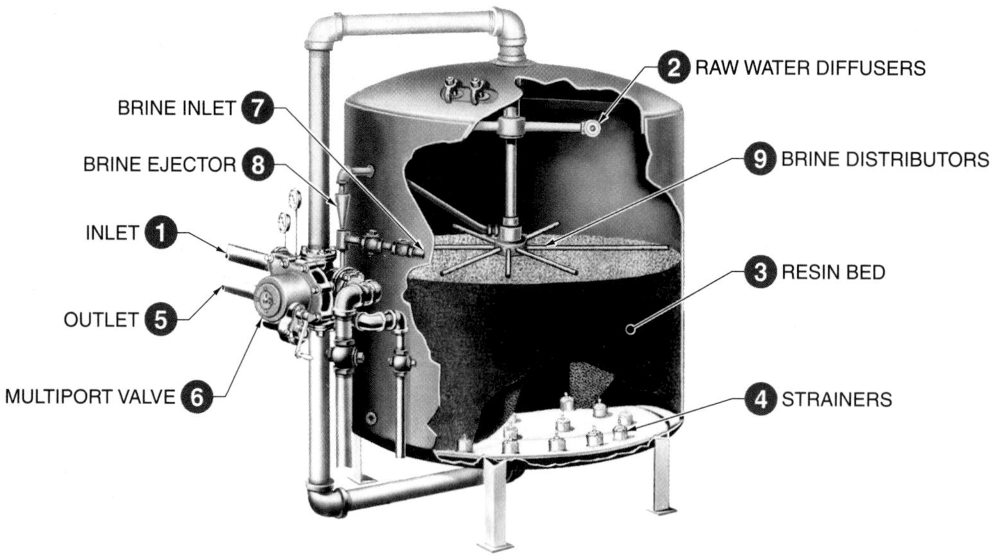

**Figure 6-16.** An ion-exchange water softener is commonly used to soften water.

### Lime-Soda Softeners

A *lime-soda softener* is a water softener that uses lime and soda ash to remove hardness from water. It is primarily used to reduce dissolved calcium, magnesium, and silica. Lime-soda softeners use chemical precipitation to remove dissolved compounds from water. *Chemical precipitation* is a process in which a chemical is added to raw makeup water to react with dissolved minerals, creating heavier particles that settle out of the solution. Lime-soda softeners use lime in the form of $Ca(OH)_2$ (calcium hydroxide) as the chemical that causes the sludge to precipitate out of the solution. The two types of lime-soda softeners are cold lime-soda and hot lime-soda.

**Cold Lime-Soda Softeners.** A cold lime-soda softener is used at ambient temperature. **See Figure 6-17.** During the process, water enters through the raw water inlet (1) and immediately blends with the chemicals introduced at the chemical inlets (2). This design causes a swirling action as the mixture passes through the catalyst bed (3). Sludge deposits onto the surface of the catalyst as the water moves toward the top. The softened water leaves at the softened water outlet (4).

As the catalytic material becomes heavily coated with sludge, it is removed through the draw-off valve (5) located at the bottom of the softener. After the sludge is removed, the water is passed through a filter for further treatment. Fresh catalyst material is added as needed at the top. The test cocks (6) are used to indicate the level of the catalyst material in the softener.

**Hot Lime-Soda Softeners.** A hot lime-soda softener is used above 212°F and is also known as a hot process water softener. This process uses exhaust or live steam to maintain the high temperatures required. The hot lime-soda softener is quicker than the cold lime-soda softener but performs the same function. It softens the water and removes sludge. **See Figure 6-18.**

In a hot lime-soda softener, water enters through the direct-contact vent condenser (1) at the top of the softener. The steam also enters at the top of the unit through the steam inlet (2). The chemical feed pump (3) delivers the chemicals to the upper section of the softener. The steam, chemicals, and raw water mix together while the noncondensable gases are removed through the vent condenser and discharged to the atmosphere.

Water flows to the bottom before rising through the sludge (4) and discharging through the treated water outlet line (5). The water then passes through filters (6) en route to the feedwater heater. The sludge is removed at the bottom blowdown valve (7).

## COLD LIME-SODA PROCESS

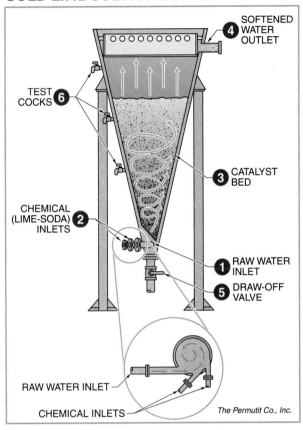

**Figure 6-17.** A cold lime-soda softener operates at ambient temperature to remove precipitated sludge.

## HOT LIME-SODA PROCESS

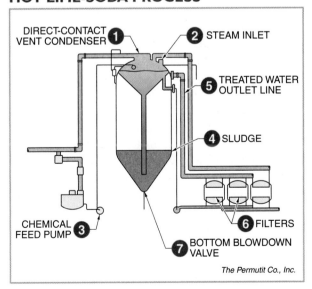

**Figure 6-18.** A hot lime-soda softener operates at a temperature above 212°F to remove precipitated sludge.

## Dealkalizers

A typical ion-exchange softener replaces hardness ions with sodium ions. However, this process does not affect alkalinity. A *dealkalizer* is an ion-exchange water conditioner that removes alkalinity from water. Dealkalizers are best used for steam boilers operating at less than 700 psi with feedwater containing more than 50 ppm of alkalinity and using 1000 gal. or more of makeup water daily. Common types of dealkalizers include chloride-ion, split-stream sodium/hydrogen, and weak-acid cation dealkalizers. A *cation* is a positively charged ion. An *anion* is a negatively charged ion.

### TECH TIP

*Water treatment must happen at the correct flow rate and for the correct period of time. The correct flow rate and time depend on the equipment and water quality.*

**Chloride-Ion Dealkalizers.** A chloride-ion dealkalizer replaces carbonate, bicarbonate, and sulfate ions with chloride anions. It uses a strong-base anion resin to remove bicarbonate, carbonate, sulfate, nitrate, and carbon dioxide from the makeup water. It regenerates the anion resin with salt plus a small amount of caustic solution blended into a brine. All the water entering the dealkalizer must go through a water softener first to remove the calcium and magnesium or fouling of the dealkalizer resin is possible. This type of dealkalizer should be used with caution because chloride ions can cause corrosion on stainless steel vessels.

**Split-Stream Sodium/Hydrogen Dealkalizers.** A split-stream sodium/hydrogen dealkalizer sends part of the water through a sodium zeolite softener and the remainder of the water through a strong-acid zeolite softener. The streams combine after passing through the softener. The sodium zeolite stream contains sodium carbonate and bicarbonate. The strong-acid zeolite stream contains free mineral acids and carbonic acid. The free mineral acids react with the carbonates and bicarbonates to form carbonic acid. Carbonic acid is unstable and converts to carbon dioxide, which can be removed in the deaerator.

**Weak-Acid Cation Dealkalizers.** A weak-acid cation dealkalizer is very similar to a strong-acid unit, but it uses different resins and only removes some of the alkanity. Cations removed include calcium, magnesium, and sodium.

## Demineralizers

Typical ion-exchange softeners may not soften water sufficiently for high pressure boilers. A *demineralizer* is an ion-exchange conditioner that consists of a strong cation-exchange resin and a strong anion-exchange resin and purifies water more than is possible with a zeolite softener. **See Figure 6-19.** A cation-exchange resin exchanges hydrogen ions for metal ions. After the water passes through the cation-exchange resin, it goes through the anion-exchange resin. An anion-exchange resin exchanges hydrogen ions for ions such as chloride and silicate.

## DEMINERALIZERS

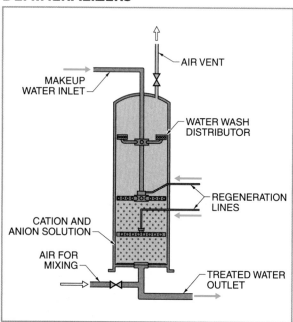

**Figure 6-19.** A demineralizer uses two types of ion-exchange resins for improved water softening.

Demineralization of makeup water is necessary for boilers operating above 1000 psi. Boilers operating at these high pressures require total dissolved solids be kept to a minimum. The possibility of carryover is reduced and the possibility of scale or sludge buildup is eliminated by removing the minerals before they enter the boiler. Demineralization systems are complex and require careful operation and accurate recordkeeping. The basic demineralizers are mixed-bed and two-vessel demineralizers.

Mixed-bed demineralizers have both types of resins within one vessel. Two-vessel demineralizers have two separate vessels for the two types of resins. Makeup water is purified as it passes through the vessel. Demineralizers are relatively expensive and only used when very pure water is needed.

## Condensate Polishers

After steam is used in a process, it forms condensate that is usually returned to the boiler. Steam and condensate can collect impurities that can cause problems in the boiler. Typical impurities in condensate include dissolved silica and suspended iron oxide particles. The concentration of impurities increases as the water is continuously reused in the boiler.

A *condensate polisher* is an ion-exchange conditioner that removes impurities from condensate. Condensate polishers usually use a mixed-bed design where two different types of resins are used. One resin is used to remove cations, like dissolved metal ions. The other resin is used to remove anions, like amines. Polishers usually include filtration within the polisher.

Multiple polishers may be placed in parallel so that some can continue to operate while others can be taken off-line for regeneration. Polishing may be used in steam plants that return condensate from process or heating applications. Polishing is almost always used when condensate is returned from a turbine.

## DEAERATION

Boiler feedwater usually contains some oxygen and carbon dioxide as dissolved gases. If the dissolved gases are not removed, they can cause severe corrosion in the boiler, steam lines, and condensate lines. This damage can be very costly and lead to dangerous conditions.

A deaerator separates oxygen and other gases from feedwater by preheating the feedwater. **See Figure 6-20.** Dissolved gases are less soluble in hot water. A deaerator heats the water to a temperature near the saturation temperature. The dissolved gases come out of solution and are vented to the atmosphere. A continuous vent from the deaerator is necessary to allow the discharge of the noncondensable gases.

A vacuum breaker is included on all deaerators to ensure that a vacuum cannot be accidently created in the deaerator. A safety valve is included to ensure that the deaerator cannot be accidently pressurized above a specified amount.

## DEAERATORS

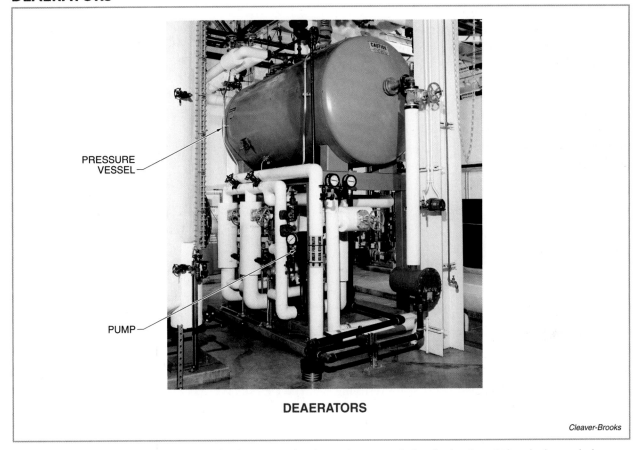

PRESSURE
VESSEL

PUMP

**DEAERATORS**

*Cleaver-Brooks*

**Figure 6-20.** A deaerator removes dissolved gases and reduces the amount of water treatment chemicals needed.

### Agitation

Feedwater is normally agitated in some way to help remove dissolved gases. Agitation increases the amount of surface area of the water to allow more of the dissolved gases to disperse. When there is more surface area, more dissolved gas can escape from the water.

The three common ways to agitate water are with a tray deaerator, a spray deaerator, or a mixed tray-spray deaerator. A tray deaerator causes water to cascade downward over staggered trays. A spray deaerator uses spray nozzles to break the water flow into a spray of water droplets. A mixed tray-spray deaerator uses trays and spray nozzles.

### Vent Condensers

A *vent condenser* is an internal element in some types of deaerators that condenses steam and uses the heat produced to warm the feedwater. **See Figure 6-21.** This type of deaerator uses spring-loaded, self-cleaning spray nozzles to break the feedwater flow into a spray of small water droplets. Steam entering the deaerator is directed into the water spray. The water is heated to within 2°F to 3°F of the steam temperature and agitated, and virtually all oxygen is released.

*Deaerators normally include a safety valve to ensure that the pressure in the deaerator cannot rise above a safe pressure.*

## VENT CONDENSERS

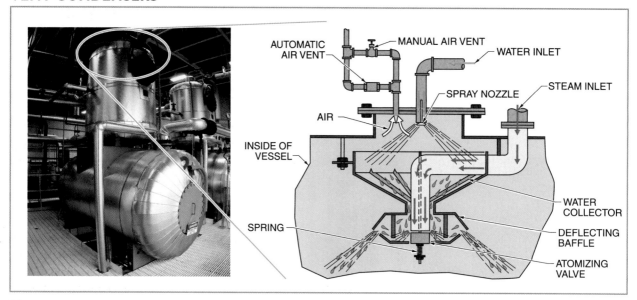

**Figure 6-21.** A vent condenser removes noncondensable gases from the deaerator and returns condensate.

### Operator Responsibilities

A boiler operator is generally responsible for verifying that the water leaving a deaerator is free of oxygen. In addition, an operator may be responsible for verifying that the deaerator is operated at the proper pressure and that the storage tank water is at saturation temperature. Deaerators normally run at about 5 psi with a saturation temperature of 227°F.

In addition to daily responsibilities, the operator should periodically inspect the internal components of the deaerator for corrosion or other damage. This is an important safety duty because a damaged deaerator can rupture. If this happens, much of the water in the deaerator can flash to steam, filling the boiler room with dangerous steam.

---

### SECTION 6.2–CHECKPOINT

1. What is external boiler water treatment?
2. What is the difference between coagulation, filtration, and reverse osmosis?
3. What are common types of water-conditioning devices?
4. How does a deaerator separate dissolved gases from feedwater?
5. What is a vent condenser?

---

### SECTION 6.3
## INTERNAL BOILER WATER TREATMENT

### CHEMICAL TREATMENT

*Internal boiler water treatment* is the treatment of boiler water through the direct addition of chemicals into the boiler to prepare the boiler water for optimum operation. Internal treatment is used to control pitting, scale, and caustic embrittlement. It is also used to condition impurities (hardness) for sedimentation, condition sludge for removal, scavenge remaining oxygen, and prevent foaming. The boiler water must be treated to prevent scale formation on tubes and heating surfaces, corrosion of the shell and tubes, caustic embrittlement at seams, and carryover of boiler water into superheater and steam lines.

Internal treatment methods depend on the condition of the boiler water and on any external water treatment equipment used. Chemicals are added to a steam boiler to prevent boiler water conditions such as scale, corrosion, caustic embrittlement, foaming, priming, or carryover. In most steam boiler plants, chemicals for boiler water treatment can be regulated and added continuously in a diluted form by using a proportioning chemical pump. **See Figure 6-22.** A daily boiler water analysis determines whether the chemical amount needs to be increased or decreased.

## CHEMICAL PUMPS

**Figure 6-22.** A chemical pump is used to add chemicals to the boiler water.

### Caustic Soda and Phosphate

Scale formation in a boiler is a result of improper boiler water treatment. Adding caustic soda and phosphate to the boiler water can prevent scale formation by changing calcium carbonate and magnesium carbonate into nonadhering sludge. The sludge is then removed from the boiler by using the bottom blowdown valves. The level of phosphate in the boiler water is maintained by the addition of phosphates. A color comparison test is used to determine the residual phosphate level in the boiler water.

With properly controlled boiler water treatment, there should be no scale formation regardless of the condition of the supply water. Preventing scale formation using chemical treatment is much easier than removing the scale after it forms on the boiler heating surface. Scale can only be removed by mechanical removal means (steam- or air-driven turbines) or by acid cleaning. Both methods can lead to thinning of the boiler tubes.

### Amines

Heating feedwater before it enters the boiler helps prevent corrosion because heating removes most of the carbon dioxide and oxygen from the water. If the carbon dioxide is not removed, it is released from the boiler water as a gas and passes out of the boiler with the steam. The carbon dioxide then condenses in the condensate lines where it mixes with the condensate and forms carbonic acid. The carbonic acid corrodes the condensate return lines. A neutralizing amine or filming amine is generally added to the boiler water to protect the condensate lines from carbonic acid.

### Sodium Sulfite

Deaerators remove almost all dissolved gases. However, trace amounts of dissolved oxygen may remain in the feedwater. Sodium sulfite is often added to feedwater to remove the remaining oxygen to prevent pitting of the boiler metal. Any oxygen still present in the water combines with sodium sulfite to form sodium sulfate. Boiler blowdown is used to remove the sulfate. When sodium sulfite is used, the boiler water should be tested to verify that there is a residual amount of the sodium sulfite remaining in the water at all times. If there is no sodium sulfite in the boiler water, the remaining dissolved oxygen will react with the metal surfaces and cause corrosion.

**TECH TIP**

*Sodium sulfate is used up to about 1000 psi. An alternative to using sodium sulfate is the chemical hydrazine. However, hydrazine can be toxic. It is rarely used in lower pressure systems and cannot be used in food processing plants.*

### BOILER BLOWDOWN

Two reasons to blow down a boiler are to remove sludge from the bottom of the boiler and to remove contaminants from the water. Removing any water from the boiler means that clean, pure makeup water must be added. Adding makeup water also reduces the overall concentration of contaminants.

During boiler blowdown, a set of valves at the bottom of the boiler are opened, and the flow of water carries the sludge out of the boiler. Contaminants are also removed from the water by opening a surface blowdown valve at the NOWL of the boiler. Surface blowdown removes any floating debris that may have collected.

## Blowdown Frequency

The most accurate means of determining the frequency of blowdown is through boiler water analysis. **See Figure 6-23.** However, a boiler should be blown down, following established procedures, at least once every 24 hr, regardless of the analysis results. Pressure must be on the boiler during blowdown. The only time there is not pressure on the boiler is when the boiler is being drained. The boiler must be cool and the boiler vent open before being drained.

## BOILER WATER TEST KITS

**Figure 6-23.** A boiler water test kit is part of a water analysis program.

Priming, carryover, and foaming are prevented by minimizing the total dissolved solids and surface impurities in the boiler water. The total dissolved solids in the boiler water must be kept within predetermined levels. The total dissolved solids present are regulated by the continuous blowdown valve and/or by using the bottom blowdown valves.

## Bottom Blowdown

*Bottom blowdown* is the process of removing water from the bottom of a boiler in order to remove impurities from the water. In addition, bottom blowdown can be used to control the concentration of contaminants in the water and to drain the boiler for cleaning and inspection.

## TECH TIP

*Foaming can be caused by high concentrations of solids, oil, or other organic contaminants. The potential for foam and carryover can be minimized by proper water treatment and boiler blowdown as required.*

All raw water contains a certain amount of scale-forming salts. The salts settle on the boiler heating surface, insulating the surface and reducing boiler efficiency. The salts can cause the boiler tubes to overheat and burn out. To prevent this, chemicals are added to the boiler water. The chemicals turn these scale-forming salts into a nonadhering sludge. The sludge stays in suspension or settles to the lowest part of the water side of the boiler. The sludge is removed by boiler bottom blowdown. **See Figure 6-24.** The bottom blowdown connection is designed for intermittent use, usually once every 8 hr to 24 hr period.

## BOILER BLOWDOWN

**Figure 6-24.** Bottom blowdown typically uses two valves to remove sludge and sediment from the bottom of a boiler.

## Surface Blowdown

Surface tension on the water in a steam drum is increased by impurities that float on the surface of the water. *Surface blowdown* is the process of removing water from a boiler near the NOWL to control the quantity of impurities in the remaining water or to remove a film of impurities on the surface of the water.

Surface blowdown systems fall into two types. The most common type of surface blowdown system uses a pipe that enters the steam drum slightly below the NOWL. This removes the water that has the highest impurity concentrations near the surface. The other type of surface blowdown system has a swivel joint and a short pipe with a funnel arrangement at the end inside the drum. The open end is maintained at the surface by a float attached to the funnel. This system is used in installations where fuel oil may collect on the surface. The ASME Code specifies the maximum size of a surface blowdown line to be 2½″.

## Continuous Blowdown

Large boilers may use a proportioning type of chemical treatment with a continuous blowdown. *Continuous blowdown* is the process of continuously draining a small, controlled amount of water from a boiler to control the quantity of impurities in the remaining water. Continuous blowdown uses a manual valve that is adjusted periodically to maintain the desired water quality. **See Figure 6-25.**

### CONTINUOUS BLOWDOWN

**Figure 6-25.** Continuous blowdown may use a manual valve that is adjusted periodically to maintain the desired water quality.

The continuous blowdown connection is made at a point just below the low water level in the steam drum. The discharge from the continuous blowdown line goes to the blowdown tank or to a flash tank. With continuous blowdown, it is possible to maintain a fairly constant total solids count in the boiler water. The advantage of this system is that solids are being removed as they are formed in the boiler water.

Continuous blowdown is used in preference to bottom blowdown. Continuous blowdown maintains more consistency in the level of chemicals within the boiler water. It is still necessary to use the bottom blowdown, but this can be done when the boiler is on a light load so that a greater concentration of solid matter can be removed. The water level must be observed, and the opened valves must be continuously manned during a blowdown.

## Automatic Blowdown

*Automatic blowdown* is the process of automatically controlling the amount of boiler blowdown in order to maintain the level of total dissolved solids at the given setpoint. Automatic blowdown controls the frequency and duration of a continuous blowdown operation. The amount of total dissolved solids in the boiler water is used for blowdown control. Conductivity meters can be installed in the continuous blowdown piping to continuously monitor the conductivity of the boiler water.

Automatic blowdown systems prolong the life of boilers, pumps, and other system components by reducing the amount of dissolved solids in the steam drum. This reduces carryover and scale buildup. In addition, compliance with environmental laws has increased the need for better control of blowdown operations. Excessive blowdown results in increased usage of fuel, makeup water, and boiler water chemicals. It also results in excessive amounts of treated boiler water being discharged into the wastewater system. Two types of automatic blowdown are timer-controlled and concentration-controlled automatic blowdown.

**Timer-Controlled Automatic Blowdown.** A timer-controlled automatic blowdown system includes a controller, a sampling assembly, and a conductivity sensor. **See Figure 6-26.** A controller is programmed for the desired blowdown interval and duration. An interval timer allows the controller to open the blowdown valve at a specified interval. A duration timer allows the controller to hold the blowdown valve open for a specified interval. The sampling assembly, in conjunction with the conductivity sensor, controls the flow of boiler water and measures the total dissolved solids present.

## TIMER-CONTROLLED AUTOMATIC BLOWDOWN

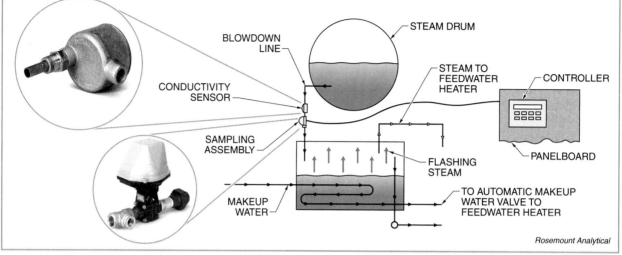

*Rosemount Analytical*

**Figure 6-26.** A timer-controlled automatic blowdown includes a controller, a sampling assembly, and a conductivity sensor.

*A conductivity controller modulates a valve to adjust the automatic blowdown.*

A disadvantage of timer-controlled blowdown is that it does not take into consideration any changes to the feedwater or quality of the returned condensate. In some systems, an instrument can be added that will sense the condition of the boiler water during the timed blowdown.

**Concentration-Controlled Automatic Blowdown.** A concentration-controlled automatic blowdown system includes a conductivity sensor inserted into the boiler water, a controller, and a modulating blowdown valve. The instrument continuously measures the conductivity of the boiler water and sends a signal to the controller. **See Figure 6-27.**

If the measured conductivity is above the desired setpoint, the controller opens the blowdown valve to discharge the boiler water. Feedwater replaces the discharged water, lowering the concentration of the dissolved solids. If the measured conductivity is less than

the desired setpoint, the blowdown valve automatically closes to minimize the loss of boiler water. At a steady state, the blowdown valve can be kept slightly open to allow blowdown to happen continuously in order to maintain the conductivity at the given setpoint. However, with changing loads and varying feedwater quality, the controller may need to modulate the blowdown valve.

### Blowdown Accessories

An *accessory* is a piece of equipment that is not directly attached to a boiler but is necessary for its safe and efficient operation. Blowdown is used to remove sludge and sediment and remove contaminants from the water. Blowdown accessories commonly include lines, tanks, separators, heat recovery systems, and flash tanks.

**Blowdown Lines.** According to the ASME Code, if the MAWP exceeds 125 psi, the blowdown lines and the fittings between the boiler and the valve must be composed of extra heavy bronze, brass, or malleable iron suitable for the temperature and pressures involved. The line must be run full size with no reducers or bushings. Each bottom blowdown line must be fitted with at least two valves, both of which must be extra heavy.

The opening in the boiler setting for the blowdown line must be arranged to provide for free expansion and contraction. In addition, when the bottom blowdown line is exposed to direct furnace heat, it must be protected by firebrick or other heat resistant material. The bottom blowdown line should be installed so that the line can be inspected.

## CONCENTRATION-CONTROLLED AUTOMATIC BLOWDOWN

**AUTOMATIC BLOWDOWN VALVE**

**CONDUCTIVITY SENSOR**

**Figure 6-27.** A concentration-controlled automatic blowdown includes a conductivity meter, a controller, and a modulating blowdown valve.

If the boiler has a heating surface of more than 100 sq ft, the blowdown lines and fittings must be at least 1″ and not more than 2½″. If the boiler has a heating surface of 100 sq ft or less, but more than 20 sq ft, the blowdown lines and fittings must be at least ¾″. If the boiler has a heating surface of 20 sq ft or less, the bottom blowdown lines and fittings must be at least ½″.

**Blowdown Tanks.** The discharge from a blowdown line cannot be connected directly to a sewer system. A direct connection exposes the sewer system to steam and water with high temperatures and pressures. This can cause damage to plumbing systems and injury to people. In order to prevent hot water and steam from entering the sewer system, a blowdown tank must be used between the blowdown line and the sewer system.

A *blowdown tank* is a vented tank that cools blowdown water and steam to protect sewer lines from the high pressures and temperatures of blowdown water. **See Figure 6-28.** If several boilers are connected in battery, multiple tanks may be used or blowdown lines from each boiler in the plant may be connected to the same tank. The hot water and steam enter at the top of the tank. The flash steam leaves through the vent to prevent the tank from pressurizing, and the hot water stays in the tank.

As the level in the tank rises, cooler water from the bottom of the tank overflows through the discharge line into the sewer. The blowdown tank equipment should be of sufficient capacity to prevent discharging water over 150°F and/or 5 psi into the city sewer. It may be necessary to have a cold-water line introduce makeup water into the blowdown tank outlet connection. After the boiler is blown down, the water remains in the blowdown tank to cool until another boiler is blown down. A vent, or siphon breaker, is installed in the overflow line to prevent the water from being siphoned from the tank.

**Blowdown Separators.** Some plants are equipped with a blowdown separator instead of a blowdown tank. A *blowdown separator* is a small tank in which makeup water is added to the boiler blowdown water after it flashes in order to reduce the discharge water temperature. **See Figure 6-29.** The cooling water enters the separator above a diffuser plate, which spreads the water into a shower. The cooling water mixes with the boiler blowdown water to cool it. The water discharged from the separator is sufficiently cooled to meet the local code requirements.

## BLOWDOWN TANKS

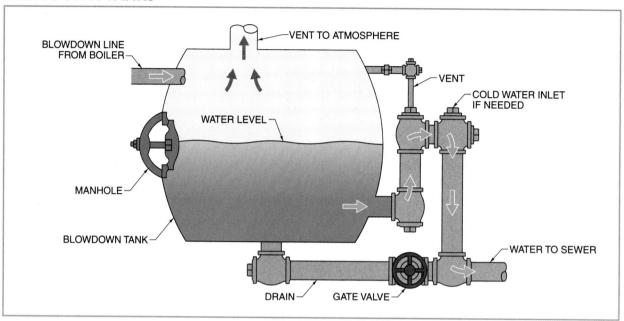

**Figure 6-28.** A blowdown tank is used to cool blowdown water before sending it to the sewer.

## BLOWDOWN SEPARATORS

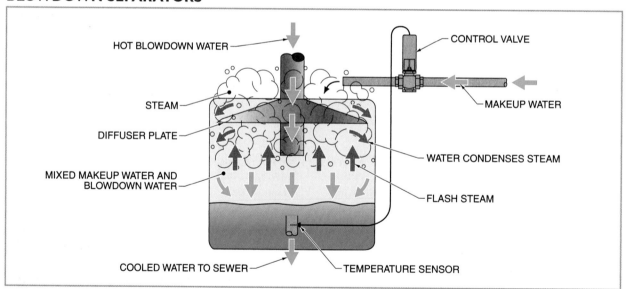

**Figure 6-29.** A blowdown separator cools blowdown water by mixing it with cold water.

**Blowdown Heat Recovery Systems.** A *blowdown heat recovery system* is equipment that is installed to reclaim heat that is normally lost during continuous blowdown. **See Figure 6-30.** As blowdown water enters the blowdown tank, its pressure drops and some of the water flashes into steam. *Flash steam* is the steam created when water at a high temperature experiences a sudden drop in pressure. In a continuous blowdown, the water flows through a heat recovery system. A flash tank is often used as well.

## BLOWDOWN HEAT RECOVERY SYSTEMS

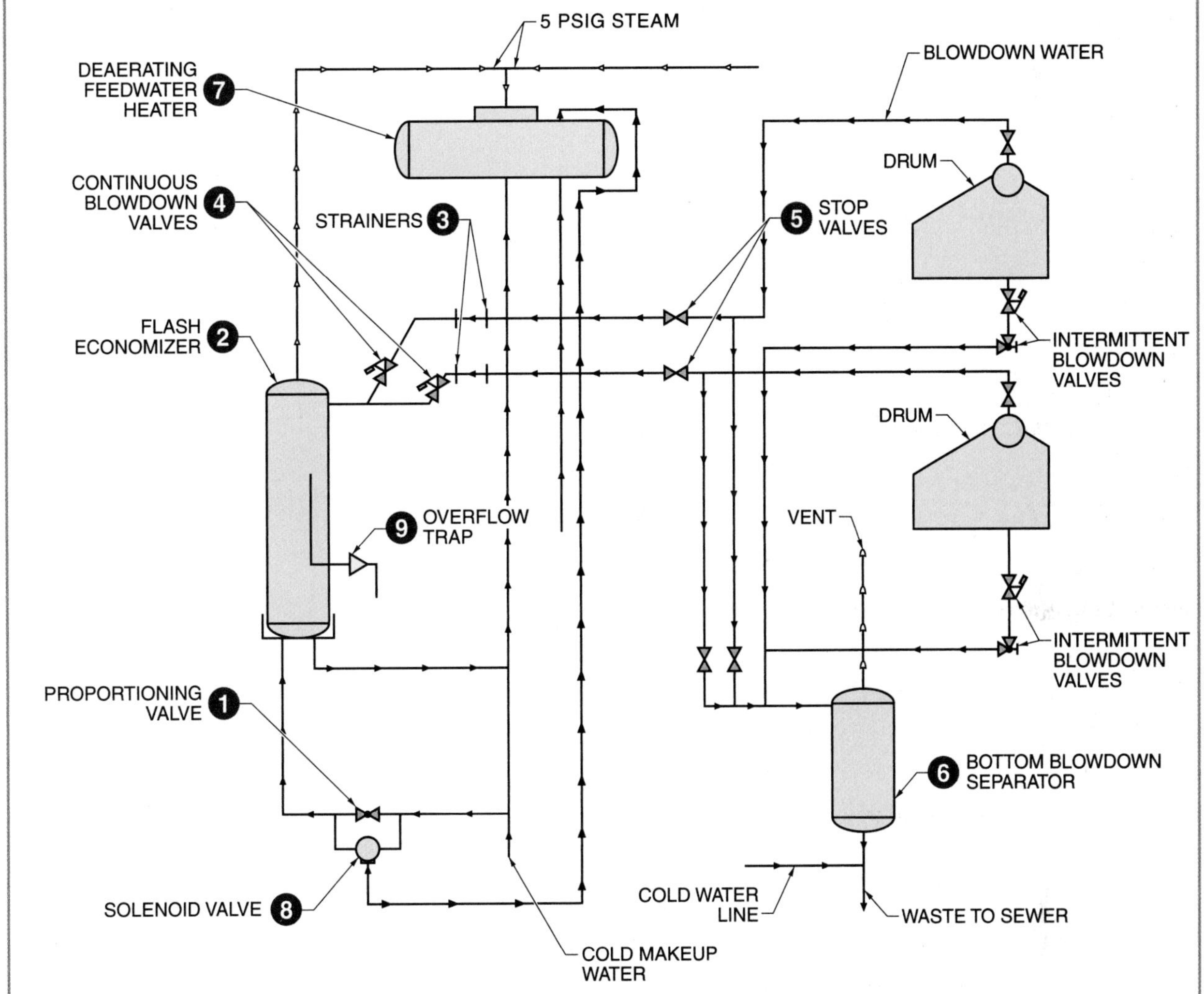

**Figure 6-30.** A blowdown heat recovery system increases the overall efficiency of a steam plant.

In the blowdown heat recovery system, continuous blowdown valves are open during normal operation. The proportioning valve (1) regulates the flow of cold makeup water to the flash economizer (2). Blowdown water flows through strainers (3) and the continuous blowdown valves (4). Stop valves (5) force blowdown water to bypass the flash economizer and flow to the bottom blowdown separator (6) instead. If there is an alarm for low water in the deaerating feedwater heater (7), the solenoid valve (8) opens to increase the amount of water to the heater. The overflow trap (9) removes excess water from the flash economizer.

**Flash Tanks.** A *flash tank* is a tank used with a continuous blowdown system to recover the flash steam from the water being removed during blowdown. **See Figure 6-31.** The heat energy in the flash steam can be recovered when it is sent to a feedwater heater. The makeup water line passes through coils below the water level in the flash tank, picking up heat. If the water level in the flash tank gets too high, the water flows through an internal overflow and a stream trap as it flows to the sewer.

### TECH TIP

*Recovering flash steam from continuous blowdown can often significantly reduce operating costs.*

## FLASH TANKS

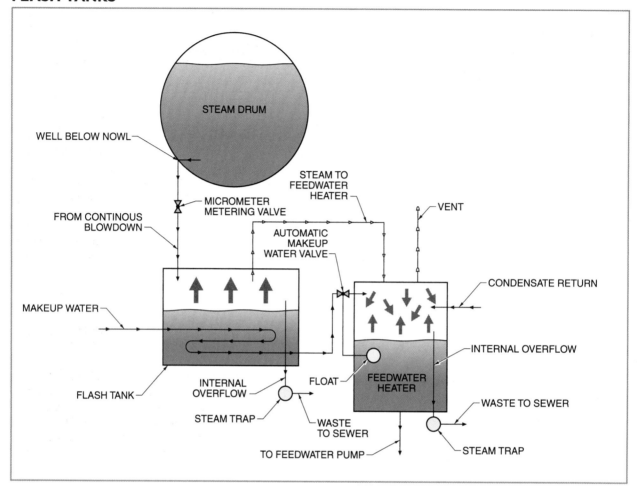

**Figure 6-31.** A flash tank is used to recover useful steam from a blowdown.

*Blowdown heat recovery systems typically use a heat exchanger to transfer heat from hot blowdown water to cooler feedwater.*

### SECTION 6.3–CHECKPOINT

1. What is internal boiler water treatment?
2. How is scale formation prevented?
3. How is corrosion prevented?
4. What is the difference between bottom, surface, continuous, and automatic blowdown?
5. What are common blowdown accessories?
6. What is the difference between a blowdown tank and a heat recovery system?

## CASE STUDY—BOILER WATER TREATMENT PROBLEMS

### Situation

The operating engineer in a nursing home and retirement center was responsible for starting, stopping, operating, maintaining, and logging the boilers, air compressors, air handlers, and HVAC equipment throughout the buildings. The two boilers for this facility were watertube boilers. The fuel for the boilers was natural gas with a heavy fuel oil backup. The fuel oil burner was steam-atomized.

### Problem

Over a long period, some of the controls started to respond more slowly when the burners were operating at a high firing rate. Over the course of a year, the controls continued to react more and more slowly. The affected controls were removed from the boiler and tested. The results were normal.

A boiler controls expert thoroughly inspected and evaluated the controls problem. The steam sensing line from the steam drum to the control panel was nearly plugged with a white residue. The piping was replaced and the controls worked properly.

That summer, the boilers were dismantled for the annual cleaning and inspection. The shift engineer ran water down the water tubes to clean them. However, one of the tubes was completely plugged. After unplugging the tube, scale buildup was found to have broken loose and plugged the tube. The scale was about ⅛″ thick and light brown in color. Chemical analysis determined it was mostly phosphate.

### Solution

A chemical consultant was called in to evaluate the problem. By studying the records for water treatment, the consultant determined that the volume of chemicals had been slowly increased over the years, with little change to the water conditions. It was determined that an inexpensive phosphate was being used as filler in the chemicals used for the boiler water.

The water treatment company was replaced with a new company. The new company was given direction to try to remove the phosphate from the boiler tubes over the following year by using a chemical treatment that would not harm the tubes. The new chemical company agreed it could be done and attempted to remove the scale.

The next annual inspection of the boilers indicated that the new water treatment company was making very little progress. The decision was made to acid clean the boiler tubes. During acid cleaning, the boiler is isolated from the system and drained. It is then refilled with a mild acidic solution and fired periodically to heat and circulate the solution. This process continues for several hours before the acidic solution is drained.

The boiler is refilled with a neutralizing solution and again periodically fired to heat and circulate the solution. The neutralizing solution is then drained, and the boiler is rinsed before being filled with chemically treated water and put back into service. The acid cleaning removes the scale without damage to the boiler tubes, but it is a long and costly process.

**Name** _____ **Date** _____

_____ **1.** A ___ is a vented tank that cools blowdown water and steam to protect sewer lines from the high pressure and temperature of the blowdown water.
    A. blowdown separator
    B. heat recovery system
    C. flash tank
    D. blowdown tank

_____ **2.** Hardness is a measure of scale-forming ___ dissolved in water.
    A. minerals
    B. pollutants
    C. oxygen
    D. carbon dioxide

_____ **3.** Boiler water having high alkalinity could develop ___.
    A. acidic corrosion
    B. oxygen pitting
    C. caustic embrittlement
    D. warping

_____ **4.** Deposits of calcium and magnesium carbonates on a boiler heating surface cause ___.
    A. scale
    B. oxygen pitting
    C. caustic embrittlement
    D. carryover

_____ **5.** Pitting and channeling of boiler metal in the water are caused by ___.
    A. scale
    B. oxygen and carbon dioxide
    C. caustic embrittlement
    D. carryover

_____ **6.** Chemicals added to the boiler water change the scale-forming salts into a(n) ___.
    A. adhering sludge
    B. oxygen scavenger
    C. nonadhering sludge
    D. corrosion inhibitor

_____ 7.  ___ is the act of small water particles and impurities being carried out of the boiler into the steam lines.
    A. Superheated steam
    B. Saturated steam
    C. Carryover
    D. Condensate

_____ 8.  Surface tension on the water in a steam drum is increased by ___.
    A. high-steam loads
    B. impurities that float on the water surface
    C. the use of very soft water
    D. low-steam loads

_____ 9.  A direct cause of overheating of boiler tubes is ___.
    A. soot formation
    B. operating at high fire
    C. insufficient air
    D. scale formation on the tubes

_____ 10. Priming is dangerous and can lead to ___ hammer.
    A. steam
    B. water
    C. gas
    D. air

_____ 11. Scale buildup on the water side of a steam boiler is caused by ___.
    A. an impure water supply
    B. scale in the raw water
    C. improper boiler water treatment
    D. dirty steam

_____ 12. Trace amounts of dissolved oxygen in the boiler water are removed by ___.
    A. adding sodium sulfite to the feedwater
    B. adding amines to the feedwater
    C. using the bottom blowdown valves
    D. using the surface blowdown valves

_____ 13. The required frequency for blowing down a steam boiler is best determined by the ___.
    A. firing rate
    B. boiler inspector
    C. boiler water analysis
    D. plant load

_____ 14. A cold lime-soda softener works at ___.
    A. temperatures of 212°F or above
    B. ambient temperature
    C. temperatures below 32°F
    D. a high, controlled pressure

_____ **15.** ___ is a filtration process in which water is pressurized and applied against the surface of a semipermeable membrane.

    A. Coagulation

    B. Zeolite filtration

    C. Hot process filtering

    D. Reverse osmosis

_____ **16.** External feedwater treatment conditions the boiler water before it enters the boiler using a(n) ___.

    A. scheduled blowdown

    B. amine treatment

    C. caustic soda solution

    D. ion-exchange softener

_____ **17.** A cold lime-soda softener removes hardness by mixing lime-soda with the raw water and passing it through a ___.

    A. catalyst bed

    B. blowdown valve

    C. strainer

    D. vent condenser

_____ **18.** An ion-exchange softener uses a resin bed to exchange ___ with ions that cause hardness in order to soften raw water.

    A. magnesium

    B. sodium

    C. calcium

    D. oxygen

_____ **19.** Alkalinity in boiler water is normally measured using a ___.

    A. steam pressure gauge

    B. conductivity meter

    C. titration test

    D. filtration test

_____ **20.** One reason to perform a bottom blowdown is to ___.

    A. use the heat energy of the blowdown water to preheat feedwater

    B. remove sludge from the bottom of the boiler

    C. reduce the concentration of dissolved oxygen

    D. minimize carryover

_____ **21.** A(n) ___ is generally added to the boiler water to protect the condensate lines from carbonic acid.

    A. sulfite

    B. zeolite

    C. condensate polisher

    D. amine

_____ **22.** A disadvantage of a timer-controlled blowdown is that it does not ___.
  A. consider changes to the feedwater
  B. permit analysis of the blowdown water
  C. remove contaminants from the water
  D. discharge treated water to the wastewater system

_____ **23.** A bottom blowdown should be performed at least once every ___.
  A. day
  B. week
  C. month
  D. year

_____ **24.** A(n) ___ is a feedwater heater that separates oxygen and other gases from feedwater by preheating the feedwater.
  A. ion-exchange softener
  B. surge tank
  C. deaerator
  D. demineralizer

_____ **25.** ___ is the rapid fluctuation of the water level that occurs when steam bubbles are trapped below a film of impurities on the surface of the boiler water.
  A. Water hammer
  B. Pitting
  C. Carryover
  D. Foaming

# Chapter 7 Objectives

### SECTION 7.1—FUEL OIL SYSTEMS

- Explain the operation of the different types of fuel oil burners.
- Describe accessories typically used with fuel oil systems.

### SECTION 7.2—NATURAL GAS SYSTEMS

- Explain the operation of the different types of natural gas burners.
- Describe accessories typically used with natural gas systems.
- Describe combination gas/fuel oil burners.

### SECTION 7.3—COAL SYSTEMS

- Explain the operation of the different types of coal burners.
- Describe ash and explain how it is handled.

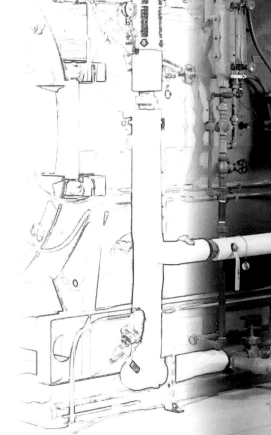

## ARROW SYMBOLS

| AIR | GAS | WATER | STEAM | FUEL OIL | CONDENSATE | AIR TO ATMOSPHERE | GASES OF COMBUSTION |
|-----|-----|-------|-------|----------|------------|-------------------|---------------------|
| ⇨ | ⇨ | ➪ | ➡ | ⇨ | ➡ | ⇨ | ➡ |

**Digital Resources**
ATPeResources.com/QuickLinks
Access Code: **472658**

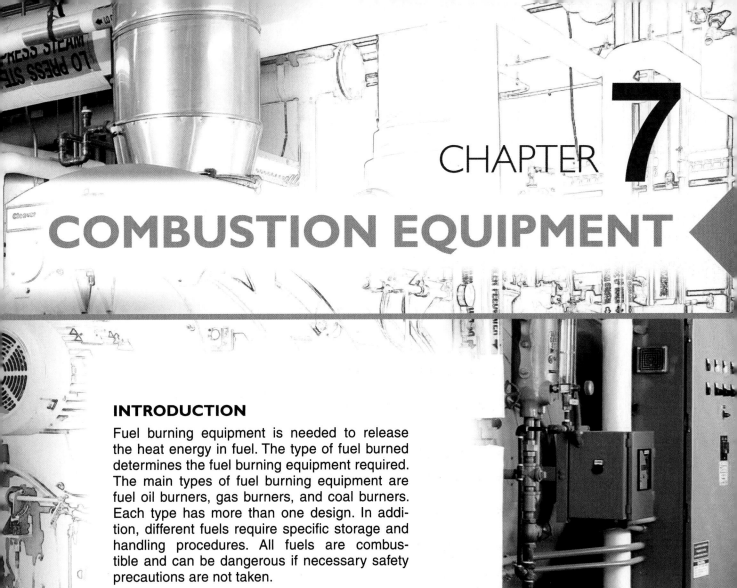

# CHAPTER 7

# COMBUSTION EQUIPMENT

## INTRODUCTION

Fuel burning equipment is needed to release the heat energy in fuel. The type of fuel burned determines the fuel burning equipment required. The main types of fuel burning equipment are fuel oil burners, gas burners, and coal burners. Each type has more than one design. In addition, different fuels require specific storage and handling procedures. All fuels are combustible and can be dangerous if necessary safety precautions are not taken.

## SECTION 7.1
## FUEL OIL SYSTEMS

### FUEL OIL BURNERS

A *fuel oil burner* is a device that delivers fuel oil to a furnace in a fine spray where it mixes with air to provide efficient combustion. When the fuel burns, it releases heat energy that is used to heat the boiler water. Different types of fuel oil burners use different methods to produce the spray. These burner types differ in their accessories, fuel oil pressure, and how they use steam and air to function. The most common types of fuel oil burners used in industrial plants are atomizing burners. Less commonly, rotary cup burners are used.

### Atomizing Burners

The most common method of burning fuel oil is with an atomizing burner. Atomizing burners use a sprayer or nozzle to break the fuel into small particles that mix easily with air for best combustion. Atomizing burner types are named for the method by which they atomize the fuel. Common types of atomizing burners include pressure, steam, and air atomizing burners. After the fuel is atomized and mixed with air, the air-fuel mixture can be ignited by a gas pilot or by a spark from a high-voltage transformer.

**Pressure Atomizing Burners.** A *pressure atomizing burner* is a fuel oil burner that uses pressure to force fuel through a nozzle or sprayer and into the furnace where the vaporized fuel mixes with air and is ignited. **See Figure 7-1.** A pressure atomizing burner is also called a pressure jet burner. The two types of atomizers used in pressure atomizing burners are the plug-and-tip type and the sprayer plate type. The fuel oil enters the burner tube (1) and is forced through the plug channel (2) or the sprayer plate channel (2). In both cases, the atomized fuel oil (3) leaves the tip at a high velocity in a rotating pattern.

## PRESSURE ATOMIZING BURNERS

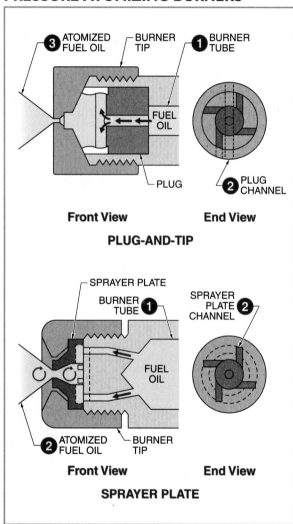

**Figure 7-1.** A pressure atomizing burner requires high temperatures and pressures to atomize the fuel oil.

Both plug-and-tip and sprayer plate atomizers require fuel oil pressures of 100 psi or higher for atomization. Other types of pressure atomizing burners can use a pressure much lower, below about 15 psi.

The amount of fuel oil going to the burner is controlled by the pressure of the fuel oil and the size of the orifice in the burner tip. The firing rate can be adjusted for small changes in boiler load with a bypass line back to the oil tank. Large changes in steam demand require changing the burner tip to a larger orifice size or the use of a multistage burner with two or three nozzles. The amount of air going to the furnace is controlled by a damper or by a variable-speed forced draft blower.

A diffuser located at the end of the burner assembly is used to help mix the air and fuel. An improper flame pattern will cause fuel oil to impinge (hit) the furnace walls, resulting in a buildup of carbon deposits and damage to the firebrick.

**Steam Atomizing Burners.** A *steam atomizing burner* is a fuel oil burner that sends steam and pressurized fuel oil through a nozzle into the furnace where the vaporized fuel mixes with air and is ignited. The two basic types of steam atomizing burners are the outside-mixing and inside-mixing types. With the outside-mixing steam atomizing burner, fuel oil and steam mix outside the burner. With the inside-mixing steam atomizing burner, fuel oil and steam mix inside the burner. **See Figure 7-2.**

## STEAM ATOMIZING BURNERS

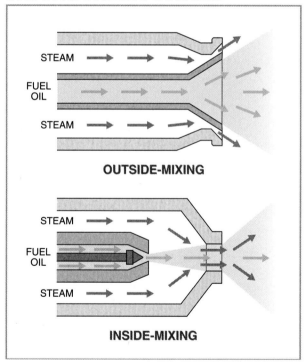

**Figure 7-2.** Steam is used to atomize fuel oil and can mix with the fuel oil outside or inside the burner.

The purpose of the steam is to atomize the fuel oil. The fuel oil is typically supplied to the burner at a pressure of about 10 psi to 50 psi and a temperature of 120°F to 150°F, depending on the grade of fuel oil and burner design. Steam at a pressure 20 psi higher than the fuel oil pressure is usually required for atomization.

Live steam that is used for atomization of fuel oil is obtained directly from the boiler. This live steam is saturated. Steam and fuel oil do not come into contact with each other until the steam is directed across the path of the fuel oil flowing outside or inside the burner nozzle. For efficient operation of a steam atomizing burner, the pressure of the fuel oil and steam must be regulated at all times. The difference in pressure between the fuel oil and steam must be maintained during the changes in steam flow rates.

**Air Atomizing Burners.** An *air atomizing burner* is a fuel oil burner that sends compressed air and pressurized fuel oil through a nozzle into the furnace where the vaporized fuel mixes with air and is ignited. Air atomizing burners are inside-mixing and use air to atomize fuel oil. Air atomizing burners must have an air compressor (air pump) to supply the air needed for atomization. This type of burner is primarily used in smaller installations. Air atomizing burners require inlet pressures ranging from 8 psi to 20 psi.

A low-pressure air atomizing burner uses low-pressure (less than 1 psi) fuel oil. Fuel oil is introduced into the burner as a thin film of oil over the outside surface of an air nozzle. As the air is forced through the nozzle at a high velocity, it induces the fuel oil from the tip and forms a mist of fuel oil in the airstream.

The high-velocity air creates turbulence, which atomizes the fuel oil and mixes it with the air. The mixture passes through a register that causes it to spin into the combustion chamber, filling the chamber for more complete mixing and combustion. The fuel oil does not pass through the nozzle, only the clean high-velocity compressed air passes through. Fuel oil contaminated with water or nonsoluble particles can be burned without plugging the nozzle.

## Rotary Cup Burners

A *rotary cup burner* is a fuel oil burner consisting of a quickly spinning cup that discharges the fuel oil with high-velocity air into the boiler, resulting in a finely atomized fuel that mixes with air and ignites. A rotary cup burner can burn a wide range of fuel oil at low temperatures and pressures. Inlet pressures of about 8 psi to 10 psi are required for rotary cup burners. Rotary cup burners are not allowed in some jurisdictions because of environmental concerns.

A rotary cup burner atomizes fuel oil using a spinning cup and high-velocity air. **See Figure 7-3.** A solenoid valve (1) controls the flow of oil through the fuel tube (2). The fuel tube is the center shaft of the motor that turns the primary air blower and the spinning cup. Fuel oil flows onto the spinning cup (3), which rotates at 3450 rpm, and forms a thin film on the inside surface of the cup.

## ROTARY CUP BURNERS

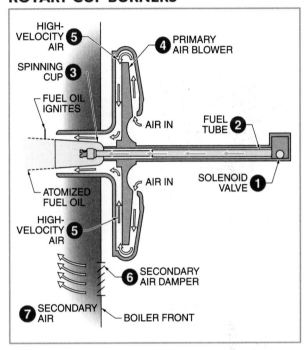

**Figure 7-3.** A rotary cup burner burns fuel oil and is readily adaptable to automatic operation.

The spinning cup increases the velocity of the fuel oil as the oil moves toward the end of the cup. The primary air blower (4) produces high-velocity air (5), which atomizes the fuel oil as it leaves the spinning cup. Air used to atomize fuel oil is primary air. The secondary air damper (6) regulates secondary air (7), which controls combustion efficiency. On smaller package units, the motor also powers the fuel oil pump through a worm gear arrangement.

## FUEL OIL ACCESSORIES

Fuel oil accessories are supporting accessories needed for the safe and efficient operation of the fuel oil burner. Fuel oil accessories store, clean, and control the temperature and pressure of the fuel oil. Common fuel oil accessories include tanks, lines, strainers, pressure gauges, heaters, and pumps.

## Fuel Oil Tanks

A fuel oil tank is used to safely and efficiently store fuel oil in required quantities. The construction, installation, and location of any fuel oil storage tank must conform to local, state, and national regulations. **See Figure 7-4.**

## FUEL OIL TANKS

**Figure 7-4.** A fuel oil storage tank must conform to local, state, and national regulations.

A fuel oil tank has a covered opening used as an entrance for inspection and cleaning. Either a fuel oil heater within the tank or a pipeline heater heats the fuel oil so it can be pumped. The fill line is located on top of the tank. A required vent line is installed on the tank with a minimum pipe size of 1¼″. The vent line prevents pressure buildup when filling the tank and also prevents a vacuum from developing in the tank when fuel oil is removed.

Other openings on the tank are for the connections used to measure the fuel oil level in the tank. A measuring stick dipped into a measuring well is also used to determine the fuel oil level. A chart is used to convert inches on the measuring stick to the number of gallons present in the tank.

## Storage Tank Regulations

An *underground storage tank (UST)* is a tank and any underground piping connected to the tank that has at least 10% of its combined volume underground. The U.S. Environmental Protection Agency (EPA) regulates hundreds of thousands of USTs.

Owners and operators of UST systems are now required to upgrade any underground storage tanks to minimize the possibility of leakage and to demonstrate the financial ability to pay for cleanups if leaks occur. The regulations apply to underground tanks and piping storing petroleum or other hazardous substances.

**TECH TIP**

*Old fuel oil tanks are a source of leaks and spills and can lead to environmental damage.*

EPA regulations specify leak detection and monitoring requirements to ensure there is no ground or water contamination from filling operations or leaking tanks and lines. All new USTs have spill, overfill, and corrosion protection. Spill protection requires the use of a spill bucket to catch spills at the fill pipe when the delivery truck hose is disconnected. Overfill protection requires the use of equipment such as an automatic shutoff device, an overfill alarm, or a ball float valve to restrict or stop the flow of fuel during delivery before the tank reaches full capacity.

Corrosion protection is required on steel USTs, which are subject to corrosion of the tank and piping. This equipment includes cathodic protection and tank interior lining protection. Cathodic protection uses the flow of electrical current to protect against corrosion.

An alternative to a UST is an aboveground storage tank (AST). An *aboveground storage tank (AST)* is a tank and any underground piping connected to the tank that has less than 10% of its combined volume underground. ASTs are also subject to federal, state, and local regulations. In addition, most ASTs need to meet the EPA requirements for spill prevention, control, and countermeasure (SPCC).

## Fuel Oil Lines

Fuel oil lines are used to connect the various parts of a fuel oil system and to transport fuel oil safely and efficiently. **See Figure 7-5.** High or low suction lines can be used to supply the burner from the tank, but the low suction line is used under normal conditions. However, if water or sludge is found in the fuel oil, the high suction line is used until the tank is pumped out and the cause of the problem is identified.

The fuel lines and equipment located on the discharge side of the fuel oil pump must be protected against excessive fuel oil pressure by relief valves. The relief valves discharge excess fuel oil, bypassing the burner to return lines feeding into the tank. Return lines are used to circulate fuel oil that is being heated or fuel oil that bypasses the burner.

## FUEL OIL LINES

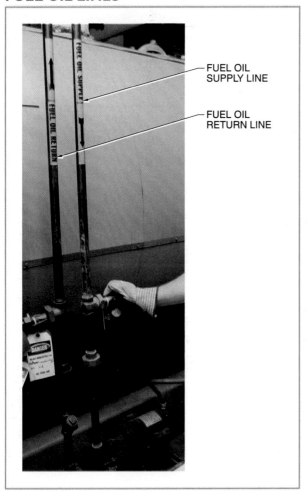

**Figure 7-5.** Fuel oil lines are used to transport fuel oil safely and efficiently.

### Fuel Oil Strainers

A fuel oil strainer is used to remove foreign matter in a fuel oil system before the fuel oil reaches the burner assembly. **See Figure 7-6.** Fuel oil, although it has been refined, still has impurities. If not removed, these impurities will settle in the fuel oil heater. Variations in the purity of the fuel oil may cause buildup of foreign matter in the lines of the burner, causing restricted flow and clogged burner tips, both resulting in boiler downtime.

Two sets of duplex fuel oil strainers are used in many installations to remove foreign matter in a fuel oil system. The first set is located on the suction side of the fuel oil pump. It consists of a machined casting that houses both strainers, each with a coarse

mesh basket screen. The fuel oil enters the top of the basket and passes through the screen before entering the suction line to the fuel oil pump. To gain access to a basket screen, the cover of the strainer must be removed. The selector lever controls a plug valve, allowing the use of one strainer while the other is removed from service for cleaning.

## DUPLEX FUEL OIL STRAINERS

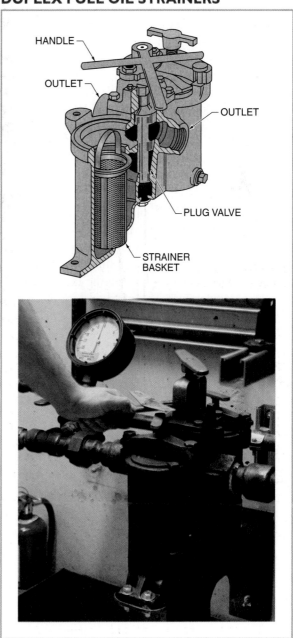

**Figure 7-6.** Duplex strainers allow the operator to keep the boiler on-line while cleaning one of the strainers.

The second set of strainers is located close to the fuel oil burner on the discharge side of the fuel oil pump. When the fuel oil is heated, fine particles of varnish can form. The second set of strainers has basket screens in each strainer with fine mesh to help remove the particles before the oil passes through to the burner. Alternatively, the strainers may consist of rotating plate strainers that grind up the impurities.

### Fuel Oil Pressure Gauges

Maintaining the correct fuel oil pressure is essential for proper atomization of the fuel oil. A fuel oil pressure gauge is used to measure the pressure of the fuel. Fuel oil pressure gauges should be located on both sides of all strainers. A large pressure drop across a strainer indicates that a strainer is dirty. On the suction line, the gauge shows an increase in vacuum when the suction strainer is dirty. On the burner line, the gauge shows a decrease in pressure when the discharge strainer is dirty. The gauges will indicate any pressure drop caused by restriction in the strainer, alerting the operator of the need to replace or clean the strainer.

Pressure gauges should also be located on the suction and discharge side of the fuel oil pump and at the fuel oil burner inlet. The suction pressure gauge changes with the fuel oil level in the tank and with the fuel oil temperature. The pressure at the discharge side does not vary much during the normal operation of the burner.

### Fuel Oil Heaters

A fuel oil heater is used to reduce the fuel oil viscosity so that the fuel oil can be pumped easily and to heat it to the temperature needed for combustion. **See Figure 7-7.** Many industrial steam plants that burn fuel oil use No. 6 fuel oil, which must be heated before it can be pumped or burned. To permit adequate flow for transport of fuel oil, No. 6 fuel oil tanks are equipped with heaters.

The two types of tank heaters commonly used are the coil heater and the bell heater. The coil heater is a pipe coil heated with steam. The fuel oil suction line passes through the center of the coil. This arrangement makes it unnecessary to heat all the fuel oil in the tank. The bell heater functions in the same manner except that it uses a closed bell-shaped vessel instead of coils. Fuel oil in the storage tank is heated only high enough to permit adequate flow.

## FUEL OIL HEATERS

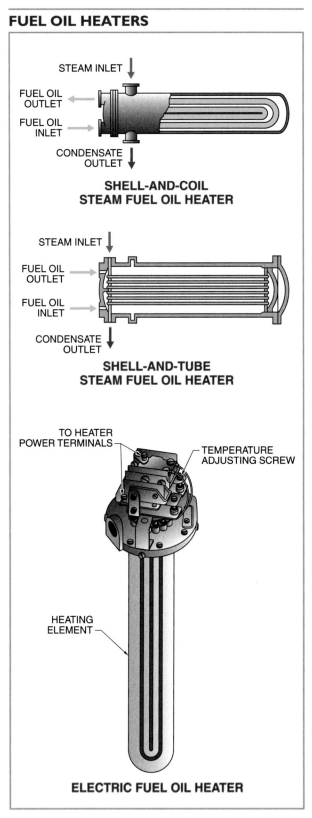

Figure 7-7. Fuel oil heaters are used to heat fuel oil so it can be pumped easily and burned at the correct temperature.

Fuel oil heaters are also used to heat fuel oil to its firing point. The temperature required for efficient combustion will vary according to fuel oil grade. Fuel oil heaters are located on the discharge side of the fuel oil pump and heat fuel oil for efficient use in the burner. The two types of fuel oil heaters used for this purpose are steam fuel oil heaters and electric fuel oil heaters.

A steam fuel oil heater is of shell-and-coil or shell-and-tube construction. In both types of heaters, the fuel oil passes through the tubes while steam and condensate are in the shell. An electric fuel oil heater consists of a shell with one or more electric heating elements inserted into the shell. The fuel oil enters at the lower section of the shell and leaves from the top. Both steam and electric fuel oil heaters are equipped with automatic temperature regulators.

A safety relief valve is installed after the fuel oil heater to protect against excessive fuel oil pressure. **See Figure 7-8.** The discharge from the safety relief valve is connected to the return line going to the tank. If the inlet and outlet valves on the heater are closed and steam is allowed to enter the heater, the fuel oil will expand, causing a pressure increase within the shell of the heater.

## FUEL OIL SAFETY RELIEF VALVES

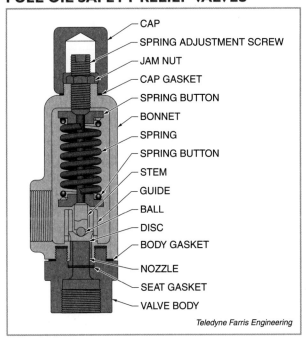

CAP
SPRING ADJUSTMENT SCREW
JAM NUT
CAP GASKET
SPRING BUTTON
BONNET
SPRING
SPRING BUTTON
STEM
GUIDE
BALL
DISC
BODY GASKET
NOZZLE
SEAT GASKET
VALVE BODY

*Teledyne Farris Engineering*

**Figure 7-8.** Because fuel oil pumps often are positive-displacement pumps and produce a constant flow of fuel oil, safety relief valves must be installed to relieve excess pressure.

## Fuel Oil Pumps

A fuel oil pump takes fuel oil from the fuel oil storage tank and delivers it to the burner at the correct pressure. **See Figure 7-9.** Reciprocating or gear-type pumps are commonly used. Relief valves are required because fuel oil pumps are positive-displacement pumps that constantly discharge fuel oil.

## FUEL OIL PUMPS

MOTOR
PUMP

**Figure 7-9.** A fuel oil pump delivers fuel to a burner at the correct pressure.

Not all of the fuel oil is delivered to the burner. The safety relief valve is located on the discharge line from the pump and discharges fuel oil to the return line, which goes back to the fuel oil tank. The amount of fuel oil returned to the fuel oil tank depends upon the rate of combustion and the pressure-regulating device at the fuel oil burner that works to maintain a constant pressure.

## Soot Blowers

*Soot* is the carbon deposits resulting from incomplete combustion. A *soot blower* is a device used to remove soot deposits from around tubes and permit better heat transfer in the boiler. Soot blowers are used where coal and fuel oil are used. Soot blowers are not needed when natural gas is used. In many locations, soot blowers cannot be used without the appropriate emission control equipment. Boiler operators must be aware of any restrictions associated with the plant operating permit relating to air pollution.

Soot acts as an insulator when deposited on the heating surface of the boiler. Insulation reduces the amount of heat that can be transferred. **See Figure 7-10.** Removing the soot increases heat transfer. Boiler efficiency increases with better heat transfer, thus reducing fuel consumption.

## EFFECT OF SOOT ON FUEL CONSUMPTION

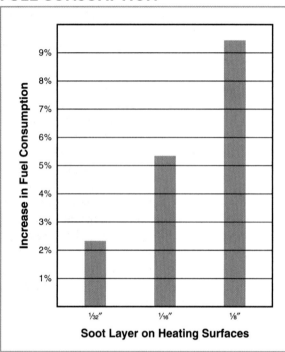

**Figure 7-10.** Soot acts as an insulator that reduces the heat transfer efficiency of a boiler.

**Soot Blower Types.** Soot blower types vary depending on boiler type, fuel type, and soot blower application. They can be fixed or retractable, and the cleaning medium can be saturated steam, superheated steam, compressed air, or water.

In watertube boilers, gases of combustion pass around the tubes leaving soot deposits, which can easily be removed using soot blowers. Most modern watertube boilers that burn fuel oil or coal are equipped with permanently installed soot blowers. **See Figure 7-11.** In a firetube boiler, gases of combustion pass through the tubes. To remove soot deposits in a firetube boiler, it is necessary to take the boiler off-line and use brushes or scrapers. Soot blowers are not very effective in firetube boilers but are found on some older HRT boilers.

## SOOT BLOWERS

**Figure 7-11.** Soot blowers are used on many watertube boilers to clean the outside of the tubes.

Soot blower elements extend into the boiler into the direct path of the gases of combustion. These elements are made of a steel alloy that can withstand the high temperature of the gases. The nozzles in the element are spaced to allow steam or air to blow between the rows of tubes. The nozzles are held in place by bearings clamped or welded to the boiler tubes. These bearings allow the element to rotate. Element alignment is important. If the element were to shift, steam impinging on the boiler tubes would cut through the tubes in a very short time.

## TECH TIP

*The temperature of the gases of combustion increases when soot builds up on heating surfaces. This is easy to recognize when the stack temperature is trended on a graph. Soot blowers are used to clean heating surfaces and restore boiler efficiency.*

**Soot Blower Operation.** In most modern plants, boiler performance information is evaluated by computers that monitor flue gas temperatures, flow, stack emissions, and draft. This information is used to automatically operate soot blowers at appropriate locations and sequences to optimize heat transfer. In some designs, a pipelike blowing element remains in the boiler at all times. Nozzles are located on this blowing element. A valve opens to allow steam to blow through the nozzles to clean the tubes. An electric motor turns a gear that rotates the blowing element. **See Figure 7-12.** As the blowing element rotates, the steam is directed toward the boiler tubes to blow off any soot.

## SOOT BLOWER OPERATION

**Figure 7-12.** Some soot blower designs have blower elements rotated by a motor and gear combination.

The steam line feeding the soot blowers must come from the highest part of the steam side of the boiler to ensure moisture-free steam. It is important that the soot blower lines be thoroughly warmed and completely drained. Any water discharged with the steam will cause the soot to cake up, and sulfur in the soot mixing with the water can cause damage to the boiler heating surface.

### SECTION 7.1–CHECKPOINT

1. What are the three common types of atomizing burners?

2. How is a rotary cup burner different from an atomizing burner?

3. What are common regulations regarding fuel oil storage tanks?

4. Where are fuel oil strainers typically located in a fuel oil system?

5. What types of burners typically require soot blowers?

## SECTION 7.2
# NATURAL GAS SYSTEMS

### NATURAL GAS BURNERS

Natural gas burners, or gas burners, may be either high-pressure or low-pressure burners, depending on the gas service available. Gas burners supply the proper mixture of air and gas to the furnace so that complete combustion can be achieved. Components required for proper gas flow include gas regulators, solenoid valves, control valves, blowers, and control mechanisms that maintain the proper air-fuel ratio. Both high and low pressure boilers can use either type of gas burner.

#### High-Pressure Gas Burners

Gas is supplied to high-pressure gas burners at a set pressure. In some modern high-pressure burners, air mixes with gas outside the combustion chamber and is forced into the burner using a blower. Premixing the air and gas allows for more complete combustion and increases efficiency. In other high-pressure gas burners, air mixes with gas on the inside of the burner register. **See Figure 7-13.** A *register* is a formed sheet-metal plate with slots where the metal is bent on an angle. A register is also known as a diffuser. As the mixture of air and fuel passes through the register, it causes the mixture to swirl turbulently into the combustion chamber.

## HIGH-PRESSURE GAS BURNERS

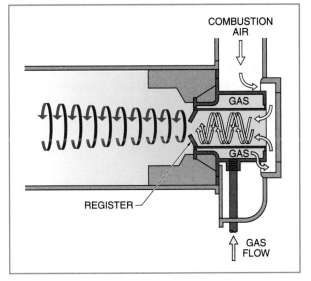

**Figure 7-13.** A high-pressure gas burner must be supplied with gas at a set pressure.

The gas entering the furnace is ignited by a gas pilot flame. The gas pilot flame is supplied with gas from a line connected to the main gas line. On the gas pilot line is a gas shutoff cock, a direct-acting solenoid valve, a gas-pressure gauge, and a gas pilot regulator.

Air for combustion is typically supplied by a forced draft blower. The two types of blowers are squirrel cage blowers and centrifugal blowers. A *squirrel cage blower* is a blower with a wheel that has blades attached at the rim and rotates in a housing. When the blower is running, air is drawn into the center and discharges at the rim of the wheel. The squirrel cage blower is typically used on large boiler installations.

A *centrifugal blower* is a blower that has a rotating impeller in a housing that throws the air to its outer edge, increasing the air's velocity and pressure. **See Figure 7-14.** The centrifugal blower is typically used on small to medium installations. With both blowers, airflow can be controlled by inlet or outlet dampers. In larger installations, a variable-frequency drive (VFD) adjusts the blower speed, controlling the volume of air supplied to the burner. Another way to maintain the correct air-fuel ratio, if a VFD is not used, is by mechanical linkage between the air damper, the butterfly valve, and the modulating motor. This is called single-point positioning control.

## AIR BLOWERS

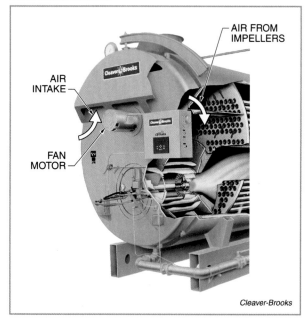

*Cleaver-Brooks*

**Figure 7-14.** Air blowers supply combustion air to a burner.

## Low-Pressure Gas Burners

Low-pressure gas burners are occasionally seen in older plants. Using a low-pressure gas system reduces the possibility of gas leaks, making it safer than a high-pressure gas system. Gas leaks in a low-pressure system usually occur up to the zero gas governor.

The main components of a low-pressure gas system are the manual reset valve, air blower, mixing chamber, regulator, solenoid valves, and air control valve. Low-pressure gas burner systems supply gas to the mixing chamber at 0 psig. The gas and primary air mix together outside of the combustion chamber and are forced by the blower through to the gas nozzle.

The main gas solenoid valve controls the gas supply to the burner. It opens fully when energized and closes tightly when deenergized. From the solenoid valve, the gas is drawn into the mixing chamber. **See Figure 7-15.**

## LOW-PRESSURE GAS BURNERS

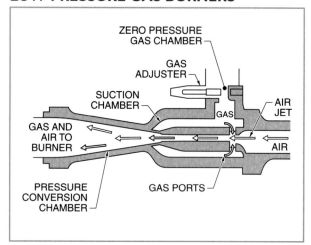

**Figure 7-15.** Air and gas combine in the mixing chamber of a low-pressure gas burner system.

The mixing chamber functions when air from the blower is forced through a nozzle and creates a venturi effect. Gas is drawn into the stream of air through the gas ports, and the air and gas mix. The mixture of air and gas leaves the nozzle and enters the furnace. The airflow from the blower to the mixing chamber is controlled by a butterfly valve. This valve allows the ratio of primary air to gas to be varied for different firing rates. The mixture of air and gas is ignited by a pilot flame. Secondary air, which is necessary to complete the combustion process, is controlled by an adjustable ring on the burner front.

## NATURAL GAS ACCESSORIES

There are many accessories that are needed to make a gas burner system operate properly and safely. **See Figure 7-16.** The accessories can be broadly classified as the gas lines and gauges, gas valves and regulators, and gas safety switches.

**TECH TIP**

*A boiler operator should understand the concepts of efficient fuel combustion. However, burner tuning is potentially dangerous and should be left to experienced, factory-trained technicians who are equipped with proper flue gas analysis equipment.*

## HIGH-PRESSURE GAS ACCESSORIES

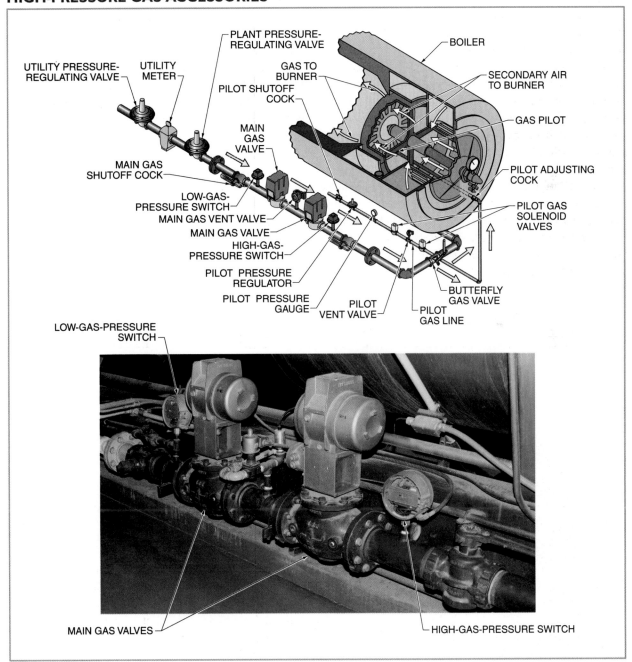

**Figure 7-16.** A gas train contains many safety and control accessories to ensure safe and efficient operation.

### Gas Lines and Gauges

The gas supply line to the burner must be large enough to prevent excessive pressure drop through the line. It is fitted with a pressure gauge that is typically calibrated with two scales. The first scale reads in inches of water column and the second in ounces per square inch. The gas line is also fitted with a manually operated shutoff valve. It is a plug cock valve that can be opened or closed with a quarter turn of the lever.

---

**TECH TIP**

*Natural gas is an odorless fuel in its natural state. A sulfur-containing compound, called mercaptan, is added to make leaks easier to detect.*

---

### Gas Valves and Regulators

A gas-pressure regulator controls the desired set pressure at the burner. **See Figure 7-17.** The regulator consists of a diaphragm connected to a globe valve with an opposing adjustable spring. By adjusting the spring tension, the pressure at the burner can be changed. The valve reduces the gas pressure to 0 psi.

The gas-flow control valve gradually allows gas to flow to the burner at startup. This valve prevents a large volume of gas from going into the furnace before ignition has taken place. The gas-flow control valve is basically a solenoid-operated diaphragm valve. When the solenoid valve opens, the gas pressure on top of the diaphragm is gradually released. The high-pressure gas under the diaphragm now moves the valve toward the open position.

Fluid-powered gas valves are slow-opening, fast-closing valves that use a small internal hydraulic pump. Hydraulic oil is pumped from a reservoir into a cylinder with a piston to slowly open the valve against spring pressure.

When the valve needs to be closed, a large internal solenoid valve opens and the spring pressure pushes the piston in the cylinder, forcing the hydraulic oil back into the reservoir. This action quickly closes the valve. The valve position (open or closed) is determined by the actuator position. The actuator position can be seen through the indicator window on the valve. Auxiliary switches can be added to prove (or verify) valve position at the combustion controller.

## GAS-PRESSURE REGULATORS

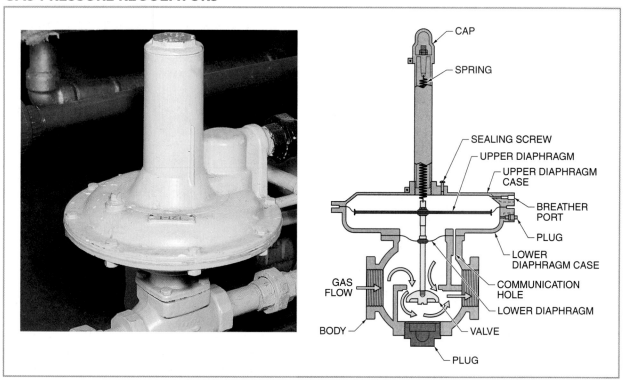

**Figure 7-17.** A gas-pressure regulator supplies the burner with gas at the correct pressure.

The gas control valve, commonly known as a butterfly gas valve, controls the volume of gas going to the burner. This valve maintains the correct air-fuel ratio. **See Figure 7-18.** The valve consists of a body, a circular disc, and a pin. The pin passes through the body and circular disc. When the pin turns, it moves the valve toward the open or closed position for accurate gas-flow control. The butterfly valve is controlled by the modulating motor. The modulating motor is connected to the combustion equipment control linkage to regulate the burner firing rate. The burner firing rate changes to meet steam demand.

## BUTTERFLY VALVES

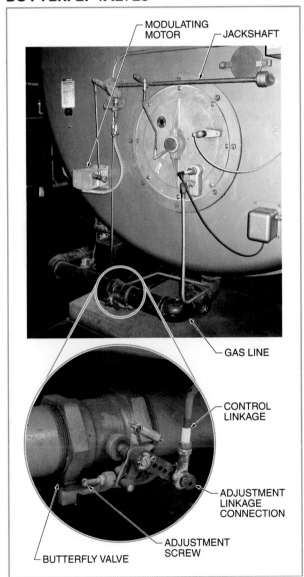

**Figure 7-18.** A butterfly gas valve controls the volume of gas going to the burner to maintain the proper air-fuel ratio.

## Gas Safety Switches

To prevent the system from operating when the gas pressure is low, a low-gas-pressure switch is connected to the gas line. **See Figure 7-19.** The low-gas-pressure switch is an electrical switch actuated by the gas pressure acting on a diaphragm. During normal operation, the switch is closed. The switch opens to secure the system if there is a drop in gas pressure. The correct gas pressure must be restored and the manual reset lever must be pushed to restore the switch to its operating position.

A high-gas-pressure switch is used to prevent the system from operating with gas pressure that is too high. This switch is similar to the low-gas-pressure switch except, during normal operating conditions, it is closed and secures the system on an increase in gas pressure. The correct gas pressure must be restored and the manual reset lever must be pushed to restore the switch to its operating position.

On smaller boilers, an electrically operated solenoid valve is used as an automatic shutoff gas valve. It is a direct-acting valve that opens when the coil is energized electrically and produces an electromagnetic field. An iron core connected to the disc of the valve is attracted to the magnetic field, and when the iron core moves toward it, the valve opens. The valve closes when the coil is deenergized.

The most important valve on a low-pressure gas system is the manual reset valve. The manual reset valve is located on the gas line to the burner and shuts off the gas supply if the pilot flame goes out or a low-water condition exists. The manual reset valve is a solenoid valve that remains open as long as the electrical coil is energized. When the coil is deenergized for any reason, the valve closes. To reset the valve, a reset button must be pressed or a manual lever must be lifted into position. Before the valve can be reset, a pilot flame is needed.

## COMBINATION GAS/FUEL OIL BURNERS

Combination gas/fuel oil burners are used in some installations. Using a combination gas/fuel oil burner is essentially the same as having a high-pressure gas burner system and a fuel oil burner connected together at the burner throat. The fuel is delivered through an orifice with gas and a nozzle with fuel oil. **See Figure 7-20.**

## LOW-GAS-PRESSURE SWITCHES

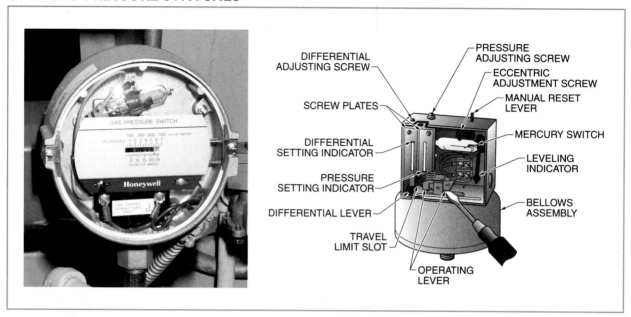

**Figure 7-19.** A low-gas-pressure switch prevents the system from operating when the gas pressure is too low.

## COMBINATION GAS/FUEL OIL BURNERS

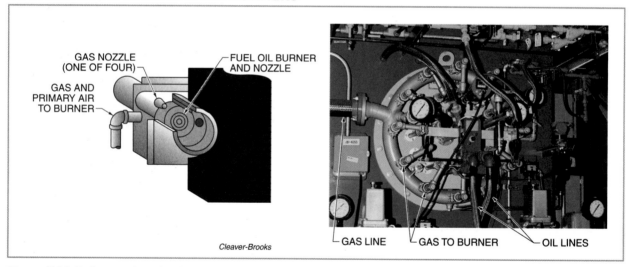

**Figure 7-20.** Boilers equipped with combination gas/fuel oil burners are more flexible because they can burn either gas or fuel oil.

Combination gas/fuel oil burners are designed to use a high-pressure gas burner in conjunction with an oil burner. Using two types of fuel allows for flexibility in operation because the operator can change from one fuel to another in the event of a mechanical failure in one of the systems. In addition, using a combination gas/fuel oil burner makes it possible to burn the fuel that is least expensive at a particular time of year. Larger, more sophisticated plants may burn gas and fuel oil at the same time for maximum efficiency and minimal air pollution.

In combination burners, the gas system is separate and includes components such as gas-pressure regulators, solenoid valves, pressure gauges, and gas-flow control valves. However, a forced draft blower supplies the necessary primary and secondary air to both the gas and the fuel oil systems. When larger quantities of gas at higher pressures are required, gas booster compressors are sometimes used.

---

## SECTION 7.2 – CHECKPOINT

1. What are the methods used to mix fuel with air in a high-pressure gas burner?
2. What is the difference between a squirrel cage blower and a centrifugal blower?
3. What is the purpose of a gas-pressure regulator?
4. What are the different types of gas safety switches?
5. How does a combination gas/fuel oil burner increase flexibility?

---

# SECTION 7.3
# COAL SYSTEMS

## COAL BURNERS

Coal is used primarily in utility boilers and some larger industrial or educational facilities. At one time coal was commonly used to fire steam boilers. However, fuel oil and gas have become more common due to price, availability, and environmental standards. Pulverized coal is commonly used in utility or very large industrial or commercial high pressure boilers. Typical methods for burning coal include using pulverized coal burners, cyclone boilers, fluidized-bed boilers, or stokers. One of the products of burning coal is flyash. Flyash is controlled by dust collectors (flyash precipitators) that trap and hold ash so it may be disposed of properly.

## Pulverized Coal Burners

*Pulverized coal* is coal that is ground into a fine powder and is then blown into the combustion chamber where it is burned in suspension. Pulverized coal is highly explosive, so care must be taken when burning it. A flame must be maintained in the furnace at all times. Also, the furnace must have a negative pressure (pressure below atmospheric pressure). If the pressure in the furnace becomes positive, the flame could travel back to the pulverizer and cause an explosion. Pulverized coal systems are most suitable for use with bituminous coals.

The main parts of a pulverized coal system are the overhead coal bunker, coal conveyor, coal scale, coal feeder, pulverizer, and exhauster. **See Figure 7-21.** In a pulverized coal burner, coal is stored in the overhead bunker to allow it to flow by gravity to the conveyor. This movement of the coal is necessary to prevent spontaneous combustion from occurring within the bunker. A belt conveyor at the base of the coal bunker moves the coal to the coal scale. The position of the scale controls the belt conveyor. When the scale is in an up position, the conveyor will run until the scale is filled with the correct amount of coal. When the scale moves down to the dump position, the conveyor stops.

## PULVERIZED COAL SYSTEM PARTS

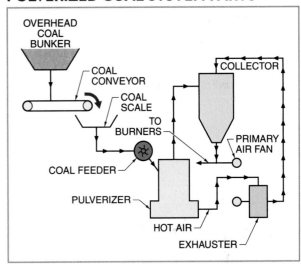

**Figure 7-21.** The parts of a pulverized coal system include an overhead coal bunker, a conveyor, a scale, a feeder, a pulverizer, and an exhauster. These feed into a collector that sends pulverized coal to the burners.

From the coal scale, the coal moves to the feeder. The speed of the rotating blade of the coal feeder controls the amount of coal being delivered to the pulverizer. **See Figure 7-22.** Once the coal is in the pulverizer, it is ground using steel balls or rollers rotating around the bowl. The coal is crushed as it passes between the surfaces, which results in the coal having the consistency of talcum powder.

Warm air is introduced into the pulverizer to dry the coal and prevent caking. Air mixes with powdered coal and passes into the exhauster. Coal and air are discharged from the exhauster to the burner throat and then into the furnace. Coal is burned in suspension and requires a furnace temperature of approximately 3000°F to achieve complete combustion.

## COAL PULVERIZERS

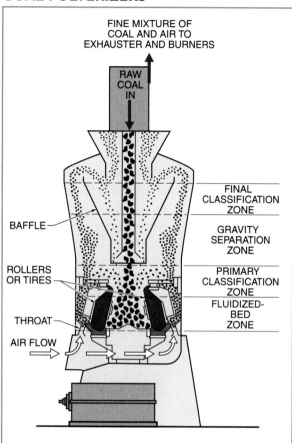

**Figure 7-22.** A pulverizer grinds coal to the consistency of talcum powder before it is blown into the furnace.

A forced draft blower supplies air for the pulverizer and windbox. Air from the windbox enters the furnace around the outer edge of the burner throat. Air introduced into the pulverizer is primary air. Air from the windbox is secondary air. Steam boilers that burn pulverized coal are usually equipped with an air preheater to supply warm air for the pulverizer and windbox.

### Cyclone Boilers

A cyclone boiler uses a burner design to produce high-temperature flames that create circulation within a boiler. Cyclone boilers typically use grades of coal not suitable for pulverizing. Crushed coal is fed at an angle, blown by high-pressure air, into an inclined cylindrical furnace. The crushed coal is thrown to the furnace wall by centrifugal force where it sticks to molten slag. The combination of coal and molten slag slowly moves down the inclined furnace. The coal burns and the ash falls into an ash

hopper at the end of the furnace. This results in cleaner gases of combustion and reduced emissions.

### Fluidized-Bed Boilers

An abundant coal supply provides an economical, predictable fuel source. However, burning coal without proper environmental equipment creates pollution. Fluidized-bed boilers were developed to help overcome pollution problems in large industrial boilers. High-sulfur-content coal can be burned without expensive flue-gas treatment equipment by injecting limestone into the bed to absorb the sulfur dioxide. Low combustion temperatures are used to minimize slag formations in the furnace and nitrogen oxides in the stack gases.

A fluidized-bed boiler is designed to have a thick bed of granular particles of limestone through which air is blown at a high enough velocity to hold the limestone particles in suspension. **See Figure 7-23.** Crushed coal is then injected into the bed and burned at a low temperature to capture the sulfur in the lime. The low temperature minimizes the formation of nitrogen oxides ($NO_x$). The waste material is drawn from the bottom of the bed as fresh limestone is injected into the bed.

## FLUIDIZED-BED BOILERS

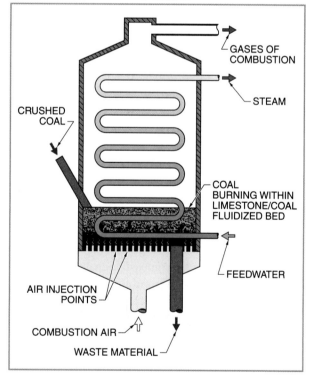

**Figure 7-23.** A fluidized-bed boiler contains a bed of granular particles.

## Stokers

A *stoker* is a mechanical device used for feeding coal into a boiler. The three types of stokers used are underfeed stokers, spreader stokers, and chain grate stokers. The properties of the coal burned determine the type of stoker required. Soft coal, which has a high volatile content, cannot be burned in the same furnace as hard coal, which has a high fixed-carbon content.

The type of stoker, the type of grate, and the furnace volume are all designed for the specific type of coal being burned. The amount of ash produced varies with the type of coal used. In some stokers, ash serves to protect the grates from excessive furnace temperatures. This minimizes the possibility of the cast iron grates warping. Coal stokers are primarily used for large industrial or utility boilers. Smaller boilers rarely burn coal. Most smaller and older stoker units have been converted or replaced with fuel oil or gas units.

**Underfeed Stokers.** An *underfeed stoker* is a stoker that feeds coal into a boiler from under the fire. **See Figure 7-24.** Coal is fed into a hopper and pulled down to a coal feeder (coal ram) by gravity. The coal ram forces the coal into a retort chamber. Pusher blocks move the coal forward and upward, distributing and leveling it over the entire grate surface. Fresh coal is continuously fed under the fire. Ash is removed from the ash pit either automatically or by hand. The lateral movement of the grate bars slowly moves the burned-out refuse toward the dump grates.

Air for combustion is supplied to the underfeed stoker by a blower that forces it into the air chamber directly under the retort. From the air chamber, the air moves through the grate openings and fuel bed. Combustion takes place from the top of the fuel bed down and over the fuel bed.

*Ash conveyors are an essential component of furnaces that burn solid fuels.*

Underfeed stokers are capable of burning either anthracite or bituminous coal. Because bituminous coal has a high volatile gas content, overfire air, which is additional air introduced above the fire, is required to prevent smoke. When bituminous coal is heated, the gases are driven off and burned above the fuel bed. Overfire air is taken from the main air chamber and its flow is controlled by a damper.

Load changes are controlled by adjusting the speed of the feeder block and the amount of air supplied. An underfeed stoker can be designed for automatic operation. The rates at which the coal is fed, the air for combustion is supplied to the furnace area, and the gases of combustion are removed are all contingent upon the steam load. In some stokers, zone dampers are installed below the grates to allow the boiler operator flexibility in distributing the combustion air to the coal.

**Spreader Stokers.** A *spreader stoker* is a stoker that feeds the coal into the boiler in suspension and on the grate. A spreader stoker is also known as an overfeed stoker or sprinkler stoker. The main parts of the spreader stoker are the coal conveyor, the coal hopper, the coal

## UNDERFEED STOKERS

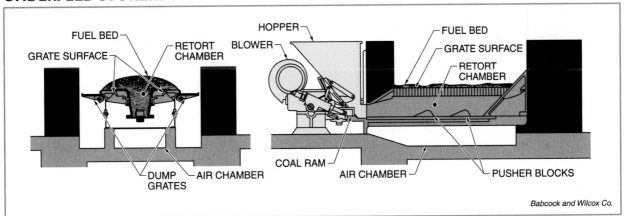

*Babcock and Wilcox Co.*

**Figure 7-24.** An underfeed stoker supplies the coal from under the fire.

feeder, and the overthrow unit. The two types of coal feeders used in spreader stokers are the reciprocating type and the conveyor type. **See Figure 7-25.** Both types function similarly.

## SPREADER STOKERS

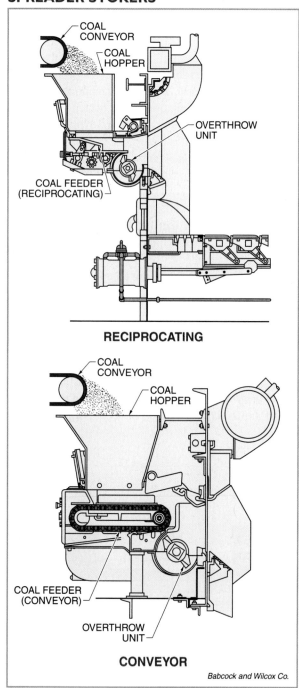

**RECIPROCATING**

**CONVEYOR**

*Babcock and Wilcox Co.*

**Figure 7-25.** Smaller particles of coal will burn in suspension when a spreader stoker is used.

The conveyor moves the coal from the coal bunker to the coal scale. When the correct amount of coal is on the coal scale, the conveyor stops and the scale dumps the coal into the coal hopper. The coal is spread onto the grates. Fine particles of coal burn in suspension above the grates. Larger particles of coal ignite while in suspension and fall to the grate to complete the combustion process. This method prevents the grates from being covered with green coal, thus keeping the fuel bed hot.

From the coal hopper, the coal drops into the coal feeder. The coal is then fed into the overthrow unit, which throws the coal out in a sprinkled form to be burned in suspension or on the grates. Air for combustion is supplied to spreader stokers by a forced draft blower. Air is forced through the fuel bed from under the grates. Overfire air is also supplied to complete the combustion of particles in suspension.

### TECH TIP

*Leaks in a pressurized furnace can result in gases of combustion leaking into the boiler room.*

The rate of combustion is controlled by the speed of the drive on the coal feeder as well as by the overfire and underfire air supplied. Spreader stokers are designed for firing small pieces of bituminous coal. Spreader stokers can use all grades of bituminous coal and operate with a very thin fuel bed.

**Chain Grate Stokers.** The *chain grate stoker* is a stoker that feeds the coal into the boiler on a traveling grate. A chain grate stoker is also known as a traveling grate coal stoker. Chain grate stokers are used for larger steam boilers. **See Figure 7-26.** Parts on a chain grate stoker include the coal conveyor, coal hopper, regulating coal gate, traveling chain grate, hydraulic drive unit, and ash hopper. The coal conveyor and coal hopper on the chain grate stoker function as in the spreader stoker. The regulating coal gate controls the thickness of the fuel bed as the sliding gate at the back of the hopper is moved up or down.

The chain grate is formed by a chain that runs around two sprockets, one of which is attached to a variable-speed drive. Coal is deposited on the front of the grate and travels past the ignition arch, which starts the combustion process. The ignition arch is formed from the refractory, which absorbs radiant heat from the combustion process and then radiates the heat back to the incoming green coal. Combustion must be completed before the coal reaches the end of the grate. Ash falls off the grate into the ash hopper.

## CHAIN GRATE STOKERS

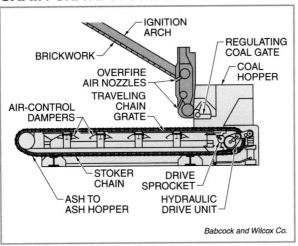

*Babcock and Wilcox Co.*

**Figure 7-26.** A regulating coal gate controls the thickness of the fuel bed in a chain grate stoker.

Coals best suited for the chain grate stoker are noncaking coals with relatively high ash content. The high ash content helps protect the grate from overheating. A forced draft blower supplies air that is fed through a windbox under the grates. The amount of coal burned on the chain grate stoker is regulated by the thickness of the fuel bed, the rate of grate travel, and the amount of air supplied.

The air supplied under the grates is controlled so that the coal is continuously burning as it moves toward the end of the traveling grate. This ensures that the coal is burned completely, so the ash is dumped into the hopper rather than on the burning green coal. The boiler operator must regulate the air to prevent the air from blowing holes through the fuel bed.

### ASH HANDLING

*Ash* is the solid material left behind in the process of burning coal. Coal-burning boiler plants must dispose of ash in an environmentally responsible manner. Ash can contain many environmental contaminants such as mercury, cadmium, and arsenic. Without proper management, these contaminants can pollute waterways, ground water, drinking water, and the air.

The amount of ash produced depends on the coal composition and equipment design. Clinker can cause problems with ash-handling equipment. *Clinker,* or slag, is the noncombustible components of coal that melt and fuse together during combustion, leaving chunks that interfere with ash handling. Coal boiler plants may use clinker grinders to reduce the size of the clinker for easier handling.

The ash created in a coal-fired boiler is categorized as fly ash or bottom ash. *Fly ash* is relatively small ash particles that are light enough to be suspended in a combustible gas stream and carried though a boiler. Fly ash and carbon particles can collect as soot on internal surfaces such as boiler tubes, economizers, air preheaters, breechings, and stacks. This can cause corrosion problems and poor heat transfer.

Soot blowers are used to clean fly ash from internal surfaces. The loosened soot particles are carried away by the gases of combustion. Modern boilers systems include baghouses and precipitators to remove the particles before they are released into the atmosphere. Soot or fly ash that is not trapped by electrostatic precipitators or baghouses can be carried out the stack and cause air pollution.

*Bottom ash* is relatively large ash particles that are heavy enough to be collected and removed from the bottom of a furnace. Bottom ash is typically removed from the boiler area by conveyors. **See Figure 7-27.** The conveyors carry the ash to a silo or dumpster where it can be disposed of properly. Some coal-fired plants dispose of ash in impoundments such as settling ponds or in abandoned mines as fill. Coal ash may also be recycled into products like concrete, wallboard, or asphalt.

### ASH CONVEYORS

**Figure 7-27.** Ash conveyors move bottom ash to a silo or a dumpster for disposal.

---

### SECTION 7.3—CHECKPOINT

1. What is the role of a pulverizer?
2. How does a fluidized-bed boiler reduce air emissions?
3. What are the main types of stokers?
4. What are the differences between handling bottom ash and fly ash?

## CASE STUDY—BOILER FURNACE EXPLOSION

### Situation
The duties of the head custodian at an elementary school include starting, stopping, and checking the building's boiler. At this particular school, the boiler room is located next to the school's cafeteria and separated by a block wall. Students line up along this wall in the morning for breakfast.

### Problem
The students had all been served and had just taken their seats when the boiler furnace exploded. The concussion from the explosion knocked down the block wall, which fell into the cafeteria. This included an area where several dozen students had been standing a few minutes previously. The explosion blew the doors, stack, and breeching off the boiler and started a small fire in the boiler room. The head custodian was in a room above the boiler room and felt the floor shake when the explosion occurred. He immediately ran down to the boiler room and saw the destruction and fire. He went to the outside gas shutoff and closed the gas valve. The fire was out when he returned to the boiler room.

### Solution
The furnace explosion incident investigation fell under the jurisdiction of the fire department. The fire department inspector concluded that there was a problem with the gas supply to the burner that caused ignition of a combustible mix after the control system had safely shut the burner off. Most probably this was from contaminants in the fuel supply that did not allow proper seating of the valves. As the gas built up in the furnace, something caused the gas to ignite (hot refractory or igniter), causing the furnace explosion.

Fortunately no one was injured. Costly repairs were needed to the boiler and building, and the school closed for nearly a month.

**Name** _____ **Date** _____

_____ **1.** Burning a fuel releases ___.
  A. solar energy
  B. hydrogen gas
  C. heat energy
  D. natural gas

_____ **2.** A vent line on a fuel oil tank prevents ___ when filling the tank with fuel oil and also prevents a vacuum from developing when fuel oil is removed.
  A. loss of suction
  B. pressure buildup
  C. loss of pressure
  D. spills

_____ **3.** A high suction line on a fuel oil tank is used when ___.
  A. the tank is full
  B. the tank is low on fuel
  C. signs of sludge and water are evident in the tank
  D. changing over tanks

_____ **4.** Coal ash can contain many environmental contaminants such as ___.
  A. nitrogen
  B. oxygen
  C. moisture
  D. increase in natural gas

_____ **5.** In both steam and air atomizing burners, atomization is accomplished by ___.
  A. live steam or air
  B. high fuel oil pressure
  C. the sprayer plate
  D. low-temperature, high-pressure fuel oil

_____ **6.** The amount of gas going to the burner at startup in a high-pressure gas system is controlled by a solenoid-controlled ___ valve.
  A. globe
  B. gate
  C. check
  D. diaphragm

_____ **7.** Pulverized coal burns ___.
A. on inclined grates
B. in suspension
C. in suspension and on grates
D. in the pulverizer

_____ **8.** An indication of a dirty fuel oil strainer would be ___ across the strainer.
A. a large pressure drop
B. an equalizing of pressure
C. a slight pressure drop
D. no change in pressure

_____ **9.** A low-gas-pressure switch opens to secure the system if there is a(n) ___ pressure.
A. drop in steam
B. increase in steam
C. drop in natural gas
D. increase in natural gas

_____ **10.** Fuel oil heaters must always be used when burning No. ___ fuel oil.
A. 1
B. 2
C. 4
D. 6

_____ **11.** A safety relief valve is used to protect against excessive fuel oil ___ that could damage the shell of a fuel oil heater.
A. temperature
B. flow
C. pressure
D. contaminants

_____ **12.** A rotary cup burner atomizes fuel oil using ___.
A. oil pressure
B. a spinning cup only
C. high-velocity air only
D. both a spinning cup and high-velocity air

_____ **13.** In a pressure atomizing burner, atomization is accomplished by ___.
A. high-pressure gas in the burner blower
B. rotating, high-pressure oil coming out of the plug or sprayer plate
C. air or steam mixing with the oil
D. the low-temperature, high-pressure oil supplied to the burner

_____ **14.** A gas pressure regulator is used to control the ___.
A. flow of the gases of combustion
B. desired gas pressure at the burner
C. airflow going to the burner
D. mixture of air and gas

_____ 15. A solenoid valve is a direct-acting valve in the gas system used ___.

    A. to regulate the flow of gas to the burner

    B. to regulate the flow of air to the burner

    C. as an automatic gas shutoff valve

    D. as a gas return valve

_____ 16. To prevent caking and to dry the coal entering a pulverizer, ___.

    A. heating coils are used

    B. electric pipe heaters are used

    C. warm air is used

    D. dryers are used

_____ 17. Blowers are used in a high-pressure gas system to ___.

    A. supply air for combustion

    B. introduce gases of combustion into the burner

    C. regulate the flow of gas

    D. cool the combustion chamber

_____ 18. If a VFD is not used, the correct ___ is maintained by mechanical linkage between the air damper, the butterfly valve, and the modulating motor.

    A. fuel-oil flow

    B. steam pressure

    C. steam temperature

    D. air-fuel ratio

_____ 19. The thickness of the fuel bed on a chain grate stoker is regulated by a coal ___.

    A. conveyor

    B. scale

    C. hopper

    D. gate

_____ 20. Boilers are equipped with a combination gas/fuel oil burner to allow ___.

    A. higher cost

    B. more flexible operation

    C. safer operation

    D. higher rates of combustion

_____ 21. Spreader stokers burn coal ___.

    A. only in suspension

    B. only on grates

    C. in suspension and on grates

    D. in retorts

_____ 22. To achieve complete combustion in a spreader stoker, air is introduced ___.

    A. under and over the fuel bed

    B. only under the fuel bed

    C. only over the fuel bed

    D. using an induced draft fan

_____ **23.** On a low-pressure gas system, the ___ is located on the gas line to the boiler and shuts off the gas supply if the pilot flame goes out or a low-water condition exists.
  A. gas-pressure regulator
  B. manual reset valve
  C. adjustable-spring valve
  D. butterfly gas valve

_____ **24.** An aboveground storage tank (AST) is a tank and any underground piping connected to the tank that has less than ___% of its combined volume underground.
  A. 10
  B. 25
  C. 50
  D. 75

_____ **25.** A fluidized-bed boiler is designed to have a thick bed of ___ through which air is blown at a high enough velocity to maintain the bed material in suspension.
  A. sulfur dioxide
  B. $NO_x$
  C. pulverized coal
  D. granular particles of limestone

# Chapter 8 Objectives

## SECTION 8.1—BOILER FUELS

- Describe fuel oil characteristics and handling procedures.
- Describe natural gas heating values and handling procedures.
- Describe coal characteristics and handling procedures.
- List and describe alternative fuels that can be burned in a boiler.

## SECTION 8.2—COMBUSTION OF FUELS

- Explain types of combustion and air supply.
- Describe the combustion reactions that happen when a fuel is burned.
- Describe the products of combustion and explain how they are analyzed.
- Describe how furnace volume affects combustion.

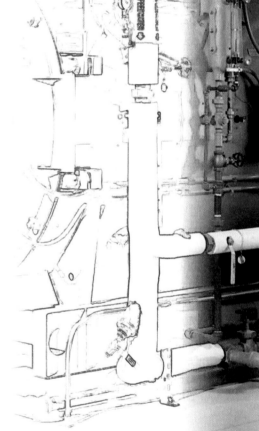

## ARROW SYMBOLS

| AIR | GAS | WATER | STEAM | FUEL OIL | CONDENSATE | AIR TO ATMOSPHERE | GASES OF COMBUSTION |
|-----|-----|-------|-------|----------|------------|-------------------|---------------------|
| ⇨ | ⇨ | ⇨ | ➡ | ⇨ | ➡ | ⇨ | ➡ |

**Digital Resources**
ATPeResources.com/QuickLinks
Access Code: **472658**

# CHAPTER **8**

# FUELS AND COMBUSTION

## INTRODUCTION

Combustion is required to generate the heat used to produce steam in a boiler. The fuels commonly used to sustain combustion in a steam boiler include fuel oil, natural gas, and coal. Each fuel is capable of producing a given amount of heat, and each fuel requires special consideration for storage, handling, and combustion procedures. Also, specific safety precautions for each fuel must be followed. The products of combustion and the type of fuel used determine the burner efficiency.

---

### SECTION 8.1
## BOILER FUELS

### FUEL TYPES

*Combustion* is the rapid reaction of oxygen with a fuel that results in the release of heat. The three common types of fuels used for combustion in a boiler are fuel oil, natural gas, and coal. Alternative fuels are becoming more common. The storage and handling procedures depend on the type of fuel being used. Federal, state, and local laws may require specific procedures. In addition, the boiler operator is responsible for understanding any potential hazards when storing and handling a fuel.

Factors to consider when choosing which type of fuel to use in a particular boiler include the design of the boiler, the price and availability of the fuel, and environmental considerations like the nitrogen and sulfur content of the fuel. An additional consideration in choosing a fuel is heating value. The *heating value* is the amount of heat that can be obtained from burning a fuel. The heating value of a fuel can be expressed in

British thermal units. A *British thermal unit (Btu)* is a measurement of the quantity of heat required to raise the temperature of 1 lb of water by 1°F. The Btu rating of a fuel indicates how much heat it can produce when burned.

## FUEL OIL

Fuel oil is a liquid fossil fuel that consists primarily of hydrocarbons (compounds of hydrogen and carbon) and is produced by distilling crude oil in a refinery. Crude oil is distilled into individual products that include gasoline, diesel fuel, fuel oil, lubricating oils, and other materials. Other elements present in fuel oil in varying amounts may include sulfur, nitrogen, arsenic, and phosphorus. Fuel oil may also be called distillate fuel oil.

### Fuel Oil Grades

ASTM International has established standards for grading fuel oils. Fuel oil is classified by grades as No. 1, No. 2, No. 4, No. 5, and No. 6 fuel oil. **See Figure 8-1.** No. 2 fuel oil is a light, clean fuel oil that is easy to handle and is often used in smaller package boilers. No. 4 fuel oil is a rarely used fuel that is heavier and darker than No. 2 fuel oil. No. 5 fuel oil is a rarely used fuel that is heavier and darker than No., 4 fuel oil. No. 6 fuel oil is a heavy, black oil that is relatively difficult to handle. It must be heated to allow it to flow and to burn properly.

**Heating Values.** The heating value of fuel oil is usually expressed in Btu/gal., although occasionally it is expressed in Btu/lb. By volume, heavier, higher-numbered fuels produce more Btu/gal. than lighter, lower-numbered fuel oils. By weight, lighter fuel oils produce more Btu/lb than heavier fuel oils.

The fuel oils most commonly used in high pressure boilers are No. 2 and No. 6. No. 2 fuel oil has a heating value of approximately 141,000 Btu/gal. It does not have to be preheated and is usually burned using a high- or low-pressure atomizing burner. No. 6 fuel oil, commonly referred to as "bunker C" fuel oil, is a residual fuel oil containing heavy elements from the distillation process, with a Btu content of approximately 150,000 Btu/gal. Tank heaters and line heaters are needed to bring No. 6 fuel oil up to the range of 220°F to 260°F. "Bunker grade" is the term used to designate the types of fuel oil used on ships.

No. 4 and No. 5 fuels are less common, but they are sometimes available. No. 4 fuel oil is heavier than No. 2 fuel oil and has a heating value of approximately 146,000 Btu/gal. In some locations, it is not available as a straight distillate but only as a blend of No. 2 fuel oil and heavier fuel oils. Normally, No. 4 fuel oil does not require preheating. However, in colder climates, preheating No. 4 fuel oil may be necessary to lower its viscosity so it can be pumped. If the fuel oil is not properly blended, it may stratify (separate) in the tank.

No. 5 fuel oil is divided into hot No. 5, which requires preheating, and cold No. 5, which can be burned as it comes from the tank. No. 5 fuel oil has a heating value of approximately 148,000 Btu/gal. In colder climates, tank heaters may be required in order to reduce the fuel oil viscosity so that pumping and atomization are possible.

**Flammability.** In addition to fuel oil grades, other characteristics of fuel oils include characteristics related to the flammability of a fuel. **See Figure 8-2.** *Pour point* is the lowest temperature at which a liquid will flow from one container to another. *Flash point* is the lowest temperature at which the vapor of a fuel oil ignites when exposed to an open flame. Fuel oils with low flash points

| FUEL OIL HEATING VALUES | | | | | |
|---|---|---|---|---|---|
| Characteristics | No. 1 Fuel Oil | No. 2 Fuel Oil | No. 4 Fuel Oil | No. 5 Fuel Oil | No. 6 Fuel Oil |
| Type | Distillate kerosene | Distillate | Very light residual | Light residual | Residual |
| Color | Light | Amber | Black | Black | Black |
| American Petroleum Institute (°API) 60°F | 40 | 32 | 21 | 17 | 12 |
| Specific gravity | 0.8250 | 0.8654 | 0.9279 | 0.9529 | 0.9861 |
| Lb/gal. | 6.87 | 7.206 | 7.727 | 7.935 | 8.212 |
| Btu/gal. | 137,000 | 141,000 | 146,000 | 148,000 | 150,000 |
| Btu/lb | 19,850 | 19,500 | 19,100 | 18,950 | 18,750 |

**Figure 8-1.** Heavier, higher-numbered grades of fuel oil have a higher Btu content per gallon than lighter, lower-numbered grades of oil.

can be dangerous and require special precautions when being handled. *Fire point* is the lowest temperature to which a fuel must be heated to burn continuously when exposed to an open flame.

## FLAMMABILITY OF NO. 2 FUEL OIL

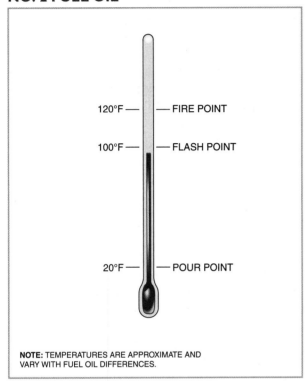

NOTE: TEMPERATURES ARE APPROXIMATE AND VARY WITH FUEL OIL DIFFERENCES.

**Figure 8-2.** Pour point, flash point, and fire point are related to how fuel oil flows and burns.

No recommended standard temperatures exist for burning the various grades of fuel oil. The fuel oil temperature required depends on the type of burners used in the plant and whether a straight distillate fuel oil or a blend of fuel oils is used. For the best results, the flash point, fire point, and pour point of the fuel oil can be obtained from the fuel oil supplier. A rule of thumb is to burn No. 4 fuel oil at 100°F, No. 5 at 150°F, and No. 6 at 220°F.

**Viscosity.** *Viscosity* is the measurement of the internal resistance of a fluid to flow. The viscosity of a fuel oil indicates how well or poorly the oil will flow during pumping and atomization. The lower the viscosity, the easier the fuel will flow. Normally the viscosity of oil is reported as the Saybolt viscosity,

which is the time it takes for 60 ml of oil to flow through a calibrated tube at a specific temperature. **See Figure 8-3.** Saybolt viscosity is measured in Saybolt universal seconds (SSU). The viscosity of fuel oil is reduced by increasing its temperature. The viscosity of No. 2 fuel at room temperature is about 35 SSU to about 50 SSU. The viscosity of No. 6 fuel oil at 160°F is about 100 SSU to about 300 SSU.

## SAYBOLT VISCOSITY

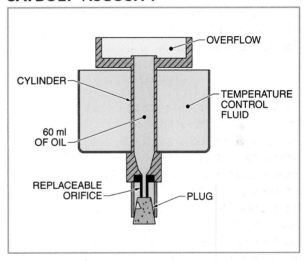

**Figure 8-3.** Saybolt viscosity is the time it takes 60 ml of oil at a specific temperature to flow through an orifice.

*The viscosity of fuel oil affects its ability to flow through pipes and strainers.*

**Specific Gravity.** *Specific gravity* is the ratio of the weight of a given volume of a substance to the weight of the same volume of water at a standard temperature of 60°F. A specific gravity less than 1 indicates that the fuel is less dense than water. In addition to specific gravity, °API is also used to describe fuels. The value °API is a value used by the American Petroleum Institute (API) to describe the density of fuel oil. When °API is more than 10, the fuel is less dense than water. The value °API is calculated from the specific gravity using the following formula:

$$°API = \left(\frac{141.5}{SG}\right) - 131.5$$

where
°API = API specific gravity
SG = specific gravity

For example, the °API gravity of a fuel with a specific gravity of 0.93 is calculated as follows:

$$°API = \left(\frac{141.5}{SG}\right) - 131.5$$
$$°API = \left(\frac{141.5}{0.93}\right) - 131.5$$
$$°API = 152.2 - 131.5$$
$$°API = \mathbf{20.7}$$

### Fuel Oil Handling

Different grades of fuel oils present different storage and handling problems. Heavier fuel oils must be heated to reduce their viscosities when pumping for transport. Care should be taken to not overheat fuel oils. As over-heated fuel oil becomes hotter, the sludge, sediment, and other impurities settle in the fuel oil tank. When residue builds up, the tank must be opened for cleaning. In addition, if the fuel oil is too hot, it can reach its flash point, which could lead to a fire.

When receiving a fuel oil delivery, care must be taken to prevent spills. Some fuel oil tanks are filled using gravity, although pumping fuel oil into the tanks is faster. The tank vent line must be clear to prevent a pressure buildup, which could result in fuel oil being blown out of the vent and fill lines.

Boiler rooms with furnaces that burn fuel oil must have the proper type of fire extinguishers placed in strategic locations. Automatic fire-suppression systems are also common in boiler rooms. **See Figure 8-4.** Foam, carbon dioxide, or dry chemical extinguishers should be used on fuel oil fires. If a fuel oil spill occurs, it should be cleaned up immediately and all fuel oil rags should be disposed of properly.

## FIRE-SUPPRESSION SYSTEMS

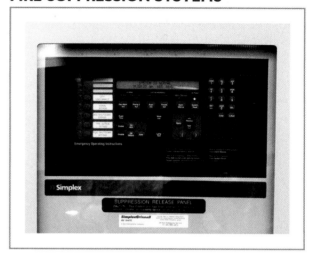

**Figure 8-4.** Fire-suppression systems help minimize fire risks.

## NATURAL GAS

*Natural gas* is a colorless and odorless combustible gas that consists mainly of methane with small quantities of ethylene, hydrogen, and other gases. An odorant is added to make leaks easy to detect. Natural gas is often simply referred to as "gas." Natural gas is found in oil fields and coal fields and can be recovered from shale fields. It burns cleaner than other fuels, causing less pollution. To comply with environmental regulations, natural gas is being used more frequently than fuel oil.

### Natural Gas Heating Values

Natural gas has a Btu content that ranges from 950 Btu/cu ft to 1050 Btu/cu ft. Because the Btu content of natural gas varies, heating values are expressed using a standard measurement called a therm. A *therm* is a unit used to measure the heat content of natural gas and is equal to 100,000 Btu. The heating value of natural gas is determined by using a gas calorimeter or by a chemical analysis of the gas.

### Natural Gas Handling

Natural gas is very explosive. Natural gas received from the supplier must be regulated for use in either low-pressure or high-pressure gas burners. All natural gas lines and controls should be tested for leaks by using gas leak detectors. Vent lines from regulators, reducing valves, or governors must be piped out of the boiler room to an area where they can be discharged safely. NFPA 54, *National Fuel Gas Code,* regulates the procedures and equipment used for handling natural gas. It requires the use of air or inert gas when purging gas lines. It also requires that the purge gases be vented outside whenever possible and that combustible gas detectors be used.

## COAL

*Coal* is a solid black fossil fuel formed when organic material is hardened in the earth over millions of years. Coal for firing steam boilers is fed using pulverized coal or stokers. There are several types and classifications of coal that must be understood. Also, since coal is a solid fuel, there are handling issues that must be managed in order to successfully use coal.

### Coal Types

Coal commonly used for boiler fuel includes anthracite coal and bituminous coal. *Anthracite coal,* commonly known as hard coal, is a geologically old coal that contains a high percentage of fixed carbon and a low percentage of volatile matter. *Volatile gas* is the gas given off when coal burns. Anthracite coal burns more cleanly than bituminous (soft) coal. *Bituminous coal,* commonly known as soft coal, is a geologically young coal that contains a low percentage of fixed carbon and a high percentage of volatile gas.

Lignite, another form of coal, is used almost exclusively as fuel for electric power generation. *Lignite coal,* commonly known as brown coal, is a geologically young coal that contains a low percentage of fixed carbon and a very high percentage of volatile gas and moisture. Lignite coal is rarely used in smaller boilers as a fuel for combustion.

**Heating Values.** The heating values for coal commonly used in boilers vary considerably depending on the type of coal. **See Figure 8-5.** Hard, anthracite coal has a heating value of about 12,700 Btu/lb to about 13,600 Btu/lb. Typical soft bituminous coal has a heating value of about 11,000 Btu/lb to about 13,800 Btu/lb. Common medium-volatile bituminous coal has a heating value of about 14,000 Btu/lb.

| COAL HEATING VALUES | |
|---|---|
| **Coal Type** | **Heating Values\*** |
| Anthracite | 12,700 to 13,600 |
| Bituminous | 11,000 to 13,800 |
| Medium-volatile bituminous | 14,000 (standard) |

\* in Btu/lb

**Figure 8-5.** The content of coal determines its characteristics and furnace requirements.

**Classification and Analysis.** Coal is classified according to rank and grade. *Coal rank* refers to the hardness of coal. *Coal grade* refers to the size, heating value, and ash content of coal. The characteristics of coal must be identified with a proximate analysis or an ultimate analysis. A *proximate analysis* is an analysis used to determine the amount of moisture, volatile gas, fixed carbon, and ash in a coal specimen. An *ultimate analysis* is an analysis used to determine the elements present in a coal specimen.

### Coal Handling

When coal is used as the fuel for combustion, it must be stockpiled to meet plant needs. Anthracite coal is safer to stockpile because it is not as susceptible as bituminous coal to spontaneous combustion. *Spontaneous combustion* is the process where a material can self-generate heat until its ignition point is reached. Bituminous coal, having a higher volatile content, must be carefully monitored when being stockpiled to prevent spontaneous combustion. Rotating the coal in the stockpile also reduces the chance of spontaneous combustion.

From the stockpile, coal is transported into overhead bunkers where the possibility of spontaneous combustion also exists. In addition, the gases given off from the coal bunker are extremely toxic and explosive. The operator should make sure the coal stays dry because wet coal can produce sulfuric acid, which corrodes the plates of the coal bunker.

The burning of coal produces residual ash that must be removed and stored for disposal. Various types of conveyors are used to transfer the ash from the point where it is generated to the storage point. One common type of conveyor consists of a chain with scrapers that drag the ash along to a point where it can be dumped.

## ALTERNATIVE FUELS

Furnaces can be designed to burn available alternative fuels in the boiler as the sole fuel source. Many furnaces are cofiring units that are capable of burning two dissimilar fuels by using separate fuel trains. The firing head may change depending on the pressure and heat value of the fuel being fired. A *cofiring furnace* is a furnace that mixes burnable material, such as biomass, with a fossil fuel for combustion. Cofiring units can be used where abundant alternative fuel is available. Many of these furnaces are designed as stoker units.

Other types of furnaces use a gasifier to remove volatile gases from alternative fuels. A *gasifier* is a device used to extract volatile gases from a solid fuel by heating the fuel in the absence of oxygen. The extracted gases are then sent to the burner. These fuels may be considered renewable. The ashes from the gasifier are handled in a similar manner as the ashes from coal burning.

The heating values of alternative fuels can vary considerably from one source to another and even at different times from the same source. Therefore, all statements of heating values are approximations only. **See Figure 8-6.**

### ALTERNATIVE FUEL HEATING VALUES

| Fuel Type | Heating Values* |
|---|---|
| Wood (hog fuel) | 10 to 30 |
| Landfill gas | 400 to 500 |
| Digester gas | 600 to 650 |
| Coke oven gas | 550 to 600 |
| Propane | 2500 |

* Btu/cu ft

**Figure 8-6.** The heating value of alternative fuels can vary considerably from one fuel to another.

One concern with some alternative fuels, especially landfill gas and digester gas, is that these fuels may contain sulfur compounds. When sulfur compounds burn, they produce sulfur dioxide. The sulfur dioxide combines with water to produce corrosive compounds, which can cause damage to the combustion equipment. Stainless steel equipment is used in the presence of sulfur to prevent this corrosion.

### TECH TIP

*Many alternative fuels are solid fuels that cannot be pulverized. Therefore, burners for these fuels use a stoker to feed fuel into the furnace. Alternatively, a gasifier may be used to remove volatile gases for combustion.*

### Biomass

*Biomass* is a biological material used as a renewable energy fuel. Many operations generate combustible byproducts. For example, agricultural operations generate a considerable amount of biomass material. During the processing of a crop, the biomass can be collected for fuel. Biomass generally contains excessive moisture and will not sustain a flame without fossil fuels in a cofiring furnace. The cofiring furnace is normally limited to about 10% biomass to maintain the flame and stay within the limits of EPA air standards.

### Wood

Some industrial operations generate scrap wood as part of their operations. For example, pulp mills and lumber mills generate wood waste, known as hog fuel, as a byproduct of making wood and paper products. **See Figure 8-7.** Rather than scrapping this material, it can be gasified or mixed with other fuels and burned to produce power and heat for an industrial process. The wood scrap can also be ground up by a chipper to make hog fuel and burned in a boiler. However, lumber mills may find it more economical to sell the wood chips for chipboard and use natural gas in their boilers.

### WOOD WASTE

**Figure 8-7.** Wood waste, a byproduct of manufacturing wood and paper products, can be used as a fuel for boilers.

### Municipal Solid Waste

*Municipal solid waste (MSW)* is the trash or garbage from residential and commercial users that is collected for disposal. Waste-to-energy (WTE) plants use specially designed combustors to incinerate MSW and produce electricity.

MSW contains a diverse mix of waste material, most of which is burnable. After metals and other nonburnable items are removed, incineration reduces the remaining material to a small fraction of its bulk. Paper and wood products included in the collected material may be considered biomass. A furnace for MSW is typically designed to cofire with coal or fuel oil to ensure a steady flame. Additional controls and equipment are required to handle MSW.

### Landfill Gas

*Landfill gas* is gas collected from a landfill that is made up of mainly methane gas and carbon dioxide with varying amounts of miscellaneous gases. Landfill gas consists of 40% to 60% methane, depending on the makeup of the constituents forming the gas from the waste. The remainder is mostly carbon dioxide with usually less than 1% miscellaneous gases, which may include nitrogen, oxygen, water vapor, hydrogen sulfide, and some heavy metals such as mercury. Decomposition in landfills causes the gases to be released into the atmosphere. Methane and carbon dioxide are considered greenhouse gases that add to global warming.

Options for managing landfill gas are flaring (burning the gas off as it is produced), using it as a fuel for boilers to produce heat, using it as a fuel for internal combustion engines, using it in gas turbines and fuel cells to produce electricity, or cleaning it sufficiently to pipe it into natural gas lines. Using landfill gas as a fuel in boilers is often the most cost-effective option and produces the cleanest emissions of all the combustion options.

According to the EPA, there were 634 landfill gas projects in the United States in 2017 that used waste methane gas. The number is increasing as more landfills install gas recovery systems. Landfill gas is often considered a renewable energy source. However, waste going into the landfill can be reduced by recycling, decreasing this renewable aspect. The EPA considers energy generated from landfill gases to be green energy.

### Digester Gas

*Digester gas* is gas produced by the biological breakdown of organic matter in the absence of oxygen. Digester gas is produced by breaking down organic material such as manure, sewage, municipal waste, plant material, and dedicated energy crops such as grass and corn in an anaerobic digester.

Digester gas can serve as a fuel in applications such as boilers, hot water heaters, reciprocating engines, turbines, and fuel cells. Digester gas is usually more than

60% methane, although some state-of-the-art plants have the potential to produce concentrations of methane up to 95%. The remainder is mostly carbon dioxide with less than 1% other gases.

In the United States, about 3500 of 16,000 wastewater treatment plants use anaerobic digestion. Most of these treatment plants use the digester gas to supply the heat needed for the digestion process. Some of these plants use excess digester gas to produce electricity.

### Other

Other gases, such as coke oven gas, blast furnace gas, butane, propane, and methane, are sometimes used in furnaces. Some of these gases may be used in boilers at locations where the gases are generated but are not typically used in utility boilers.

---

### SECTION 8.1–CHECKPOINT

1. What are the differences in how No. 2 fuel oil and No. 6 fuel oil are used?
2. Why does No. 6 fuel oil need to be heated before use?
3. What is a therm?
4. What are common procedures when handling natural gas?
5. What are common types of coal?
6. What is spontaneous combustion?
7. What are common alternative fuels that can be burned in a boiler?

---

## SECTION 8.2
## COMBUSTION OF FUELS

### COMBUSTION

Combustion of fuels occurs when the fuel, such as fuel oil, natural gas, or coal, combines with oxygen to produce heat. **See Figure 8-8.** Achieving the proper combustion can be complex. The ratio of the air to the fuel and the mixing of the two determine the type of combustion that occurs. To understand the reaction taking place in a furnace, a gas analyzer must be used.

## COMBUSTION OF FUELS

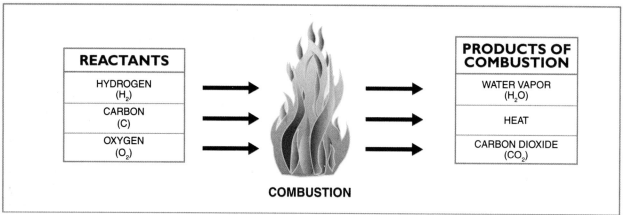

**Figure 8-8.** Fuels and oxygen react during combustion to generate water vapor, heat, and carbon dioxide.

### Types of Combustion

The three types of combustion are perfect, complete, and incomplete. *Perfect combustion* is combustion that occurs when all the fuel is burned using only the theoretical amount of air. The theoretical amount of air is the amount of air needed to completely react with all the fuel without any extra air or fuel. The theoretical amount of air is often called the stoichiometric amount of air. Perfect combustion cannot be achieved in a boiler because of imperfect mixing of the fuel and air and the varying environmental conditions that adversely affect the combustion conditions.

*Complete combustion* is combustion that occurs when all the fuel is burned using a minimum amount of excess air. Complete combustion is the boiler operator's goal. If complete combustion is achieved, the fuel is burned at the highest combustion efficiency with minimum pollution. *Incomplete combustion* is combustion that occurs when all the fuel is not burned, resulting in the formation of soot and smoke.

Hot gases of combustion are produced as any fuel is burned. The gases of combustion are sometimes called flue gases because they flow through a flue. *Flue* is the general term for the path used by the gases of combustion as they flow from the point of combustion to the point where they are released to the atmosphere.

### Air Supply

Air is necessary for the combustion of fuel. Air contains approximately 21% oxygen and 79% nitrogen. Oxygen supports combustion, but it is not combustible. It will not burn without the introduction of other elements. Nitrogen is not combustible nor does it support the combustion process.

The types of air used in the combustion process are primary air, secondary air, and excess air. *Primary air* is the air supplied to a burner to atomize fuel oil or convey pulverized coal and control the rate of combustion, thus determining the amount of fuel that can be burned. *Secondary air* is the air supplied to the furnace by draft fans to control combustion efficiency by controlling how completely the fuel is burned. *Excess air* is air supplied to the boiler that is more than the theoretical amount of air needed for combustion.

The amount of excess air needed will vary depending on the type of fuel used. Current industry practice is to monitor the oxygen level in the gases of combustion. The oxygen level in the gases of combustion should be kept low to achieve the highest combustion efficiency. **See Figure 8-9.** A minimum amount of excess air is always used in burners. However, reducing excess air while still completely burning the fuel reduces the heat lost to the atmosphere through the stack.

### Combustion Reactions

Carbon, hydrogen, and sulfur are the combustibles in fuel that combine with oxygen in the air. A combustion reaction takes place in the furnace of the boiler. In combustion, the fuel combines with the oxygen from the air to form carbon dioxide, water, and traces of other compounds. Incomplete combustion simply means that not all of the fuel combines with the oxygen and some unburned or partially burned fuel leaves the furnace. This unburned fuel may be in the form of raw fuel, carbon monoxide, or soot and smoke.

## COMBUSTION EFFICIENCY

**Figure 8-9.** Combustion control systems can automatically calculate combustion efficiency.

**Elements and Reactions.** Letters of the alphabet are used to represent elements. For example, the letter C represents carbon, H represents hydrogen, O represents oxygen, and S represents sulfur. Subscripts list the number of atoms of a particular element that occur in a molecule. For example, an oxygen molecule occurs naturally in the atmosphere as two oxygen atoms and is shown as $O_2$.

Most fuels contain hydrogen. Natural gas is primarily methane ($CH_4$), which consists of one carbon and four hydrogen atoms with small amounts of ethane ($C_2H_6$). Fuel oil and coal are more complex combinations of hydrocarbons (carbon and hydrogen), with trace amounts of other elements.

Fuels containing hydrocarbons react in the presence of heat and oxygen to form carbon dioxide ($CO_2$) and water ($H_2O$). The term "reaction" is used to describe the process where elements or molecules chemically combine with each other. For example, natural gas goes through the following chemical reaction during complete combustion:

Fuel + oxygen → carbon dioxide + water + heat

$$CH_4 + 2O_2 \rightarrow CO_2 + 2H_2O + \text{heat}$$

Heat from combustion breaks apart the complex fuel molecules into elements that can burn. Some of the energy contained in the fuel is used to break apart the fuel, and some of the energy is available as heat. With incomplete combustion, there is not enough oxygen for the amount of fuel present. For example, a fuel containing carbon may not react completely to form carbon dioxide. Some of the carbon forms carbon monoxide instead. When more air is added, the carbon monoxide can react with oxygen to form carbon dioxide.

$$C + O \rightarrow CO + \text{heat}$$
$$CO + O \rightarrow CO_2 + \text{heat}$$

Some fuels contain small amounts of sulfur. When sulfur combines with oxygen, it forms sulfur dioxide. This is undesirable because sulfur dioxide is the pollutant that causes acid rain.

$$S + O_2 \rightarrow SO_2 + \text{heat}$$

### TECH TIP

*Sulfur compounds are classified as a pollutant because they react with water vapor to form sulfuric acid, which is harmful to the environment. The level of sulfur oxides emitted depends on the sulfur content of the fuel.*

**MATT.** When firing a boiler, the boiler operator's goal is to achieve complete combustion. This involves burning all fuel using the minimum amount of excess air. To achieve complete combustion, the proper mixture of air and fuel, proper atomization, proper fuel temperature, and sufficient time are required. These requirements can easily be recalled by thinking of the term "MATT."

*MATT* is an acronym that stands for "mixture, atomization, temperature, and time." The M stands for the mixture of air and fuel. The proper air-fuel ratio must be maintained at all firing rates. High firing rates burn the maximum amount of fuel and require more air than low firing rates. The A stands for the atomization of the fuel. The T stands for temperature. The proper temperature of the air, fuel, and furnace zone must be maintained to achieve complete combustion. The second T stands for the time needed to achieve complete combustion. The combustion process must be completed before the gases of combustion come in contact with the heating surfaces.

## Products of Combustion

In the combustion process, fuel combines with oxygen to form heat and the products of combustion. This reaction results in the formation of products of combustion such as carbon monoxide, carbon dioxide, sulfur dioxide, nitrogen oxides, oxygen, and water. In addition, soot, steam, smoke, and flyash are discharged. The products of combustion will vary based on the fuel burned and the boiler design. The operator is responsible for ensuring the most efficient combustion of the fuel in the boiler. By properly controlling the combustion process, the operator can reduce emissions.

**Combustion Gas Analysis.** Combustion gas analysis is used to determine boiler combustion efficiency. Carbon dioxide in the gases of combustion indicates burned fuel. Carbon monoxide in the gases of combustion indicates incomplete combustion. Oxygen in the gases of combustion indicates excess air.

**Electronic Gas Analyzers.** Gases of combustion can be checked quickly and accurately for temperature, gases, draft, and smoke with an electronic gas analyzer. **See Figure 8-10.** An electronic gas analyzer typically uses a probe inserted in the breeching as close to the boiler as possible. An electronic gas analyzer may have a small fan or pump inside to draw in a sample of the gases of combustion. Sensors in the instrument analyze the gases for temperature, types and amounts of gases present, amount of draft, and amount of smoke.

*Cleaver-Brooks*
*Analysis of the gases of combustion helps when tuning a burner to maximize efficiency.*

Data from the sensors is sent to a controller to determine combustion efficiency, amount of excess air, and amount of carbon dioxide. The results are displayed and can be printed or plotted when logging shift activities. This information can also be sent to remote locations and used as a tool to help control plant operating costs. Many gas analyzers are Fyrite® analyzers. Fyrite is a trademarked name used by a manufacturer of combustion gas analyzers.

**Portable Gas Analyzers.** An electronic gas analyzer can also be a portable type with a handheld probe. **See Figure 8-11.** The handheld probe is temporarily inserted into the breeching to measure the gases of combustion exiting the boiler. The probe is held in place while the reading is taken. The data is sent to a processor. A printout displays an analysis of the data, which can be used for troubleshooting the combustion process.

**Orsat Analyzers.** A gas analysis can also be done with an Orsat analyzer. An *Orsat analyzer* is an old method of measuring the carbon dioxide, carbon monoxide, and oxygen in the gases of combustion by selectively absorbing the gases and measuring the volume removed from the gases of combustion. Orsat analyzers are rarely used anymore for analyzing combustion gases from boilers.

## ELECTRONIC GAS ANALYZERS

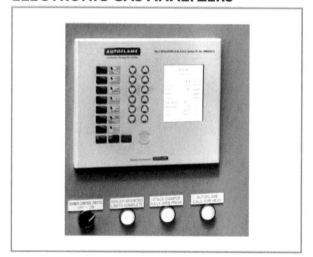

**Figure 8-10.** An electronic gas analyzer can show analysis results on a display panel.

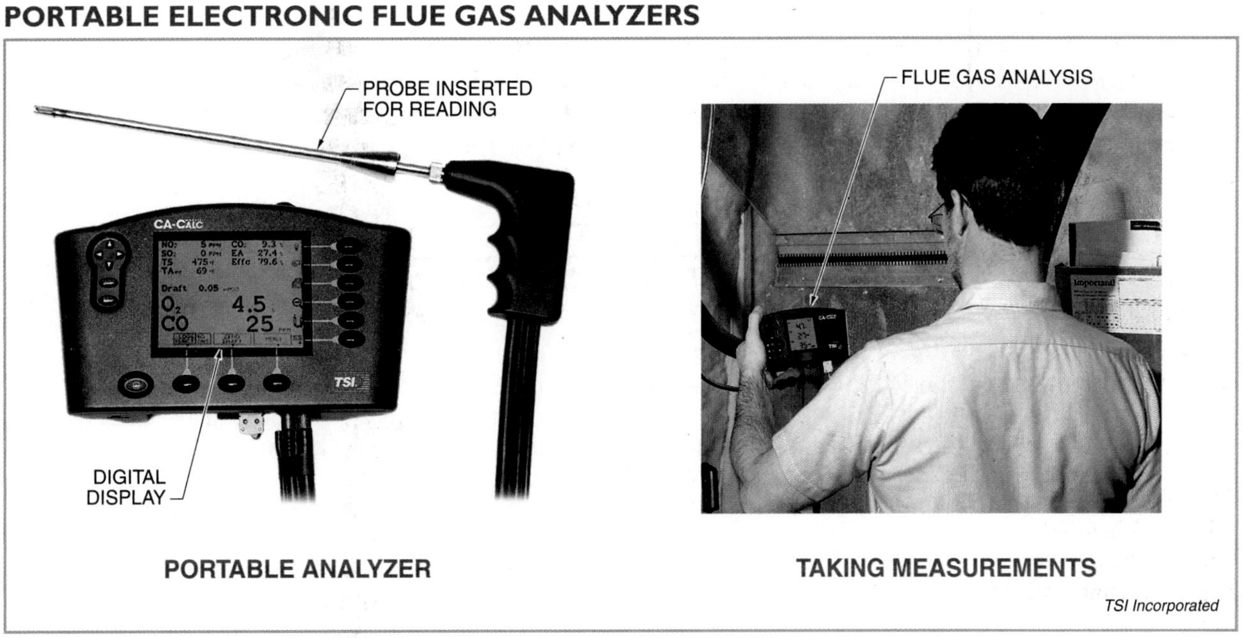

## PORTABLE ELECTRONIC FLUE GAS ANALYZERS

**PORTABLE ANALYZER**          **TAKING MEASUREMENTS**

*TSI Incorporated*

**Figure 8-11.** A portable electronic flue gas analyzer can be used on individual boilers.

### Furnace Volume

Furnace efficiency is affected by furnace volume and the amount of heat that can be released within that volume. Furnace volume is the amount of space available in a furnace where the fuel can burn before it comes in contact with the boiler heating surface. The design, shape, and size of a furnace depend on the type of fuel being burned. Another factor considered in determining the required furnace volume includes the type of draft used. The volume of the furnace must be large enough to achieve complete combustion of all the fuel under all load conditions before the fuel enters the first pass or touches the heating surfaces of the boiler to prevent the formation of soot and smoke.

### SECTION 8.2—CHECKPOINT

1. What is the difference between complete combustion and incomplete combustion?
2. What is the difference between primary air and secondary air?
3. What does MATT mean?
4. Why is it undesirable to burn fuels containing sulfur?
5. What are common products of combustion?
6. How does a gas analyzer sample the gases of combustion?
7. How does furnace volume affect furnace efficiency?

## CASE STUDY—AIR-FUEL RATIO AND SOOT DEPOSITS

### Situation

Field experience was required for students enrolled in a building maintenance and operation program. The college maintenance supervisor allowed the class to assist in the annual maintenance on the campus boilers to provide hands-on experience with boiler maintenance.

With one exception, all the boilers on the campus were small, hot water boilers used to heat a single building. The oldest section of the campus used steam from a gas-fired low pressure steam-heating boiler, which had a fuel oil backup. A low pressure steam-heating boiler is a boiler operated at pressures not exceeding 15 psi. This boiler was a firebox boiler that operated at 11 psi.

### Problem

When the front doors of the boiler were opened, exposing the front tube sheet, the students found about an inch of soot on all the exposed surfaces. According to the maintenance crew, the boiler had not used fuel oil for many years. It had been firing on natural gas. Natural gas does not normally cause a soot buildup.

The exposed surfaces of the boiler were cleaned with wire brushes. Then, the tubes were cleaned with a wire brush attached to a long pipe. The breeching was also cleaned with wire brushes. The breeching was about 4' on each side and about 8' long. Once all of the soot was removed, the students working from the inside could see daylight through hundreds of pinholes in the breeching. The firebox was then cleaned, and the students noticed damage to the firebrick. The water side was then dumped and rinsed clean.

### Solution

The college contracted an outside firm to adjust the burner for proper operation with gas as a fuel. The college also hired a contractor to replace the breeching and repair the firebrick. The boiler passed its annual inspection after all repairs were made.

In all, about 100 gal. of soot was removed from the boiler. When the class checked the boiler later, it was burning clean. It was a valuable experience on the importance of properly operating and maintaining a boiler. When the air-fuel ratio is out of balance, and the burner goes fuel-rich, heavy soot deposits can form even with a natural-gas-fired boiler.

**Name** _____ **Date** _____

_____ 1. The lowest temperature at which the vapor of fuel oil ignites when exposed to an open flame is its ___ point.
    A. flash
    B. fire
    C. pour
    D. viscosity

_____ 2. Rank refers to the ___ of a coal.
    A. ash content
    B. Btu content
    C. clinker formation
    D. hardness

_____ 3. Grade refers to the ___ of a coal.
    A. size, heating value, and ash content
    B. proximate analysis
    C. clinker formation
    D. hardness

_____ 4. Air supplied to the burner that is more than the theoretical amount needed to burn a fuel is ___ air.
    A. control
    B. excess
    C. secondary
    D. primary

_____ 5. When the temperature of fuel oil increases, its viscosity ___.
    A. remains the same
    B. increases
    C. decreases
    D. is not affected

_____ 6. The lowest temperature at which fuel oil will flow is its ___ point.
    A. flash
    B. fire
    C. pour
    D. viscosity

_____  **7.** Fuel oil with a low flash point would be ___.
    A. easy to pour
    B. used only in high pressure plants
    C. used in either high or low pressure plants
    D. dangerous to use

_____  **8.** A ___ fire extinguisher is the correct type of extinguisher to use on fuel oil fires.
    A. foam or dry chemical
    B. soda water
    C. carbon tetrachloride
    D. feedwater

_____  **9.** Air used to atomize fuel oil is ___ air.
    A. control
    B. excess
    C. secondary
    D. primary

_____  **10.** Air that controls how completely a fuel is burned is ___ air.
    A. control
    B. forced-draft
    C. secondary
    D. primary

_____  **11.** ___ combustion is when all the fuel is burned using only the theoretical amount of air.
    A. Complete
    B. Incomplete
    C. Perfect
    D. Theoretical

_____  **12.** ___ combustion is when all the fuel is burned using the minimum amount of excess air.
    A. Complete
    B. Incomplete
    C. Perfect
    D. Theoretical

_____  **13.** Gases of combustion that cool on contact with the boiler heating surface before combustion is completed cause ___.
    A. a loss of draft
    B. an increase in chimney temperature
    C. the formation of soot and smoke
    D. an increase in heat transfer

_____  **14.** Air contains approximately ___% oxygen.
    A. 5
    B. 21
    C. 50
    D. 79

_____ **15.** Anthracite coal is safer to stockpile than ___ because it is not as susceptible to spontaneous combustion.

       A. bituminous coal

       B. medium-volatile bituminous coal

       C. lignite

       D. brown coal

_____ **16.** Soot and smoke are the result of ___ combustion.

       A. incomplete

       B. complete

       C. perfect

       D. theoretical

_____ **17.** Carbon monoxide in the flue gas indicates ___.

       A. complete combustion

       B. incomplete combustion

       C. a high chimney temperature

       D. perfect combustion

_____ **18.** The lowest temperature at which fuel oil will burn continuously when exposed to an open flame is its ___ point.

       A. flash

       B. fire

       C. pour

       D. viscosity

_____ **19.** ___ must be preheated to reach the required temperature for combustion.

       A. No. 2 fuel oil

       B. No. 6 fuel oil

       C. Coal

       D. Natural gas

_____ **20.** Higher numbered fuel oils produce ___ lower numbered fuel oils.

       A. more Btu/gal. than

       B. fewer Btu/gal. than

       C. the same amount of Btu/gal. as

       D. one-half the Btu/gal. as

_____ **21.** The viscosity of a fuel is expressed in ___.

       A. specific gravity (SG)

       B. fire point units (FPUs)

       C. Saybolt universal seconds (SSUs)

       D. British thermal units (Btu)

_____ **22.** An ultimate analysis of coal provides information regarding ___.

       A. elements present in the coal

       B. moisture content

       C. volatile gases

       D. fixed carbon and ash content

_____  **23.** ___ is biological material used as a renewable energy fuel.
A. Lignite
B. Biomass
C. Peat
D. Coal

_____  **24.** ___ will support combustion but is not itself combustible.
A. Nitrogen
B. Mercury
C. Oxygen
D. Hydrogen

_____  **25.** The letter "M" in the acronym "MATT" stands for ___.
A. mixture of air and fuel
B. monitor the air-fuel ratio
C. measurement of quantity of heat
D. methane fuel

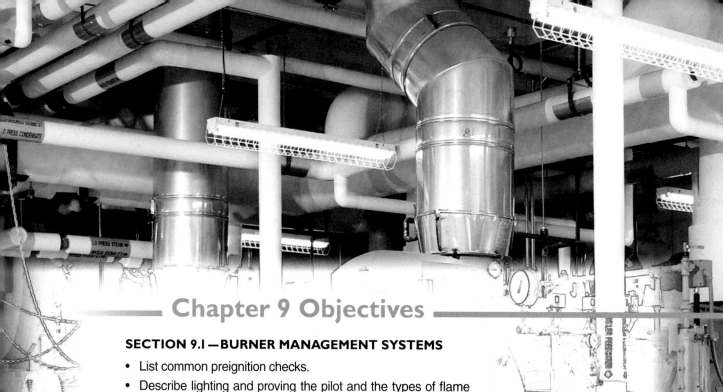

# Chapter 9 Objectives

### SECTION 9.1—BURNER MANAGEMENT SYSTEMS

- List common preignition checks.
- Describe lighting and proving the pilot and the types of flame scanners used.
- Describe lighting and proving the main flame and releasing to automatic control.

### SECTION 9.2—COMBUSTION CONTROL SYSTEMS

- Define combustion control systems and explain how steam pressure and plant masters are used to control boilers.
- Explain how ON/OFF controls systems operate.
- Explain how modulating control systems operate.
- Explain how metering controls systems operate.
- Describe the types of control signals commonly used with combustion control systems.

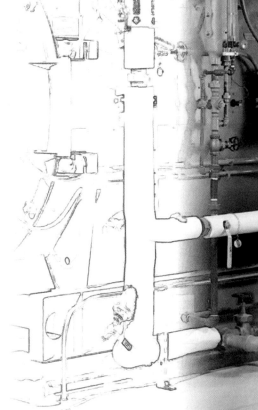

## ARROW SYMBOLS

| AIR | GAS | WATER | STEAM | FUEL OIL | CONDENSATE | AIR TO ATMOSPHERE | GASES OF COMBUSTION |
|-----|-----|-------|-------|----------|------------|-------------------|---------------------|
| ⇨ | ⇨ | ⇨ | ➡ | ⇨ | ➡ | ⇨ | ➡ |

**Digital Resources**
ATPeResources.com/QuickLinks
Access Code: **472658**

# CHAPTER 9

# COMBUSTION AND BOILER CONTROLS

## INTRODUCTION

A burner management system is responsible for managing the purging operations and firing cycles of a burner. A burner management system is also called a sequencing system. A combustion control system provides control through the burner management system by determining when to fire and how much fuel to burn in order to produce enough steam for the loads while maintaining the pressure at the setpoint.

## SECTION 9.1
## BURNER MANAGEMENT SYSTEMS

### BURNER FIRING SEQUENCE

Boiler burners can be dangerous. This is especially true when conditions change quickly during startup and shutdown. Therefore, a burner management system is used to ensure safe operation. A *burner management system* is a system of control devices and control logic used to ensure safe burner operation. A burner management system may also be known as a programmer, flame safeguard system, or sequencing system.

A burner must be started in a particular sequence to ensure a safe startup. The burner management system controls the operation and sequence of the startup. Burner status information can be shown on a flame monitor display. **See Figure 9-1.** Many manufacturers use programmable logic controllers (PLCs) for the platform for the burner management system.

An important part of the burner management system is the safety interlocks that affect the burner firing sequence. Common checks and procedures used by the burner management system include preignition checks, purge cycles, lighting the pilot, lighting the main flame, and postpurge during shutdown.

## FLAME MONITOR DISPLAYS

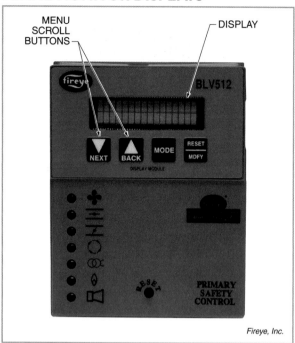

**Figure 9-1.** Burner status information can be shown on a flame monitor display.

### Preignition Checks

A burner management system checks many boiler and burner components before starting a firing cycle. The burner management system proves, or verifies, that the fuel valves are closed; the operating and high-limit controls are closed; the fuel pressure proving switch is closed; air dampers are in their low-fire, open position; and no false flame is detected. One type of limit switch is a gas-pressure proving switch. **See Figure 9-2.**

Other checks may be performed, depending on the details of the boiler and fuel system design. An interlock is provided to prove airflow during the firing cycle. The final preignition step is to pre-purge the furnace with the damper in the full open position and then return it to the low-fire position for ignition.

### Purge Cycles

A *purge cycle* is a process during which air is blown through a furnace for a predetermined period of time to remove any combustible vapors. *Pre-purge* is a purge that occurs before a burner is allowed to fire. The length of time needed for a purge cycle depends on the furnace size, standard procedures, and insurance company requirements. An airflow proving switch is used to verify

that air is moving through the furnace and that a purge has actually occurred. The first step in a purge cycle is to start the motor for the combustion or secondary air blower. **See Figure 9-3.** The secondary air blower then blows air into the furnace to purge any unburned fuel.

## GAS-PRESSURE PROVING SWITCHES

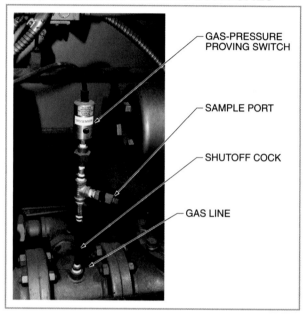

**Figure 9-2.** A gas-pressure proving switch is used to verify that the fuel is at the correct pressure before the burner management system allows the flame to be started.

### Lighting the Pilot

The next step in the firing cycle is to automatically open a valve in the fuel pilot line to allow fuel to flow through the pilot train to the pilot. At the same time, the circuit to the ignition transformer is energized, causing a spark in front of the fuel pilot tube and igniting the fuel. **See Figure 9-4.**

**Proving the Pilot.** *Proving the pilot* is a process in which a flame scanner is used to observe the pilot to verify that it is lit. A *flame scanner* is a safety device with a flame sensor that detects the presence of a flame and signals to the burner management system whether it is safe to operate the boiler. **See Figure 9-5.** The flame scanner proves the pilot to allow the firing cycle to continue when steam demand initiates the combustion cycle. The burner management system has a preprogrammed time to prove the pilot. If the burner management system cannot prove that the pilot has been lit, the burner management system shuts down the system, purges the furnace, and initiates a flame failure alarm.

## PURGE CYCLES

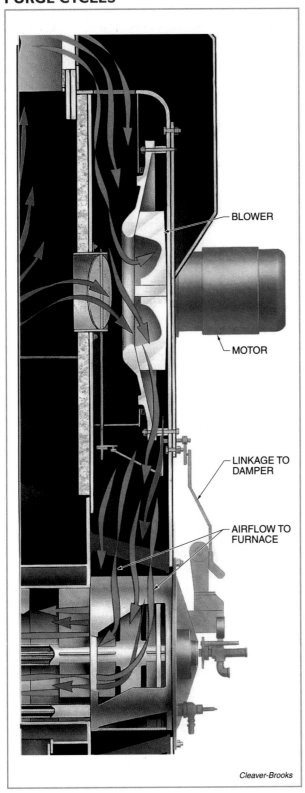

Figure 9-3. A purge cycle flushes unburned fuel vapors out of a burner.

Labels: BLOWER, MOTOR, LINKAGE TO DAMPER, AIRFLOW TO FURNACE

*Cleaver-Brooks*

## IGNITION TRANSFORMERS

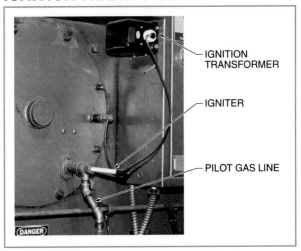

Labels: IGNITION TRANSFORMER, IGNITER, PILOT GAS LINE, DANGER

Figure 9-4. An ignition transformer provides the spark used to ignite fuel.

## FLAME SCANNERS

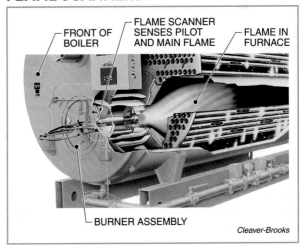

Labels: FRONT OF BOILER, FLAME SCANNER SENSES PILOT AND MAIN FLAME, FLAME IN FURNACE, BURNER ASSEMBLY

*Cleaver-Brooks*

Figure 9-5. A flame scanner uses a flame sensor to sense whether the pilot light and/or main flame is lit.

**Flame Sensors.** A *flame sensor* is a sensing device in a flame scanner that senses the pilot and main flame in the burner. **See Figure 9-6.** Flame sensors typically use a semiconductor device to sense the flame. The flame scanner sends a signal to the burner management system. The burner management system monitors this signal and shuts down the burner on flame failure when the signal is lost. A *flame failure* is a situation where the pilot or main flame fails to light properly or goes out unintentionally during normal operation. The most common flame sensor types include photocell flame sensors, infrared flame sensors, ultraviolet flame sensors, and flame rod sensors.

## FLAME SENSORS

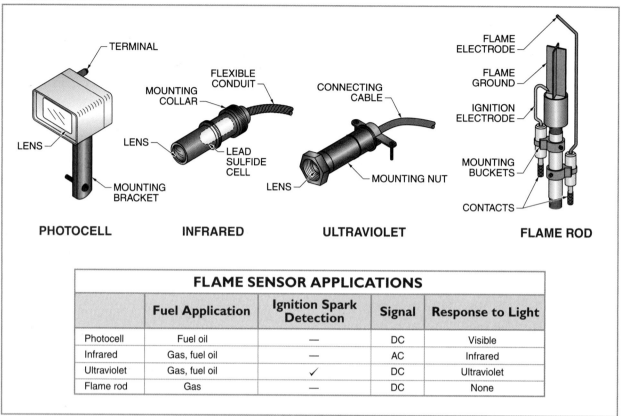

| FLAME SENSOR APPLICATIONS | | | | |
|---|---|---|---|---|
| | Fuel Application | Ignition Spark Detection | Signal | Response to Light |
| Photocell | Fuel oil | — | DC | Visible |
| Infrared | Gas, fuel oil | — | AC | Infrared |
| Ultraviolet | Gas, fuel oil | ✓ | DC | Ultraviolet |
| Flame rod | Gas | — | DC | None |

**Figure 9-6.** A flame sensor in a flame scanner is selected for the specific fuel and boiler design.

A *photocell* is a device in a flame scanner that generates a small amount of electric current when exposed to visible light. When a flame is present, the electric current flows from the photocell. This current indicates to the burner management system that a flame is present. When a flame is not present, the lack of electric current indicates to the burner management system that a flame is not present.

An *infrared flame sensor* is a device in a flame scanner that generates a small amount of electric current when exposed to infrared light. Infrared light is beyond the red end of the light spectrum and is not visible to the eye. The devices are often designed to detect multiple wavelengths of infrared light to improve reliability. As with a photocell, this current indicates to the burner management system whether a flame is present. Older flame sensors used a lead sulfide sensor.

An *ultraviolet flame sensor* is a device in a flame scanner that generates a small amount of electric current when exposed to ultraviolet light. Ultraviolet light is beyond the violet end of the light spectrum and is not visible to the eye. As with a photocell, this current indicates to the burner management system whether a flame is present. Many modern burner management systems use combination flame scanners that use a combination of two or more infrared, visible, and/or ultraviolet sensors in a single unit.

A *flame rod* is a sensing device in a flame scanner that generates a small electric current when one end of the sensor is placed directly in a flame. A flame rod must be installed so that the flame impinges on the flame rod. Flame rods are older technology and are rarely used in boilers today.

### Lighting the Main Flame

After proving the pilot, the burner management system completes a circuit to allow the main fuel valve to open. This allows the fuel to enter the furnace where it is ignited by the pilot. The valve to the pilot is closed after a predetermined time so that the only flame in the furnace is the main flame. The flame scanner is then used to prove the main flame.

**Proving the Main Flame.** *Proving the main flame* is a process in which a flame scanner observes the main flame to verify that it is lit. The flame scanner must be properly positioned to be able to sense both the pilot and main flames. The flame scanner signals to the burner management system that the main flame is lit. The burner management system allows combustion to continue as long as the signal is present. The burner management system has a preprogrammed time to prove the main flame. If the burner management system cannot prove the main flame, it shuts down the system and initiates a flame failure alarm. **See Figure 9-7.**

## FLAME FAILURE ALARMS

FLAME FAILURE ALARM

**Figure 9-7.** A flame failure initiates a flame failure alarm.

**Release to Automatic Control.** After the burner management system proves the main flame, the burner management system releases to automatic control. **See Figure 9-8.** The burner management system continues to monitor the signal from the flame scanner. As long as the flame scanner sends the signal, the burner management system allows the automatic control of the boiler.

If there is a flame failure while the burner is firing, the signal stops. The burner management system shuts down the burner and purges the furnace. This prevents furnace explosions caused by the ignition of the accumulated fuel in the furnace. After a flame failure, the burner management system must be manually reset to reinitiate the start cycle.

## Postpurge

*Postpurge* is a purge that occurs after burner shutdown. After a normal shutdown, the burner management system keeps the combustion blower running for a predetermined time to purge the furnace and prepare for the next startup.

## RELEASE TO AUTOMATIC CONTROL

**Figure 9-8.** After the burner management system proves the main flame, it releases control to the combustion control system.

---

### SECTION 9.1—CHECKPOINT

1. What are some common preignition checks?
2. What is a purge cycle?
3. Why is a flame scanner used to prove a pilot?
4. What are common types of flame sensors?
5. What must occur before the main flame is lit?
6. What happens if there is a flame failure during normal operation?

---

**SECTION 9.2**

# COMBUSTION CONTROL SYSTEMS

## BURNER FIRING RATE

A *combustion control system* is an automatic boiler control system that regulates fuel supply, air supply, air-fuel ratio, and draft in a boiler in order to deliver the required amount of steam to a load. **See Figure 9-9.** A combustion control system takes control of a boiler after the burner management system completes a firing cycle. Combustion control systems reduce the risk of human error and aid in overall plant efficiency.

## COMBUSTION CONTROL SYSTEMS

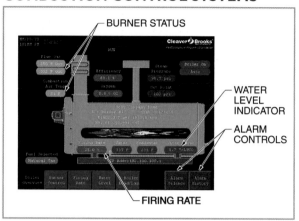

**Figure 9-9.** A combustion control system performs all the functions necessary to control the steam production rate of a boiler.

A burner should normally be started and stopped in low fire. Low fire is safer because it reduces the possibility of a flareback (minor furnace explosion) when excess fuel accumulates in the furnace. Starting in low fire minimizes thermal stress on boiler heat transfer surfaces.

The burner should normally fire for longer periods than it is idle. For example, it is better for the burner to fire for approximately 30 min and be off for 5 min instead of firing for 5 min and idling for 30 min. This firing schedule improves boiler efficiency by maintaining a relatively constant furnace temperature. Maintaining a constant furnace temperature also helps to reduce refractory maintenance caused by repeated heating and cooling of the brickwork.

### Steam Pressure

Steam pressure is the most common measurement used to determine the balance between steam generation and steam use. The steam pressure increases if steam is generated faster than it is used by the load. The steam pressure decreases if steam is used by the load faster than it is generated.

The burner firing rate is adjusted to balance the amount of steam generated by the boiler with the steam load demanded by the users. The fuel flow rate and airflow are adjusted to maintain the proper steam pressure in the burner. The fuel flow rate is controlled by a valve. The airflow rate is controlled by a damper or a combination of dampers, adjustable vanes at the blower intake, or variable-speed blowers. The control system maintains the proper ratio of fuel and air in the burner.

### Plant Masters

Many facilities have several boilers that operate together to deliver the total steam load to the steam header. A *plant master* is a master controller that calculates and distributes steam production requirements across several boilers to maintain the steam pressure at the setpoint. **See Figure 9-10.** The plant master sends a signal to each boiler telling the boiler how much steam to produce. Each boiler has a combustion control system that works with the burner management system to automatically deliver the required amount of steam to the load.

## PLANT MASTERS

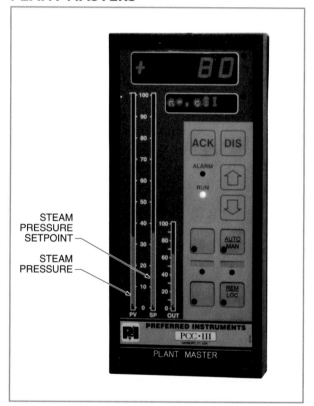

**Figure 9-10.** Steam pressure is used by a plant master to determine the firing rate.

## ON/OFF CONTROL SYSTEMS

*ON/OFF control* is a combustion control strategy that is used to start and stop a burner without any modulation of the flame. **See Figure 9-11.** ON/OFF control can be used in plants where the steam load is steady and allows the burner to be started and stopped. Simple ON/OFF control consists of an operating pressure control that signals a programmer to start and stop the burner and a high-pressure limit control with a manual reset.

## ON/OFF CONTROLS

**Figure 9-11.** ON/OFF control is the simplest type of control.

### Operating Pressure Controls

An *operating pressure control* is an ON/OFF control with an adjustable differential that regulates the operating range of the boiler between the cut-in pressure and the cut-out pressure by opening and closing electrical contacts. The contacts connect to circuits that start or stop the firing cycle in the burner management system. Operating pressure controls are connected to the highest part of the steam side of a boiler and are protected from live steam by a siphon. The three settings that can be used with an operating pressure control are cut-in (starting) pressure, differential pressure, and cut-out (shutoff) pressure. **See Figure 9-12.**

---

### TECH TIP

*ASME* Controls and Safety Devices for Automatically Fired Boilers *(CSD-1) was originally developed to ensure that operating controls and systems on boilers were checked regularly by operators and maintenance technicians to ensure safe operation.*

---

## OPERATIING PRESSURE CONTROL SETTINGS

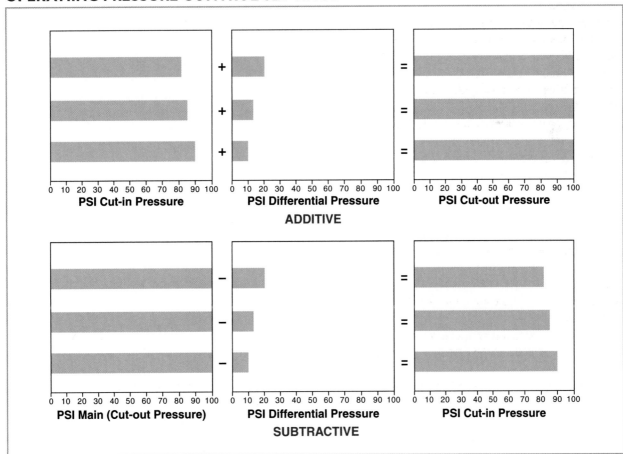

**Figure 9-12.** Operating pressure controls can be additive or subtractive.

An operating pressure control typically uses cut-out pressure and differential pressure settings to control the burner. The operating pressure control is set at the cut-out pressure that, once reached, opens a switch to shut off the burner. The differential pressure is subtracted from the cut-out pressure. The differential pressure indicates the amount the steam pressure must decrease before the burner needs to be started.

On some operating pressure controls, the differential pressure is additive. One setting is set at the cut-in pressure, where the burner needs to be started to increase the steam pressure. The other setting is set at the differential pressure, which indicates the amount the steam pressure must increase before the burner needs to be shut down. When adjusting the settings, cut-in pressure plus differential pressure equals cut-out pressure. When the cut-out pressure is reached in the boiler, the burner shuts off.

### High-Limit Pressure Controls

A *high-limit pressure control* is an ON/OFF control with a manual reset that sends a signal to the combustion control system to shut down the burner in the event of a high pressure condition that exceeds the operating pressure control cut-out setting. The presence of a reset switch means the high-limit pressure control acts as a safety switch. **See Figure 9-13.**

### HIGH-LIMIT PRESSURE CONTROLS

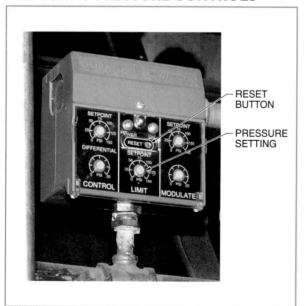

RESET BUTTON

PRESSURE SETTING

SETPOINT
DIFFERENTIAL
CONTROL
POWER
RESET
SETPOINT
LIMIT
SETPOINT
MODULATE

**Figure 9-13.** A manual reset is used to alert an operator that the cause of a shutdown must be identified and remedied before restarting the burner.

The high-limit pressure control is set slightly above the operating pressure control. In the event that the operating pressure control fails, the pressure will continue to rise to the setpoint of the high-limit pressure control, and then the high-limit control will shut the burner off. The boiler will not restart until the boiler operator pushes the reset switch. The operator must find and repair the cause of the failure before restarting the boiler.

### High/Low/OFF Control

*High/low/OFF control* is a combustion control strategy that senses the steam pressure and sends a control signal to start and stop the burner to maintain the steam pressure. High/low/OFF control is used in plants that have a variable steam demand large enough to require the burners to switch between high and low fire.

## MODULATING CONTROL SYSTEMS

A *modulating pressure control device* is a pressure control device that provides local control of the firing rate proportional to the steam pressure. **See Figure 9-14.** One type of modulating pressure control device consists of a bellows that moves a potentiometer to vary the electrical output. A modulating pressure control device can be connected directly to a modulating motor with a matching potentiometer that can be used to adjust the dampers and fuel flow.

There is both a minimum and maximum flame possible in a modulating burner. There are limitations on controlling gas and fuel oil flows with valves. In addition, there are limitations on controlling airflow with dampers and variable speed drives. These limit the amount of adjustment possible in a burner. A *turndown ratio* is the ratio of the maximum firing rate to the minimum firing rate. For example, if the maximum firing rate is five times the minimum firing rate, the turndown ratio if 5:1.

A modulating control system can be used instead of a modulating pressure control device on newer or larger boilers. A *modulating control system,* or positioning control system, is a combustion control system that is used to adjust the firing rate to a degree that is proportional to the steam load through the use of a PLC or digital control system. A modulating control system can also start or stop the burner automatically when necessary by using an operating pressure control.

## MODULATING CONTROL SYSTEMS

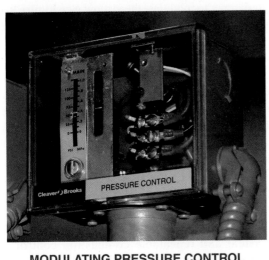

**MODULATING PRESSURE CONTROL**

LINKAGE DRIVE
TO COMBUSTION
CONTROLS

**MODULATING MOTOR**

**Figure 9-14.** A modulating pressure control device sends an electrical signal to a modulating motor to control the air-fuel ratio.

Modulating control is known as firing rate control when it is connected to a PLC-based platform. The PLC receives a signal from the pressure sensors, uses those signals to calculate what the appropriate output signals should be, and sends those signals to the final elements, which could be a modulating motor, fuel valve, or air damper. For example, a boiler may have an operating range of 90 psi to 100 psi. When the pressure in the boiler drops to 90 psi, the burner management system puts the burner through a firing cycle. The burner ignites in low fire and the modulating control system gradually brings the burner up to high fire by energizing the modulating motor.

The burner modulates between high and low fire to maintain the pressure within the operating range. If the boiler reaches its cut-out pressure of 100 psi, the operating pressure control switch opens, stopping the power to the control circuit, and shutting off the burner. The cycle is repeated when the boiler steam pressure drops to the cut-in point of 90 psi. It is not unusual for a boiler to modulate the controls to meet the steam demand for extended periods of time and not shut off for days, weeks, or months.

### Flow Characteristics

Fuel valves and air dampers are used to modulate the flow of fuel and air. They are usually nonlinear. This means that the percent that the valve or damper is open does not necessarily correspond to the percent of flow. For example, if a fuel valve is 25% open, it may have 48% total fuel flow, while at 50% open, it may have 80% total fuel flow. **See Figure 9-15.** Air dampers have different flow characteristics. Therefore, boiler designers must use cams or actuators to maintain the proper air-fuel ratio throughout the entire operating range.

## FLOW CHARACTERISTICS

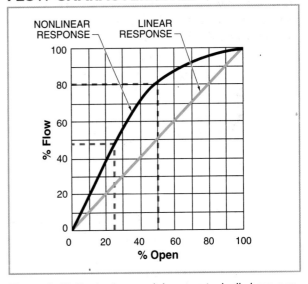

**Figure 9-15.** Fuel valves and dampers typically have nonlinear responses. Cams or actuators are used to adjust the final elements to maintain the proper air-fuel ratio.

### TECH TIP

*A boiler operator should always assume that a safety device has functioned correctly when a boiler shutdown occurs. It is always best to assume that an unsafe condition exists than to assume a safety device is faulty.*

## Single-Point Positioning

*Single-point positioning* is a modulating control strategy that uses steam pressure as the input signal and outputs a signal to a modulating motor that turns a jackshaft to modulate the air and fuel flow. Single-point positioning sends one signal only. **See Figure 9-16.** The fuel valve and air damper have mechanical linkages that connect to the jackshaft. The modulating motor turns the jackshaft and the air and fuel flow are changed at the same time. In addition, a jackshaft often has a linkage to a position sensor that verifies the movement of the jackshaft. The jackshaft arrangement puts mechanical constraints on the air-fuel ratio in order to minimize excess air.

## Parallel Positioning

*Parallel positioning* is a modulating control strategy that uses steam pressure as the input signal and outputs one signal to a fuel valve to modulate the fuel flow and another signal to a variable-speed blower motor or to an air damper to modulate the airflow. **See Figure 9-17.** Typically, the fuel valve and dampers are directly connected to motorized actuators.

Parallel positioning sends separate signals to two outputs. An important benefit of parallel positioning over single-point positioning is that these separate output signals allow the air-fuel ratio to be adjusted to different values across the complete firing range. The flow rates need to be properly adjusted to match the respective air-fuel ratios required as the burner modulates up and down. A low firing rate needs a slightly higher percentage of excess air than a high firing rate to ensure that all the fuel is burned and removed from the furnace. This results in very consistent and repeatable combustion control and efficiency readings.

## SINGLE-POINT POSITIONING

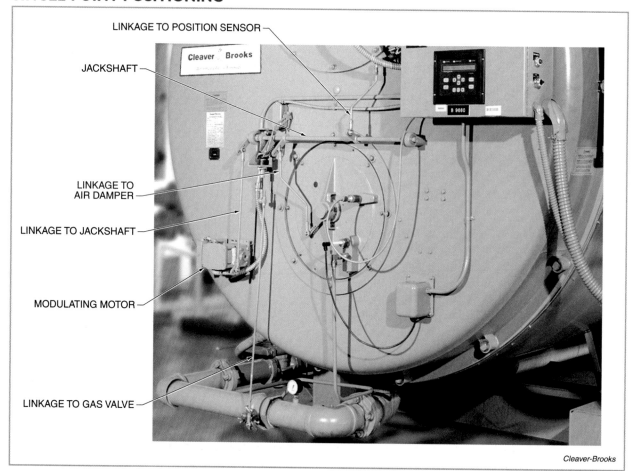

*Cleaver-Brooks*

**Figure 9-16.** Single-point positioning regulates the air and fuel flow using a single output signal to rotate a jackshaft and move linkages.

## PARALLEL POSITIONING

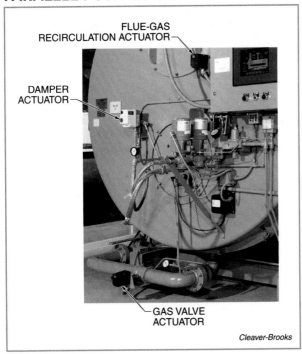

*Cleaver-Brooks*

**Figure 9-17.** Parallel positioning regulates airflow and fuel flow using two separate signals to separate actuators.

## METERING CONTROL SYSTEMS

A *metering control system* is a combustion control system in which the flow of fuel and air are measured and adjusted in correct proportions based on steam pressure. Changes in steam pressure result in control signals calling for changes in the flow of fuel and combustion air. Metering control systems maintain a constant steam pressure and high combustion efficiency and are used in generating stations, oil refineries, and large chemical plants.

A metering control system is more accurate and sensitive to steam pressure variations than a simple modulating control system and responds more quickly to air and fuel requirements. A metering control system is used in plants that require a constant steam pressure.

### Combination Gas/Fuel Oil

In automatic operation of a combination gas/fuel oil system, the metering control system functions in the following manner. **See Figure 9-18.** The gas flow control valve (1) on the gas line is controlled by signals from the gas pressure transmitter (2) and the steam pressure

transmitter (3). The most common instrument used to measure the flow of steam is an orifice installed in the steam line. The difference in pressure across the orifice varies according to the amount of steam flowing through it.

## METERING CONTROL SYSTEM—COMBINATION GAS/FUEL OIL

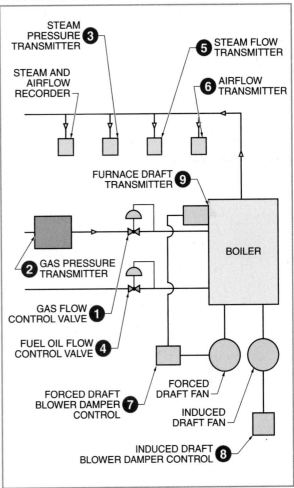

**Figure 9-18.** A metering control system in a combination gas/fuel oil system controls gas flow or fuel oil flow to the burner based on the flow and pressure of the steam generated.

When using fuel oil, the fuel oil flow control valve (4) is controlled by signals from the steam pressure transmitter, steam flow transmitter (5), and airflow transmitter (6). The forced draft blower damper control (7) is also controlled by the steam pressure transmitter, steam flow transmitter, and air flow transmitter. The induced draft blower damper control (8) is controlled by a signal from the furnace draft transmitter (9).

## Coal

In the automatic operation of a coal-fired system, a metering control system functions in the following manner. **See Figure 9-19.** The stoker feed control drive (1) regulates the amount of coal going to the stoker and is controlled by a signal from the steam pressure transmitter (2). The forced draft damper control drive (3) is controlled by a signal from the steam flow transmitter (4), steam pressure transmitter (2), and airflow transmitter (5). The outlet draft damper (6) is controlled by the outlet draft damper control drive (7), which receives its signal from the furnace draft controller (8) to regulate the flow of the gases of combustion.

## METERING CONTROL SYSTEM— COAL STOKER

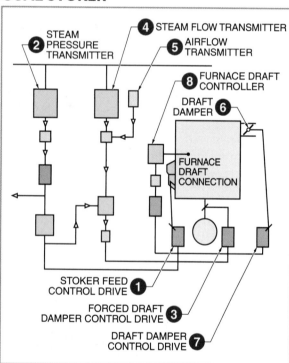

**Figure 9-19.** A metering control system in a coal-fired system is used to regulate the amount of coal going to the stoker.

## Cross-Limited Control

*Cross-limited control* is a control strategy in which increases in airflow lead increases in fuel flow, and decreases in airflow lag behind decreases in fuel flow. If the excess air is limited, any change in steam load can temporarily reduce the excess air to zero or less. This is not a safe condition because it can result in pockets of explosive combustible gases.

Cross-limited control increases safety by adding a linking system that prevents the airflow from going too low for the amount of fuel flow during load changes. This means that extra air is always provided during load changes.

## Oxygen Trim

*Oxygen trim* is a control strategy used to adjust the air-fuel ratio of a burner according to the results from a flue gas oxygen analyzer. **See Figure 9-20.** The oxygen analyzer output goes to a controller, which is programmed to maintain a specified percentage of excess oxygen. The excess oxygen setpoint depends on the steam load. For example, the excess oxygen may need to be 5% at low fire and 3% at high fire to ensure that all the carbon monoxide is consumed.

The excess air is typically measured with an oxygen sensor in the stack, which sends a signal to the combustion controls to adjust the air-fuel ratio based on varying environmental conditions such as ambient temperature, relative humidity, and atmospheric pressure.

A boiler is typically started up with more excess air than normal. Therefore, the controller should be programmed to avoid making changes based on the oxygen levels during startup. The oxygen levels can be used again once the boiler control has been released to modulation.

## OXYGEN TRIM

**Figure 9-20.** An oxygen trim strategy measures the excess oxygen in the gases of combustion and adjusts the airflow to maintain the excess air at the optimum level.

## Variable-Speed Drives

A *variable-speed drive (VSD)* is a motor controller used to vary the frequency of the electrical signal supplied to an AC motor in order to control its rotational speed. **See Figure 9-21.** A VSD may also be called a variable-frequency drive (VFD) or an adjustable-speed drive (ASD). The AC motor that drives an air blower can be controlled by a VSD.

Typical blower motors operate at a constant speed at 60 Hz. A VSD varies the frequency of the supply voltage to the value a blower motor needs to maintain a specific air pressure or flow. This is accomplished in the VSD with a rectifier that converts the supplied 60 Hz AC voltage to DC voltage. The DC then is fed to an inverter that converts the DC voltage back to AC voltage at a different frequency. The changed AC voltage is then supplied to the blower motor.

For example, a constant-speed blower motor may operate at 1800 rpm at 60 Hz, but would operate at 900 rpm when the voltage supplied to it through a VSD is 30 Hz. In addition to controlling motor speed, VSDs can control motor acceleration time, deceleration time, motor torque, and motor braking.

## CONTROL SIGNALS

Combustion controls may use data received from sensors located throughout a plant. The signals need to be sent from measuring instruments to a controller and from the controller to a final element that changes flow. Outputs that respond to this data are programmed in the combustion control system. **See Figure 9-22.** If the data received is outside the programmed limits, an annunciator or alarm is energized. An *annunciator* is an audible electronic alarm that alerts the operator of an operating condition that needs immediate attention.

Modern control systems use digital or electrical signals to communicate between the controller, the plant instruments, and the flow control valves, dampers, and variable-speed blowers. The valves and position-control linkages have actuators that receive the signals and modulate the final elements to continuously adjust the firing rate. Older control systems use pneumatic systems (air pressure) as the signal.

### TECH TIP

*PLCs use the binary system because it is easy to measure the presence or absence of voltage using binary numbers.*

## VARIABLE-SPEED DRIVES (VSDs)

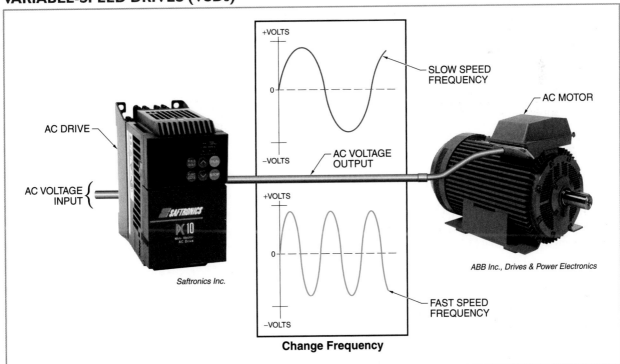

+VOLTS

0

−VOLTS

SLOW SPEED FREQUENCY

AC VOLTAGE OUTPUT

+VOLTS

0

−VOLTS

FAST SPEED FREQUENCY

AC DRIVE

AC VOLTAGE INPUT

AC MOTOR

*Saftronics Inc.*

*ABB Inc., Drives & Power Electronics*

**Change Frequency**

**Figure 9-21.** Variable-speed drives are used to change the supply voltage frequency to a motor to modulate the speed of a pump or blower.

## COMMUNICATIONS MODULE

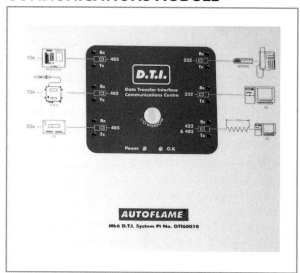

**Figure 9-22.** Data sent from sensors can be sent through a communications module to a controller.

### Digital Signals

Digital signals consist of binary (ON/OFF) signals in the form of low-voltage DC pulses sent over a network, or fieldbus. The binary signals are typically combined together in groups, called words or bytes, that can be used to communicate complex information. Networks follow a protocol to implement the transmission and reception of data. Fieldbus networks can transmit large amounts of information to a control system. For example, a fieldbus-compatible flowmeter can provide information about the flow rate in a process as well as diagnostic information about the meter itself.

### Electric Signals

Electric signals consist of continuous current signals in the form of electrical current in a loop over the range of 4 mA to 20 mA or 0 V to 10 V. The signals are scaled so that a signal of 4 mA represents the low end of the variable range and a signal of 20 mA represents the high end of the range. For example, a thermocouple may be used to measure furnace temperature. The output of the thermocouple can be scaled in a transmitter where 4 mA represents 0°F and 20 mA represents 2500°F. A signal of 12 mA would be the middle of the range and would represent a temperature of 1250°F.

### Pneumatic Signals

Older styles of control systems may use pneumatic controls. Pneumatic controls use compressed air to control combustion in the furnace. Pneumatic control systems require an air compressor, special filters, and driers so that only clean, dry air will enter the control air lines. A standby compressor or a cross-connection is used to tie in boiler room air in the event of a maintenance problem. A master control unit relays the pressure signals to relay units that control the air and fuel supplied to the furnace.

Pneumatic controls require clean, dry, compressed air for efficient operation. Compressed air tanks must be drained, air filters must be cleaned, and dehumidifiers must be functioning properly in order for pneumatic controls to operate efficiently.

---

### SECTION 9.2 – CHECKPOINT

1. What is a combustion control system?
2. What is a plant master?
3. What is an ON/OFF control?
4. What is the difference between operating pressure control and high/low/OFF control systems?
5. How do flow characteristics of a valve or damper affect control?
6. What is the difference between single-point and parallel positioning?
7. Why is cross-limited control safer than standard metering control?
8. What is oxygen trim?
9. What are three common types of control signals?
10. What is the normal range of current flow when using electric signals?

---

## CASE STUDY—FURNACE EXPLOSION

### Situation

The operating engineer in a shipyard was responsible for starting, stopping, operating, and logging the boilers, air compressors, cooling towers, and generators in the shipyard. This included any equipment that was on land or on one of the piers. If a boiler was located on a ship, barge, or floating dry dock, it was not the engineer's responsibility.

During a long four-day holiday weekend, it was decided to put a portable rental boiler on the deck of a dry dock to supply steam to a ship that was in for repairs. If the boiler was placed on the pier, the local code required the boiler to be checked every two hours by a licensed boiler operator. It was decided that putting the boiler on the dry dock meant that the security guard could check the boiler on his normal rounds and the cost of paying overtime to an engineer would be avoided.

### Problem

During the guard's normal rounds on Saturday night, the boiler was not operating, and there was a red indicator light indicating flame failure. When the guard pushed the reset button on the burner management system, the boiler went through its start-up sequence and shut down again due to flame failure. After pushing the reset button several times, the guard asked for the assistance of one of the riggers working in the area.

The burner management system on this boiler was an older style that had a mechanical timer with a thumbwheel to advance the timing sequence. The rigger told the guard that he had seen boiler repair technicians turn the timer's thumbwheel while working on the boiler. He proceeded to show the guard how to do it.

*Note:* The thumbwheel should not be used to advance the timer by anyone except experienced boiler repair technicians. When this is done, the burner management system purge cycle is bypassed. A modern burner management system has electronic timing and cannot be advanced by hand.

On the second or third attempt to restart the boiler, the buildup of unpurged fuel in the boiler's combustion chamber ignited and caused a furnace explosion. Fortunately, the explosion was not enough to blow the doors or the burner off of the boiler, but the concussion out of the stack knocked the guard into the water.

### Solution

Management decided that all of the boilers used for shipbuilding and repair should be placed on the pier so that only qualified, trained operators and the operating engineer would have jurisdiction over the boiler operations.

**Name** _____ **Date** _____

_____ 1. The primary function of a burner management system is to regulate the ___ of a boiler.
    A. fuel supply
    B. air supply
    C. draft
    D. safe operation

_____ 2. ON/OFF control is used to start and stop a burner ___.
    A. and modulate the flame to match steam load
    B. without modulating the flame to match steam load
    C. based on the measured fuel flow
    D. based on the measured draft

_____ 3. A siphon is used to protect an operating pressure control from ___.
    A. live steam
    B. water
    C. air
    D. gas

_____ 4. An operating pressure control must be connected to the ___.
    A. main steam line
    B. main feedwater line
    C. highest part of the steam side of the boiler
    D. water side of the boiler below the NOWL

_____ 5. The settings found on the boiler operating pressure control are used to ___.
    A. set the operating steam pressure range of the boiler
    B. eliminate low water
    C. maintain high and low fire
    D. warn the operator of flame failure

_____ 6. A combustion blower continues to run after fuel is shut off to the burner during ___.
    A. postpurging
    B. pre-purging
    C. synchronization
    D. ignition

_____   **7.** The burner management system in an ON/OFF control system is used to ___.
     A. control the burner firing cycle
     B. control the boiler operating range
     C. secure the burner in the event of low water
     D. regulate the steam flow

_____   **8.** To eliminate the danger of a furnace explosion during startup, the burner management system must allow ___.
     A. main flame ignition
     B. proving of the main flame
     C. a pre-purge period
     D. a varying steam load

_____   **9.** The purpose of the flame sensor in a flame scanner is to prove ___.
     A. only the pilot
     B. only the main flame
     C. both the pilot and the main flame
     D. the proper boiler water level

_____   **10.** A flame sensor is sensitive to ___.
     A. heat
     B. temperature
     C. pressure
     D. infrared, visible, or ultraviolet light

_____   **11.** In the event of a flame failure, the burner management system ___.
     A. automatically goes through a new firing cycle
     B. shuts the burner down
     C. vents the fire tubes
     D. starts the induced draft fan

_____   **12.** A modulating control system is sensitive to changes in the ___.
     A. boiler water level
     B. firing rate
     C. steam flow
     D. steam pressure

_____   **13.** A(n) ___ is a controller that distributes steam production requirements across several boilers to maintain the steam pressure at the setpoint.
     A. plant master
     B. operating pressure control
     C. ON/OFF control
     D. programmer

_____   **14.** A preignition check includes ___.
     A. verifying that the fuel pressure proving switch is closed
     B. lighting the pilot
     C. proving the pilot
     D. proving the main flame

_____ **15.** After a burner management system proves the main flame, it ___.

      A. verifies fuel pressure

      B. verifies airflow through the furnace

      C. releases to automatic control

      D. postpurges the furnace

_____ **16.** If steam is being used faster than it is being generated, the ___.

      A. steam pressure increases

      B. steam pressure decreases

      C. combustion control closes the fuel valve

      D. combustion control closes the inlet damper

_____ **17.** A high-limit pressure control ___.

      A. has a modulating controller

      B. uses a high/low/OFF controller

      C. uses differential pressure settings

      D. has a manual reset

_____ **18.** The flow of fuel through valves and air through dampers is usually ___.

      A. nonlinear

      B. linear

      C. not adjustable

      D. set at maximum

_____ **19.** Single-point positioning uses the ___ as the controller input and has only one output.

      A. fuel flow

      B. airflow

      C. steam pressure

      D. steam temperature

_____ **20.** A low firing rate needs ___ than at a high firing rate.

      A. more excess air

      B. less excess air

      C. more single-point positioning

      D. less single-point positioning

_____ **21.** Parallel positioning uses steam pressure as the controller input and ___.

      A. has ON/OFF fuel flow

      B. bypasses the postpurge

      C. uses one output signal

      D. uses two output signals

_____ **22.** A modulating control system adjusts the firing rate proportional to the ___.

      A. steam load

      B. amount of excess air available

      C. flue gas analysis

      D. water level

_____ **23.** Nonlinear flow means that the percent a valve or damper is open does not necessarily correspond to ___.
   A. percent of flow
   B. controller output
   C. steam pressure
   D. water level

_____ **24.** ___ control increases safety by adding a linking system that prevents the airflow from going too low for the amount of fuel flow during load changes.
   A. Cross-limited
   B. Modulating
   C. Single-point positioning
   D. ON/OFF with modulating

_____ **25.** ___ is used to adjust the air-fuel ratio based on the results of a flue gas analysis.
   A. Metering control
   B. Burner management
   C. Purging
   D. Oxygen trim

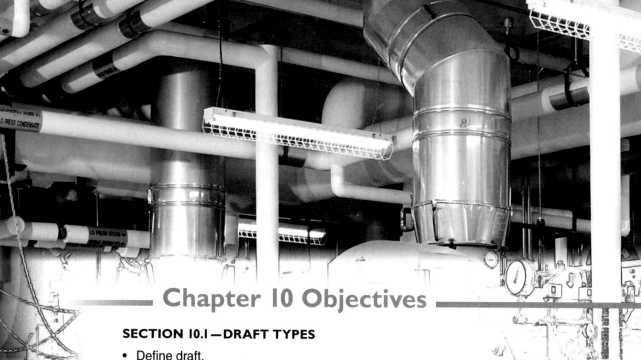

# Chapter 10 Objectives

### SECTION 10.1—DRAFT TYPES

- Define draft.
- Explain natural draft and mechanical draft.

### SECTION 10.2—DRAFT DESIGN AND CONTROL

- Identify methods for bringing outside air into a boiler room.
- Explain types of air heaters.
- Describe natural gas and fuel oil draft systems.
- Describe coal draft systems.
- Explain the principles of draft control.

### SECTION 10.3—ENVIRONMENTAL REGULATIONS

- Describe the Clean Air Act (CAA) and the six pollutants included in National Ambient Air Quality Standards (NAAQS).
- Describe the classification of emissions sources as major and area sources.
- Explain New Source Performance Standards (NSPS), hazardous air pollutants (HAPs), and the four types of control technology.
- Explain sulfur, particulate, and mercury removal methods.
- Explain low $NO_x$ burners.

## ARROW SYMBOLS

| AIR | GAS | WATER | STEAM | FUEL OIL | CONDENSATE | AIR TO ATMOSPHERE | GASES OF COMBUSTION |
|-----|-----|-------|-------|----------|------------|-------------------|---------------------|
| ⇨ | ⇨ | ➡ | ➡ | ⇨ | ➡ | ⇨ | ➡ |

**Digital Resources**
ATPeResources.com/QuickLinks
Access Code: 472658

CHAPTER **10**

# DRAFT SYSTEMS

## INTRODUCTION

Every fired boiler has a draft system. Draft provides the air needed to complete the combustion process when a fuel is burned. Draft must be strong enough to overcome the resistance caused by baffles, tube passes, dampers, economizers, air heaters, superheaters, and breechings. Draft removes the products of combustion and moves them through the boiler and up the stack.

Draft systems are designed for the type of fuel being burned. In addition, boiler rooms must be designed to permit the proper flow of outside air to the furnace. An airtight room creates a vacuum that starves the furnace of air. Outside air passes by the heated boiler shell or through an air heater where it is warmed prior to entering the furnace. This improves combustion efficiency.

Air emissions are regulated by the Environmental Protection Agency (EPA). The Clean Air Act (CAA) provides direction and requirements for the regulations. The National Ambient Air Quality Standards (NAAQS) and the EPA's list of hazardous air pollutants (HAPs) serve as the primary focus of environmental protection efforts.

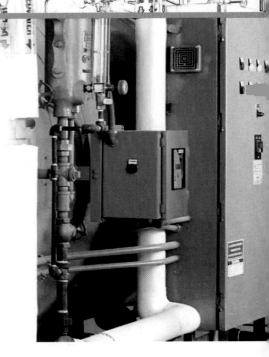

## SECTION 10.1
# DRAFT TYPES

### DRAFT

*Draft* is the flow of air or gases of combustion caused by a difference in pressure between two points. To burn a fuel in a furnace, sufficient quantities of air at the correct pressure must pass through the boiler. Without the proper draft, complete combustion of fuels cannot be achieved. Draft must overcome the flow resistance caused by equipment such as baffles, tube passes, dampers, economizers, air heaters, superheaters, and breechings that are in the path of the gases of combustion. The type of draft system used will depend on the requirements of the steam plant. Draft is classified as natural or mechanical.

The gases of combustion move through the breeching to a stack, or chimney. The terms "stack" and "chimney" are often used to mean the same thing. However, sometimes "stack" is preferred for a structure made of steel, and "chimney" is preferred for a structure made of brick or masonry.

The unit used to measure draft is inches of water column (in. WC). When the pressure is below atmospheric pressure, it is designated with a minus sign (–). When the pressure is above atmospheric pressure, it is designated without a sign or with a plus sign (+).

*Note:* Atmospheric pressure, or barometric pressure, is measured in inches of mercury (in. Hg). To convert measurements in inches of mercury to inches of water column, multiply the measurement in inches of mercury by 13.6.

## NATURAL DRAFT

*Natural draft* is draft produced by the natural action resulting from temperature differences between air and gases of combustion. Natural draft is the oldest form of draft. It is produced by a stack without the use of fans. **See Figure 10-1.**

### NATURAL DRAFT

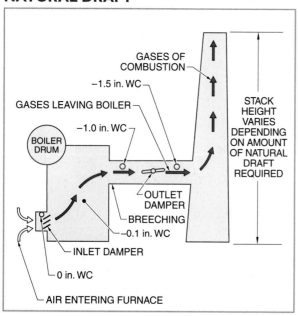

**Figure 10-1.** Natural draft occurs as hot gases leave a boiler, condense, and flow out a stack.

The amount of draft produced by a stack depends on its height and the difference in temperature between the outside air and the gases of combustion. Gases of combustion rise because they are less dense than the outside air. The denser outside air replaces the gases of combustion leaving the furnace area of the boiler.

Natural draft is controlled by dampers. An inlet damper regulates the flow of air to the burners. An outlet damper regulates the gases of combustion leaving the boiler. The amount of natural draft available is determined by measuring the difference in pressure between two points, such as the atmosphere and the inside of the boiler. The average draft pressures within a natural draft system vary depending on boiler design. Typical draft pressures are at-mospheric pressure before the inlet damper; about –0.1 in. WC (below atmospheric pressure) inside the furnace area; about –1.0 in. WC before the outlet damper; and about –1.5 in. WC on the stack side of the outlet damper.

## MECHANICAL DRAFT

Natural draft cannot be used in all applications. Modern boilers are designed to permit higher rates of heat transfer, which in turn requires more draft. A higher stack would be required to produce the amount of natural draft needed. Since higher stacks are often physically and economically impractical, mechanical draft was developed to produce the amount of draft needed for efficient combustion.

*Mechanical draft* is draft produced by power-driven fans. Mechanical draft is controlled by the speed of the draft fans and/or by dampers. The two types of mechanical draft are forced and induced draft. These two types may both be used in one boiler. This is referred to as balanced draft.

### Forced Draft

*Forced draft* is mechanical draft produced by a fan supplying air to a furnace. Forced draft is produced by a fan or blower located in front of the boiler that forces air into a furnace. **See Figure 10-2.** Forced draft was developed to provide the amount of air for combustion that is required by the increased firing rates of coal stokers, pulverized coal feeders, and fuel oil and gas-fired furnaces. A relatively high stack is usually used with forced draft fan systems.

Forced draft fans supply both primary air and secondary air. Large forced draft blowers can move combustion air to more than one boiler using windboxes and tunnels. A *windbox* is a chamber surrounding a burner assembly or coal stoker that allows the pressurized air from the forced draft fan to enter the burner. Small forced draft blower units, such as those used with firetube boilers, supply air to only one burner. The amount of draft is controlled by a combination of inlet dampers, outlet dampers, and variable-speed forced draft fans.

Draft pressures at the discharge side of the forced draft fan are normally about 2 in. WC to 3 in. WC pressure. This creates a pressurized furnace. A *pressurized furnace* is a furnace that operates at slightly above atmospheric pressure. The pressure in these furnaces is kept slightly higher than atmospheric pressure so that the air and gases of combustion can be pushed through the furnace. This prevents the infiltration of excess air into the combustion process and results in greater combustion efficiency. In addition, pressurized furnaces provide greater control of the amount of air introduced into the furnace.

## FORCED DRAFT

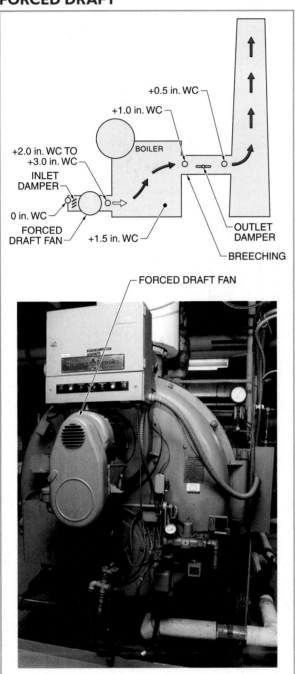

**Figure 10-2.** A forced draft system has a draft fan at the inlet damper of a boiler.

## TECH TIP

*Burners that use natural draft are often known as atmospheric burners, and burners that use a forced draft fan are often known as power burners.*

### Induced Draft

*Induced draft* is draft produced by pulling air through the boiler furnace with a fan. Induced draft is produced by a fan or blower located in the breeching between the burner and stack. **See Figure 10-3.** Induced draft fans remove gases of combustion from the boiler and breeching and prevent back pressure from developing. Additionally, the use of induced draft fans has largely replaced tall stacks, which have become prohibited by modern architectural standards and codes.

## INDUCED DRAFT

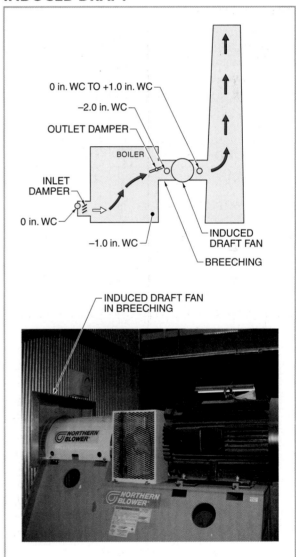

**Figure 10-3.** An induced draft system uses a draft fan at the outlet damper of a boiler.

The pressure in induced draft furnaces is kept at slightly less than atmospheric pressure so the air and gases of combustion can be pulled though the furnace. This prevents the gases of combustion from leaking into the boiler room. The amount of draft is controlled by a combination of inlet dampers, outlet dampers, and variable-speed induced draft fans. Induced draft fans must be designed to withstand increased wear caused by components in the gases of combustion.

### Balanced Draft

*Balanced draft* is draft produced from one or more forced draft fans located before the boiler and one or more induced draft fans after the boiler. **See Figure 10-4.** Large industrial and utility plants use balanced draft to overcome air and gas-flow resistance. Balanced draft may be necessary to maintain high combustion rates within a boiler. The amount of draft is controlled in the same manner as it would be if the systems were separate. The pressure in balanced draft furnaces can be kept at any desired pressure.

## BALANCED DRAFT

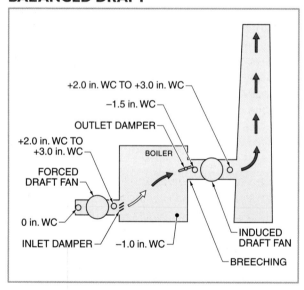

**Figure 10-4.** A balanced draft system uses draft fans at both the inlet damper and outlet damper of a boiler.

Care must be taken to maintain a balance between the two fans in order to maintain constant pressure in the boiler. The induced fan is generally designed to move more gas than the forced draft fan. Because the combustion reactions generate carbon dioxide and water

vapor from the fuel and oxygen, there is a greater volume of the gases of combustion than of combustion air. In addition, the gases of combustion are typically hotter than the combustion air. Gases increase in volume when they are heated.

---

### SECTION 10.1 – CHECKPOINT

1. What is draft?
2. What is natural draft?
3. What is the difference between forced draft and induced draft?

---

## SECTION 10.2
# DRAFT DESIGN AND CONTROL

### DRAFT SYSTEM DESIGNS

A draft system has specific design requirements for different boilers and fuels. Design is primarily based on the size of the boiler and the type of fuel used. The amount of fuel that can be burned depends on the amount of air supplied to the furnace. In some cases, the draft system supports the delivery of the fuel into the boiler. Draft system design includes bringing in outside air to replace air used for combustion and heating the air before it enters the boiler. It also includes the specific designs for burning different types of fuel.

### TECH TIP

*Although panel-mounted draft gauges do not contain water, the term "manometer" is often used to refer to all draft-measurement devices.*

### Outside Air

Combustion requires a flow of outside air. If the boiler room is airtight, a vacuum is created that starves the furnace of air. Therefore, boiler rooms must be designed to permit the proper flow of outside air to the furnace. Common methods of bringing in outside air include wall dampers that allow air to enter the boiler room and ductwork that brings air into the room near the boilers. **See Figure 10-5.**

## OUTSIDE AIR

**WALL DAMPERS**

**DUCTWORK**

**Figure 10-5.** Wall dampers and ductwork can be used to bring in outside air.

### Air Heaters

An important part of a draft system is a method to recover heat that may be lost up the stack. The boiler system loses that heat and burns extra fuel to make up for the loss. Air heaters are used in most large installations. An *air heater* is a heat exchanger that is used to heat combustion air

for a furnace and is located in the breeching between the boiler and the stack. The air heater is typically mounted in the path of the gases of combustion as they exit the boiler.

Air heaters increase combustion efficiency by warming the air supplied to the furnace. In addition, air heaters prolong the life of the firebrick and furnace refractory. Cold air entering the furnace can cause spalling (hairline cracks) in the brickwork.

The two basic types of air heaters are convection air heaters and regenerative air heaters. Both types of air heaters employ a counterflow principle. This means that the gases of combustion and airflow move in opposite directions to maximize heat transfer. The hotter gases of combustion pass by the warmer air leaving the heater. The cooled gases of combustion pass by the cooler air entering the heater.

Some heaters are designed with bypass dampers. This allows part of the air to bypass around the heater for more accurate control. Air heaters can be used most successfully in plants with a fairly constant steam load.

**Convection Air Heaters.** *Convection* is a type of heat transfer that occurs due to the circulation of a fluid or gas. A *convection air heater* is an air heater that uses convection to transfer heat from the gases of combustion to the combustion air. Convection air heaters may be of tubular or plate design. **See Figure 10-6.** A tubular air heater consists of tubes through which the gases of combustion pass. Air flows over the outside of the tubes. The air typically makes several passes across the tubes to increase the heat transfer.

A plate air heater uses plates instead of tubes. The plates are assembled and spaced in a structural steel frame and form air and gas chambers. Heat transfer occurs as the air and gases of combustion move past each other on opposite sides of the plates.

**Regenerative Air Heaters.** A *regenerative air heater* is an air heater that is divided into zones: the zone containing the gases of combustion, the sealing zones, and the air zone containing the air for combustion. Regenerative air heaters are used primarily in large boiler installations. Regenerative air heaters contain many moving parts. Therefore, they often require more care and maintenance than other types of air heaters. A slow-moving rotor consists of a honeycomb plate element that rotates through the zones of the air heater. **See Figure 10-7.**

First, the plate element rotates through the zone containing the hot gases of combustion. It absorbs heat from the hot gases. This reduces the temperature of the hot gases and increases the temperature of the plate element. The gases of combustion then go through the induced draft fan and out the stack.

## CONVECTION AIR HEATERS

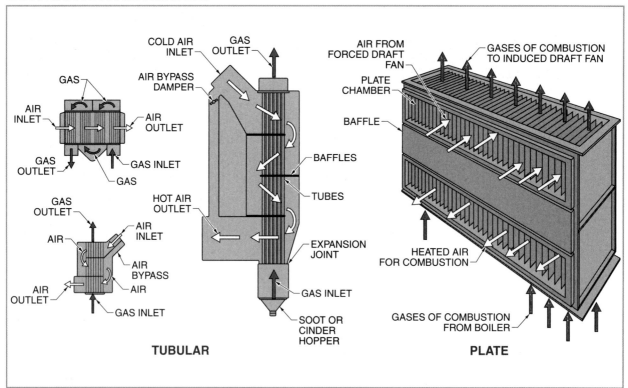

**Figure 10-6.** Convection air heaters move heat from the gases of combustion to the combustion air before the gases enter the stack.

## REGENERATIVE AIR HEATERS

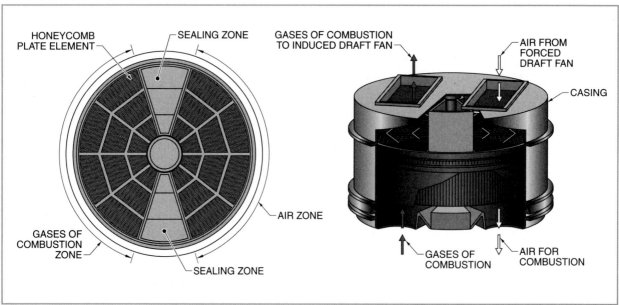

**Figure 10-7.** The plate element in a regenerative air heater rotates continuously through the zones to absorb and release heat from the gases of combustion and the combustion air.

The plate element rotates into the first sealing zone. The sealing zone is a barrier area that prevents the gases of combustion from mixing with the air for combustion.

The plate element then rotates into the air zone. It releases heat into the relatively cool combustion air. This reduces the temperature of the plate element and increases the temperature of the combustion air. The combustion air then moves through the air heater to the furnace. The plate element rotates into the second sealing zone to finish the cycle.

**Air Heater Maintenance.** Air heaters are normally made of a steel alloy and must be externally insulated to prevent heat loss. The temperature of the air entering the air heater must be monitored. If the temperature gets too low, the acidic vapor in the gases of combustion will condense. This results in corrosion when soot and moisture combine. Air heaters are located in the direct path of the gases of combustion. They are typically exposed to soot and other components of the gases of combustion. They can be cleaned with soot blowers.

---

**TECH TIP**

*The use of an air heater typically increases overall efficiency 2% to 10%. The amount of efficiency increase depends on the unit location, steam capacity, and whether an economizer is installed.*

---

Air heaters should be examined about once a year for signs of corrosion. If the tubes or plates are corroded through, some of the air will mix with the gases of combustion leaving the stack. This loss of air for combustion could lead to incomplete combustion of the fuel and create smoke. In addition, heat could be lost, resulting in lower boiler efficiency.

### Natural Gas and Fuel Oil Draft Systems

The draft system for smaller package boilers that burn natural gas and fuel oil typically consists of a forced draft fan and a windbox. As boilers increase in size and complexity, an air heater may be incorporated. **See Figure 10-8.**

A forced draft fan draws in ambient air and forces it through a metal duct to the air heater. As the air passes through the air heater, the temperature of the air increases and the temperature of the gases of combustion decreases. The air moves through an insulated duct to the windbox. The air then mixes with the fuel in the burner.

## GAS AND FUEL OIL DRAFT SYSTEMS

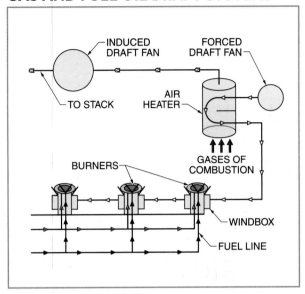

**Figure 10-8.** A natural gas and fuel oil draft system uses a forced draft fan to supply the fuel to the boiler and an induced draft fan to remove the gases of combustion.

From the burner, the air-fuel mixture enters the furnace where the combustion process takes place. The gases of combustion then pass through the boiler before entering the air heater where the heat energy is transferred to the air. The gases of combustion leave the air heater to enter the suction of the induced draft fan. The induced draft fan then discharges the gases of combustion to the stack.

### Coal Draft Systems

The draft systems for coal boilers are more complex than those for natural gas and fuel oil boilers. Coal can be burned on a grate or as a pulverized fuel blown into the furnace. The designs for these systems are different because the coal burns differently.

**Chain Grate Stoker Draft Systems.** A chain grate stoker draft system typically consists of a forced draft fan, an air heater, a windbox under the chain grate, an economizer, and an induced draft fan. **See Figure 10-9.** The forced draft fan pulls in ambient air and forces it through the air heater to increase the air temperature.

From the air heater, the air passes to the windbox under the chain grate stoker. The air is used for combustion and to help prevent the grates from warping because of overheating. The air is distributed through the windbox so that it passes through zones into the fuel bed. The gases of combustion pass through the boiler, economizer, and air heater. After they leave the air heater, the gases of combustion are pulled through the induced draft fan and discharged to the stack.

## CHAIN GRATE STOKER DRAFT SYSTEMS

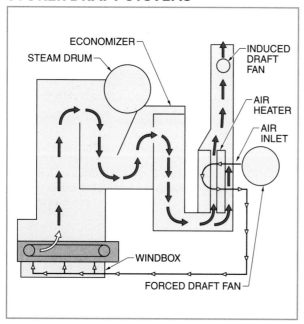

**Figure 10-9.** The air supplied for combustion in a chain grate stoker is uniformly distributed through the windbox.

**Pulverized Coal Draft Systems.** A pulverized coal draft system typically consists of a forced draft fan, an air heater, an air-tempering damper, a pulverizing mill, an exhauster, a mechanical dust collector, an electrostatic precipitator, and an induced draft fan. **See Figure 10-10.** A forced draft fan takes in ambient air and discharges it into the air heater. Heated air flows from the air heater through an insulated duct toward the windbox and the air-tempering damper. At the windbox, the air is ready to mix with the incoming pulverized air-fuel mixture in the furnace.

The air from the forced draft fan that flows to the air-tempering damper blends with ambient air to maintain an air temperature of about 200°F. This air is primary air, whereas the air at the windbox is secondary air. The primary air enters the pulverizing mill. Its purpose is to dry the coal and carry the pulverized coal, in suspension, to the burner.

The primary air and pulverized coal are removed from the pulverizing mill by the exhauster. The exhauster discharges the air-fuel mixture through a pipe to the burners. Coal is burned in suspension with sufficient secondary air to achieve complete combustion. The gases of combustion then pass through the boiler and enter the air heater.

## DRAFT CONTROL

Proper draft control is essential for improving heat transfer and combustion efficiency. Draft control reduces heat loss and minimizes operating costs. It improves flame stability to minimize pilot light failure. Draft must be proportional to fuel flow for proper operation and efficient fuel consumption in boilers. Draft control systems maintain the draft during varying load demand by balancing the forces caused by changes in draft temperature and pressure.

Forced draft fans are usually used to supply combustion air and remove the gases of combustion. Induced draft fans are usually incorporated into large, complex boilers to help remove the gases of combustion. The volume of gases removed is typically controlled by boiler outlet dampers, draft positioning controllers, or a variable-speed drive (VSD) that controls fan speed.

### Outlet Dampers

Outlet dampers are located at the outlet side of a boiler in the breeching before the stack. **See Figure 10-11.** These dampers can be controlled by electrical, electronic, or pneumatic control systems. A damper control system adjusts the dampers to maintain constant pressure in the furnace. During automatic operation, the selector relay (1) sends a signal containing the measured furnace pressure (2) to the draft positioning controller (3). The draft positioning controller will position the outlet damper (4) to open or close as the combustion rate increases or decreases.

### Draft Positioning Controllers

Draft positioning controllers set the dampers at predetermined positions. The dampers are automatically positioned during the purge cycle, burner ignition, and shutdown process when the boiler is shut off. The damper position is adjusted during operation to maintain a predetermined pressure within the furnace.

## PULVERIZED COAL DRAFT SYSTEMS

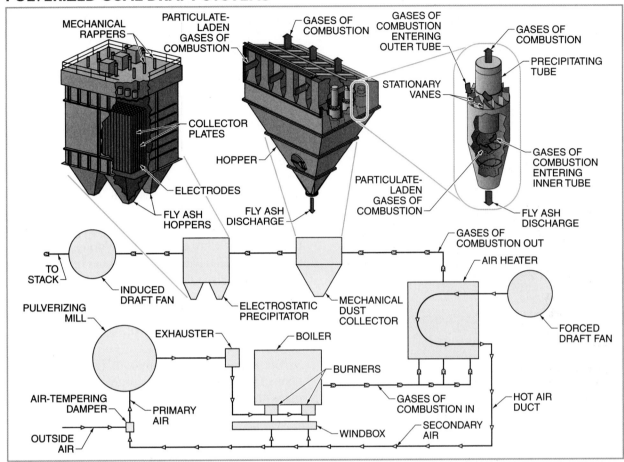

**Figure 10-10.** Heated primary air carries pulverized coal to the burner where it mixes with secondary air for efficient combustion.

## DRAFT CONTROL

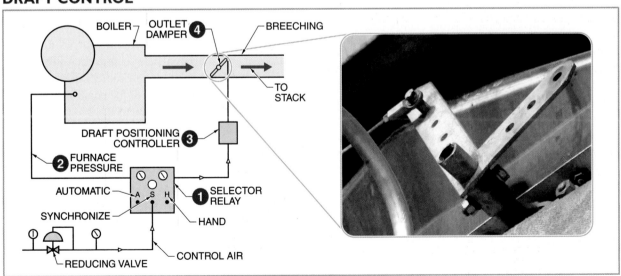

**Figure 10-11.** Draft in a furnace can be controlled by restricting the flow of the gases of combustion through an outlet damper.

## Variable-Speed Drives

It takes energy to move air and gases of combustion. Fixed-speed fans always use the same amount of energy while they are running. Dampers work by changing the amount of resistance to the flow of air and gases of combustion. Resistance to flow wastes energy. A variable-speed drive (VSD) adjusts the speed of the fan and can be used as part of a damper control system. **See Figure 10-12.**

## VARIABLE-SPEED DRIVES (VSDs)

**Figure 10-12.** A VSD can be used to change the speed of a fan motor.

A VSD saves energy. As the speed of the fan changes, the amount of air or gases of combustion that can be moved also changes. This allows the combustion control system to maintain the correct amount of air for the amount of fuel without wasting energy because of dampers that restrict the flow. This type of draft control is the most energy efficient.

### SECTION 10.2 – CHECKPOINT

1. What are two common methods of bringing outside air into a boiler room?
2. Why is an air heater used?
3. What are two common types of air heaters?
4. How do draft systems operate for furnaces burning natural gas and fuel oil?
5. How do draft systems operate for furnaces burning coal?
6. What are the three main components of draft control?

## SECTION 10.3
# ENVIRONMENTAL REGULATIONS

### CLEAN AIR ACT

Air pollutants cause an adverse effect on humans and the environment. These adverse effects may include health problems, global warming, acid rain, heavy metal poisoning, destruction of forests, and acidic deterioration of buildings and surfaces. Because of these adverse effects, governments worldwide regulate air emissions. In the United States, the Environmental Protection Agency (EPA) regulates the emissions from boilers through the Clean Air Act.

The *Clean Air Act (CAA)* is a law that was passed by Congress in 1963 to allow for monitoring and controlling environmental air emissions. It has been amended several times since then. It gives the EPA administrative enforcement capabilities and sets up a permitting program for air emissions. One of the first goals of the CAA was to establish National Ambient Air Quality Standards (NAAQS).

The EPA developed emissions requirements and plans. The EPA asked the states to develop their own plans that meet or exceed the EPA requirements. These plans are called State Implementation Plans (SIPs). The federal government has avoided imposing uniform regulations for the states because of the differences in weather, demographics, and geography. Boiler plant operations staff must know local requirements. The Clean Air Act Amendments of 1990 (1990 CAAA) expanded the requirements to include many more emissions called hazardous air pollutants (HAPs).

The EPA classifies emissions sources as major sources or area sources to help set priorities. In addition, the EPA developed rules for new sources of emissions to prevent further deterioration of the environment.

### NAAQS

The CAA requires that the EPA prioritize six pollutants that are responsible for much of the concern over air emissions. The CAA established NAAQS, which focuses on six types of criteria pollutants. These six pollutants include carbon monoxide (CO), lead (Pb), nitrogen dioxide ($NO_2$), ozone ($O_3$), sulfur dioxide ($SO_2$), and particulate matter (PM). **See Figure 10-13.** The EPA periodically reviews the scientific literature and modifies the allowed emissions of the criteria pollutants.

## NATIONAL AMBIENT AIR QUALITY STANDARDS (NAAQS)

| Criteria Pollutant | Primary Source Related to Combustion of Fuels |
|---|---|
| Carbon monoxide | Incomplete combustion |
| Lead | Generally not emitted by combustion |
| Nitrogen dioxide | High-temperature combustion |
| Ground-level ozone | Nitrogen oxides and volatile organic compounds |
| Sulfur dioxide | Combustion of fuels containing sulfur |
| Particulate matter | Fly ash and particulates from combustion |

**Figure 10-13.** NAAQS focuses on the six primary emissions regulated under the CAA.

**Carbon Monoxide.** Carbon monoxide (CO) consists of a molecule containing one carbon atom and an oxygen atom. Carbon monoxide is a chemical that is absorbed in the body and impairs hemoglobin in the blood. This reduces the oxygen supplied to the organs. Carbon monoxide is produced from incomplete combustion due to poor burner design, maintenance, or firing conditions. Complete conversion of carbon to carbon dioxide is seldom achieved in a boiler and some of the carbon ends up as carbon monoxide. High flame temperatures and good air-fuel mixing in the furnace are essential for achieving low carbon monoxide emissions.

**Lead.** Lead (Pb) consists of elemental lead and lead oxide emissions. Boilers are not a large contributor to lead emissions. Standard boiler fuels contain little or no lead. Alternate fuels, such as waste oil, that may contain lead are subject to federal, state, and local regulations. Waste oil burned as a fuel in an industrial boiler must contain less than a specified lead content. This is regulated under the federal Resource Conservation and Recovery Act (RCRA). As a result, lead emissions from industrial boilers are limited.

**Nitrogen Dioxide.** Nitrogen dioxide ($NO_2$) consists of a molecule containing one nitrogen atom and two oxygen atoms. At high furnace temperatures, oxygen and nitrogen can combine to form nitrogen oxides. *Nitrogen oxides ($NO_x$)* is a general term used to include all possible forms of molecules containing nitrogen and oxygen that result from combustion at high temperatures. Any type of combustion is a contributor to $NO_x$ emissions, but higher furnace temperatures greatly increase $NO_x$ formation.

**Ozone.** Ozone ($O_3$) consists of a molecule containing three oxygen atoms. Ozone in the upper atmosphere is beneficial. It prevents ultraviolet radiation from reaching the Earth's surface. However, ozone in the lower atmosphere is a regulated pollutant. It is a key component of smog and has harmful effects on plants and animals.

Ozone is not produced directly from the combustion process. Ozone forms from the products of combustion as a result of a reaction between $NO_x$ and volatile organic compounds (VOCs) in sunlight. Since both sunlight and hot weather precipitate its formation, ozone is primarily known as a summer pollutant. The amount of $NO_x$ that forms depends on the amount of nitrogen in the fuel and the furnace temperature. The forms of $NO_x$ typically generated by boilers are nitric oxide (NO) and nitrogen dioxide ($NO_2$).

**Sulfur Dioxide.** Sulfur dioxide ($SO_2$) consists of molecules containing a sulfur atom and two oxygen atoms. The combustion of fuels that contain sulfur (primarily oils and coals) results in $SO_2$. Sulfur compounds react with water vapor to form sulfuric acid. This is corrosive and harmful to the environment. The level of $SO_2$ emitted depends on the sulfur content of the fuel.

**Particulate Matter.** Particulate matter (PM), commonly known as particulate pollution, consists of fine particles emitted from a boiler that become individually dispersed in the air. PM is also often referred to as particulates. PM from combustion sources consists of many different types of compounds from combustion and any noncombustible elements in the fuel. These pollutants can be corrosive, toxic to plants and animals, and harmful to humans. The level of PM emitted depends on the type of fuel used. A common type of PM from combustion is fly ash. *Fly ash* is a type of PM consisting of noncombustible material found in the gases of combustion. Fly ash is a product of the combustion of coal or other solid fuels.

PM emissions are generally classified as PM, $PM_{2.5}$, and $PM_{10}$. PM is the general term for all types of particulate emissions. $PM_{2.5}$ consists of PM with diameters of less than 2.5 microns. $PM_{10}$ consists of PM with diameters of less than 10 microns. The greatest concern is $PM_{2.5}$ because its small size allows it to easily enter the body.

## Classification of Emissions Sources

The EPA has different regulations for different emissions sources. A *major source* is a facility that emits 10 or more tons per year of any single HAP or 25 tons or more per year of any combination of HAPs. Most of these boilers are in industrial facilities such as refineries, chemical plants, manufacturing plants, and paper mills. Major sources have more stringent regulations than area sources.

An *area source* is a facility that emits less than 10 tons per year of any single HAP and less than 25 tons per year of any combination of air toxics. Area source boilers emit small amounts of HAPs, primarily because most burn natural gas. Most of these boilers are in commercial and institutional facilities. A smaller number are in industrial facilities. Commercial facilities include stores, malls, laundries, apartments, restaurants, hotels, and motels. Institutional facilities include hospitals, clinics, nursing homes, schools, churches, amusement parks, courthouses, and prisons.

### New Source Performance Standards (NSPS)

The federal government has established national standards called the New Source Performance Standards (NSPS). The *New Source Performance Standards (NSPS)* are health-based environmental standards for emissions from major new sources of emissions. These standards include the best technology currently available and account for cost and energy impact. Over time, these standards have come to include best practices as well as the original technologies.

The standards differ for an installation depending on the size of the boiler and fuel type being used. Boilers are classified in three size ranges: burner input rating more than 250 million Btu/hr (MMBtu/hr), between 100 MMBtu/hr and 250 MMBtu/hr, and between 10 MMBtu/hr and 100 MMBtu/hr. All small boilers built after June 9, 1989, must meet NSPS requirements. New source facilities must test their emissions to verify compliance. **See Figure 10-14.**

### HAPs

The CAA requires the EPA to publish a list of industry groups that use, manufacture, or emit HAPs. A *hazardous air pollutant (HAP)* is any material from a list of approximately 200 pollutants that has significant health effects when it is emitted into the air. The EPA established the National Emission Standards for Hazardous Air Pollutants (NESHAP) to regulate HAPs.

## NEW SOURCE PERFORMANCE STANDARDS (NSPS)

**Figure 10-14.** New sources of emissions must be tested to verify compliance with environmental regulations.

There are many locations where EPA standards for emissions have been attained (attainment areas). There are also many locations where EPA standards have not been attained (nonattainment areas). In nonattainment areas, environmental regulations place significant restrictions on boiler emissions to help improve the air quality. The CAA identifies about 200 HAPs and has directed the EPA to regulate their emissions.

In addition to creating the list of HAPs, the EPA has identified several types of technologies and emissions rates to control emissions for different situations. These include Maximum Achievable Control Technology (MACT), Best Available Control Technology (BACT), Lowest Achievable Emission Rates (LAERs), and Reasonably Available Control Technology (RACT).

**MACT.** *Maximum achievable control technology (MACT)* is technology that meets a level of control the EPA uses to regulate HAP emissions based on the level of emissions control for the best-performing technology used in similar major sources. This technology may include control equipment, clean processes, and clean work practices. This level of control provides a baseline for minimum allowed emissions for an industry.

---

**TECH TIP**

*Mercury (Hg) is a toxic material emitted from boilers. It is a regulated contributor to boiler emissions. In 2011, the EPA issued a new regulation called the Mercury and Air Toxics Standards (MATS) to regulate mercury emissions from power plants.*

---

**BACT.** *Best available control technology (BACT)* is technology that meets the environmental requirement for emissions from major new or modified sources in an attainment area. These technologies are intended to limit emissions in attainment areas and prevent significant deterioration of the environment. A building permit application for a new or modified source must include a description of the BACT to be used. The EPA maintains a database of BACTs for different industries.

**LAERs.** A *lowest achievable emission rate (LAER)* is a rate for emissions from a new source in nonattainment areas that meets the environmental requirement. For new sources in nonattainment areas, environmental regulations may require owners to implement an LAER regardless of the cost. For boilers, the LAER requirement for $NO_x$ in some areas is 9 ppm or lower. The limits may be lower in the future. New boilers may have emissions requirements below 9 ppm.

**RACT.** *Reasonably available control technology (RACT)* is technology that meets the environmental requirement for emissions from an existing source in a nonattainment area. The RACT requirement is intended to allow greater flexibility for states to account for the cost of control technologies when setting emissions requirements while still considering air quality requirements. RACT requirements vary from state to state.

## EMISSIONS CONTROL EQUIPMENT

When regulated chemicals are generated in a boiler, equipment needs to be installed to remove the chemicals from the gases of combustion. Owners of boilers are required to know the regulations and install appropriate equipment to remove the chemicals. In addition, the EPA may require continuous emissions monitoring to prove that systems are working properly at all times.

### Continuous Emissions Monitoring Systems

A *continuous emissions monitoring system (CEMS)* is a monitoring system used for continuous measurement of emissions from gases of combustion or from industrial processes. **See Figure 10-15.** A CEMS consists of all the equipment necessary to determine the gas and PM concentration or emission rate as required by regulations. The system uses analyzer measurements and a conversion equation, graphs, or computer program to display the measured emissions. A CEMS may be required by the EPA for demonstrating continual compliance with pollution standards for regulated boilers.

**CONTINUOUS EMISSIONS MONITORING SYSTEMS (CEMSs)**

**Figure 10-15.** A CEMS uses sensors to continuously monitor the gases of combustion.

The EPA has established requirements for continuous monitoring of volumetric flow, diluent gases, sulfur oxides, nitrogen oxides, and opacity for regulated units. In addition, EPA procedures require monitoring or estimating the carbon monoxide discharge as part of the recordkeeping and reporting requirements. State and local requirements may include additional monitoring requirements.

## Sulfur Removal

Scrubbing is the method of choice for removing sulfur oxides from the emissions from coal-fired plants. Scrubbing can have removal efficiencies of 90% or higher. **See Figure 10-16.** Sulfur oxides are neutralized through direct contact with an alkaline solution, such as a limestone slurry, in an absorption tower. The slurry is produced by crushing limestone in a ball mill and mixing it with water. This converts the limestone to calcium hydroxide. The calcium hydroxide is alkaline and neutralizes the acidic sulfur oxides.

## SCRUBBERS

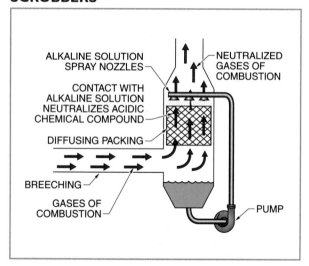

ALKALINE SOLUTION SPRAY NOZZLES
NEUTRALIZED GASES OF COMBUSTION
CONTACT WITH ALKALINE SOLUTION NEUTRALIZES ACIDIC CHEMICAL COMPOUND
DIFFUSING PACKING
BREECHING
GASES OF COMBUSTION
PUMP

**Figure 10-16.** Scrubbers are used to neutralize acidic chemical compounds in the gases of combustion in order to reduce air pollution.

The surface area of the calcium hydroxide particles is the critical factor for scrubber efficiency. The use of hot water at 110°F to 120°F supplements the heat of reaction. This improves efficiency and reduces limestone consumption. Future control requirements are likely to be more stringent.

## Particulate Removal

PM control varies for different types and sizes of boilers. Baghouses, dust collectors, and/or electrostatic precipitators are typically used for large utility or industrial boilers. A *baghouse* is a device that separates PM from the gases of combustion using cloth bags, similar to vacuum cleaner bags.

A *mechanical dust collector* is a device that separates PM from the gases of combustion using centrifugal force to spin the PM out of the gas stream. **See Figure 10-17.** Gases of combustion are directed in a circular path in the outer tubes of the precipitator tubes. The swirling motion causes the PM to discharge out the bottom, and the clean gas exits through the inner tubes to the atmosphere. The PM collects in a hopper below the tubes.

An *electrostatic precipitator* is a device that removes finer PM from the gases of combustion using electric charge. Collector plates are grounded, and the gases of combustion pass through an electrically charged grid between electrodes. The PM becomes charged as it passes through the grids and is attracted to the collector plates. The PM collects on the collector plates and is shaken off into hoppers by mechanical rappers.

## Mercury Removal

Mercury (Hg) is a toxic heavy metal emitted from boilers. It is a regulated contributor to boiler emissions. In 2011, the EPA issued a new regulation called the Mercury and Air Toxics Standards (MATS) to regulate mercury emissions from power plants. In addition, this regulation addresses other emissions such as arsenic and metals.

Mercury is found naturally in coal. It is released into the environment when coal is burned. Coal-burning power plants are the largest source of mercury stack emissions in the United States. The EPA estimates that coal-burning power plants account for over 50% of all domestic human-caused mercury emissions. Equipment used to remove mercury from boiler emissions includes selective catalytic reduction with flue-gas desulfurization, activated carbon injection, and electrostatic precipitators.

Power plants have taken steps to install controls or update operations to meet the MATS regulation. This regulation has been controversial because of the cost of compliance. The EPA has ruled that it is appropriate and necessary to issue this regulation. The EPA continues to review public comments on some sections of the regulation. Congressional or court actions to modify this regulation are possible.

## MECHANICAL DUST COLLECTORS

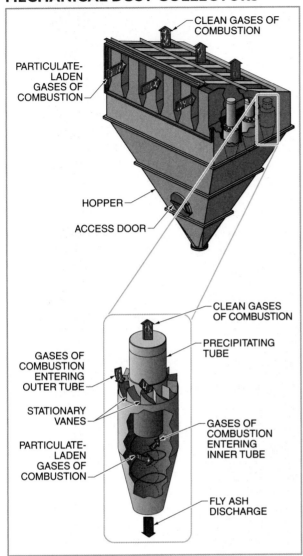

**Figure 10-17.** Mechanical dust collectors are used to remove particulate matter from the gases of combustion.

### Low NO$_x$ Burners

Boiler and burner manufacturers have developed new technologies to reduce NO$_x$ emissions. Low NO$_x$ burners usually burn natural gas since most types of fuel oil cannot be burned cleanly enough to achieve the required NO$_x$ levels. The amount of NO$_x$ produced decreases as the combustion temperature decreases. Low NO$_x$ burners reduce the formation of NO$_x$ primarily by reducing the furnace flame temperature. Processes such as flue-gas recirculation, premixing, combination premixing and flue-gas recirculation, and selective catalytic reduction have been developed to reduce the amount of NO$_x$ emitted into the air.

**Flue-Gas Recirculation.** To reduce NO$_x$ levels, low NO$_x$ burners may be designed to recirculate some of the gases of combustion. *Flue-gas recirculation (FGR)* is an emissions control method for boilers in which moderate amounts of flue gas are captured from the exhaust and recirculated back through the burner along with the secondary air. Recirculating some of the gases of combustion reduces the furnace temperature and the amount of NO$_x$. **See Figure 10-18.**

## FLUE-GAS RECIRCULATION (FGR)

*Cleaver-Brooks*

**Figure 10-18.** FGR is an emissions control method for boilers in which moderate amounts of flue gas are captured from the exhaust and recirculated back through the burner along with the secondary air.

FGR can be used with any forced draft boiler burner. A burner with 5% to 10% of the flue gas recirculated back through the burner can reduce NO$_x$ to below 80 ppm, and 10% to 20% of the flue gas recirculated back through the boiler can reduce NO$_x$ emissions to below 30 ppm. When a lower NO$_x$ emissions rate is required (less than 30 ppm), more FGR as a percentage of combustion air is required.

**Premixing.** Manufacturers have developed burners that premix fuel and air. Premixing fuel and air reduces the temperature of the combustion process and reduces $NO_x$ emissions. The principle advantage of premixing is the elimination of the expensive FGR process. A potential disadvantage of premixing is that fuel efficiency may be affected. In addition, the premixing process requires a larger furnace. This may require a larger or longer boiler.

To achieve extremely low levels of $NO_x$ emissions, the premixing process is used with high rates of FGR. A burner that premixes fuel and air and has FGR rates of approximately 25% to 30% can achieve $NO_x$ levels of less than 10 ppm. The major disadvantages of using the premixing process with FGR are increased flue-gas volume and reduced modulating range (turndown).

**Selective Catalytic Reduction.** *Selective catalytic reduction (SCR)* is an emissions control method for boilers in which ammonia gas is introduced over a catalyst located in a module that is installed in the boiler exhaust stack. **See Figure 10-19.** Some new boiler designs use SCR to treat the gases of combustion to reduce $NO_x$ emissions. SCR can reduce $NO_x$ levels to below 4 ppm. The ammonia reacts with the $NO_x$ to minimize the concentration of the $NO_x$ in the flue-gas emissions. SCR can be used with low $NO_x$, gas, or fuel oil burners. This allows maximum flexibility in the choice of fuel used.

## TECH TIP

*Nitrogen oxides ($NO_x$) are common pollutants from high-temperature combustion processes. $NO_x$ contributes to acid rain, ozone, and smog in our cities. The largest source of $NO_x$ emissions is power plants.*

A major disadvantage of SCR is that it can significantly increase the cost of new boiler installation. Other disadvantages include sensitivity to contamination and plugging, short catalyst life, catalyst degradation due to the presence of toxins, and poor gas metering.

## SELECTIVE CATALYTIC REDUCTION (SCR)

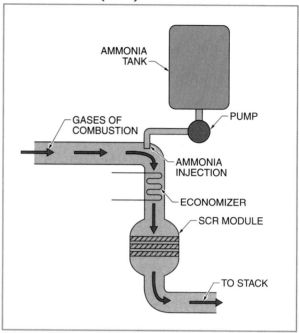

**Figure 10-19.** SCR is an emissions control method for boilers in which ammonia gas is introduced over a catalyst located in a module that is installed in the boiler exhaust stack.

## SECTION 10.3 – CHECKPOINT

1. What is the CAA?
2. What are the six regulated emissions included in the NAAQS?
3. What is the difference between major sources and area sources?
4. What are the NSPS?
5. What are HAPs?
6. What are the four common control requirement or technology categories?
7. What is the purpose of a CEMS?
8. What methods are used to remove sulfur, particulates, and mercury?
9. What equipment or procedures are used to minimize $NO_x$ emissions?

## CASE STUDY—LOSS OF AIR AND EXHAUST FLOW THROUGH BOILER BREECHING

### Situation

A maintenance engineer who works on a small college campus takes care of the day-to-day problems and occasionally checks the operating equipment. The operating equipment includes the eight boilers on campus. The boilers are also monitored by the campus energy-management system that sends an alarm to the maintenance engineer's phone when a problem arises. All the boilers are hot water boilers for heating the campus buildings.

The boilers are four-pass firetube boilers firing on natural gas. Each has a forced draft fan supplying air to the boiler and a 10″ breeching running diagonally to a 10″ stack to exhaust the gases of combustion.

### Problem

A boiler used for heating classroom and shop space at the far north end of the campus developed a problem. It started with an occasional alarm that the burner failed during ignition. When checked, everything would appear to be operating as normal. As time went on, the engineer would receive several alarms for that boiler each week, and finally the boiler would not start at all. Finding nothing wrong, the engineer informed the maintenance supervisor of the situation and recommended a service company be called to determine the problem with the boiler.

### Solution

The stack and breeching of the boiler consisted of a double-wall duct with a ½″ air gap between the walls. The technician from the service company discovered the inner lining of the breeching had collapsed and was restricting the flow of the combustion air and the products of combustion. The boiler combustion air-proving switch would sense insufficient airflow and shut down the boiler during the preignition sequence. It was determined that the inner lining was gradually collapsing, causing the intermittent failures until it nearly completely closed off the breeching. To solve the problem, the breeching was replaced.

**Name** _____ **Date** _____

_____ **1.** Draft is measured in ___.
  A. pounds per square inch
  B. inches of water column
  C. tenths of an inch of mercury
  D. ounces per square inch

_____ **2.** ___ draft is created by a draft fan that pulls the gases of combustion through a furnace.
  A. Forced
  B. Natural
  C. Induced
  D. Stack

_____ **3.** Pressure at the discharge side of a forced draft fan is ___ atmospheric pressure.
  A. balanced at
  B. less than
  C. the same as
  D. greater than

_____ **4.** ___ compounds react with water vapor to form sulfuric acid.
  A. Carbon
  B. Ozone
  C. Nitrogen
  D. Sulfur

_____ **5.** An air heater employs a counterflow principle in which the gases of combustion ___.
  A. and air move in opposite directions
  B. and air move in the same direction
  C. bypass the flowmeter
  D. and air mix together

_____ **6.** If the temperature of air entering a heater gets too low, it will cause ___.
  A. condensation on the gas side of the air heater
  B. thermal shock to the heater
  C. incomplete combustion
  D. oxygen pitting

_____ **7.** Air heaters can be cleaned by ___ to ensure good heat transfer.
  A. wire brushing monthly
  B. reversing air and gas flow
  C. using soot blowers
  D. using high-pressure water on the gas side

_____ **8.** Air heaters can most successfully be used in plants that ___.
   A. have a fluctuating steam load
   B. have a fairly constant steam load
   C. have high stacks
   D. burn only coal or fuel oil

_____ **9.** If air supplied for combustion mixes with the gases of combustion, it could lead to a(n) ___.
   A. loss of air for combustion and incomplete combustion
   B. increase in stack temperature
   C. increase in excess oxygen
   D. higher boiler efficiency

_____ **10.** Mechanical draft is produced by ___.
   A. the height of the stack
   B. the diameter of the stack
   C. power-driven fans
   D. steam jets

_____ **11.** ___ can be classified as forced, induced, or balanced.
   A. Complete combustion
   B. Air heaters
   C. Natural draft
   D. Mechanical draft

_____ **12.** The ___ is a law that was passed to allow for monitoring and controlling environmental air emissions.
   A. Ozone Emissions Act (OEA)
   B. Particulate Matter Plan (PMP)
   C. New Source Performance Standards (NSPS)
   D. Clean Air Act (CAA)

_____ **13.** The gases of combustion leaving a natural draft boiler are controlled by ___.
   A. dampers
   B. adding excess oxygen
   C. higher firing rates
   D. an induced draft fan

_____ **14.** An induced draft fan is located ___.
   A. between the burner and the stack
   B. inside the air preheater
   C. before the burner
   D. in the first pass of the gases of combustion

_____ **15.** NAAQS do not include ___.
   A. sulfur oxides
   B. nitrogen oxides
   C. lead
   D. mercury

_____ **16.** Large steam boilers that are equipped with air heaters and/or economizers typically use ___ to overcome the associated flow resistance.
   A. natural draft only
   B. balanced draft
   C. air heater diverters
   D. counterflow economizers

_____ **17.** ___ is technology that meets the environmental requirement for emissions from an existing source in a nonattainment area.
   A. MACT
   B. BACT
   C. LAER
   D. RACT

_____ **18.** In a furnace, the amount of $NO_x$ decreases as the ___.
   A. combustion air flow decreases
   B. amount of FGR decreases
   C. combustion temperature increases
   D. combustion temperature decreases

_____ **19.** The oldest form of draft used in a boiler is ___ draft.
   A. mechanical
   B. natural
   C. forced
   D. combination

_____ **20.** ___ is a type of PM consisting of noncombustible material found in the gases of combustion.
   A. Combustion air
   B. Pulverized coal
   C. Induced carbon
   D. Fly ash

_____ **21.** A ___ can be used as part of a damper control system to adjust the speed of a fan.
   A. variable-speed drive
   B. damper linkage
   C. modulating motor
   D. jackshaft

_____ **22.** ___ is the method of choice for removing sulfur oxides from emissions from coal-fired plants.
   A. FGR
   B. A baghouse
   C. Scrubbing
   D. An electrostatic precipitator

_____ **23.** The amount of draft produced in a natural draft system depends on the ___.
     A. diameter of the stack
     B. height of the stack
     C. amount of fly ash in the gases of combustion
     D. types of fans used

_____ **24.** A(n) ___ is a heat exchanger used to heat combustion air for the furnace and is located in the breeching between the boiler and the stack.
     A. open feedwater heater
     B. superheater
     C. air heater
     D. baffle

_____ **25.** ___ is produced from incomplete combustion due to poor burner design or firing conditions.
     A. Carbon monoxide
     B. Nitrogen oxide
     C. Sulfur
     D. Mercury

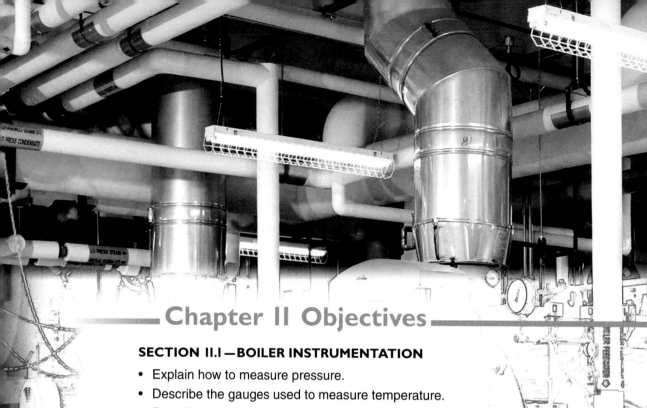

# Chapter II Objectives

### SECTION II.I — BOILER INSTRUMENTATION

- Explain how to measure pressure.
- Describe the gauges used to measure temperature.
- Describe the gauges used to measure level.
- Explain the methods used to measure flow.
- Describe the analyzers used in a boiler plant.

### SECTION II.2 — BOILER CONTROL SYSTEMS

- Explain how the three elements of a control loop work together to control a process.
- Describe the elements of a digital control system.

## ARROW SYMBOLS

| AIR | GAS | WATER | STEAM | FUEL OIL | CONDENSATE | AIR TO ATMOSPHERE | GASES OF COMBUSTION |
|---|---|---|---|---|---|---|---|
| ⇨ | ⇨ | ➩ | ➡ | ⇨ | ➡ | ⇨ | ➡ |

**Digital Resources**
ATPeResources.com/QuickLinks
Access Code: **472658**

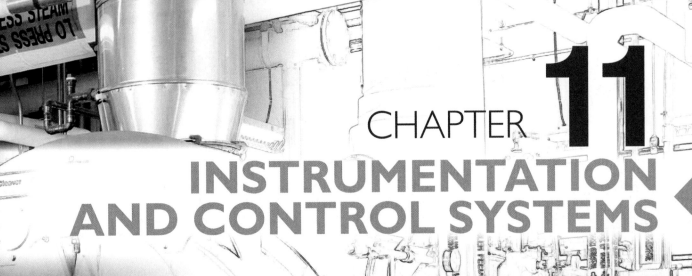

# CHAPTER 11

# INSTRUMENTATION AND CONTROL SYSTEMS

## INTRODUCTION

Instruments in a boiler room indicate the status of a boiler system to an operator. In order to operate a plant safely and efficiently, the operator must use instruments such as pressure gauges, temperature gauges, level gauges, flow meters, and analyzers. Data from these instruments are usually recorded to document the operation of a steam plant in a given period. Smaller steam plants require fewer instruments than larger, more sophisticated steam plants.

Automatic control systems streamline the operation of boilers. These systems include control loops that regulate boiler operations and plant automation control systems. The use of plant automation control systems improves the efficiency of boiler operation as part of a total building operation or production process and helps meet environmental emission requirements.

## SECTION 11.1
## BOILER INSTRUMENTATION

### INSTRUMENTS

Instruments are used to measure the operating conditions of a boiler. Readings from instruments can be viewed at the point of measurement, viewed at a local control panel near a boiler, or sent to control rooms. Instruments are set to operate under normal boiler operating conditions and must be able to operate over the range necessary for upset conditions.

As instruments and control systems age, older pneumatic and analog instruments are being replaced by digital instruments. Digital instruments are preferred because they allow for easier data collection, transmission, and storage. However, pneumatic instruments are still sometimes used in hostile or corrosive environments, and analog instruments are still used in smaller, older facilities.

Instruments are located at various points in a process to measure and transmit local readings to the controller. The controller compares the reading, performs a calculation, and sends a signal to a control device that controls the flow of a material or energy. The control device adjusts the process to regulate the system for normal operating conditions.

Some processes require continuous measurement. For example, the level of fuel oil in a tank should be known at all times. The measurement is used to determine when more fuel should be ordered or to make sure the tank will not overflow when being filled. With other processes, it is only necessary to know that the measurement is within the desired limits, such as for water level in a boiler. This ensures that water level will not go too low or too high.

## PRESSURE GAUGES

Boiler plants use pressure gauges to measure pressure at various points within a boiler system. **See Figure 11-1.** Pressure gauges measure air pressure, gas pressure, steam pressure, vacuum pressure, and discharge pressures on fuel and water lines. Common types of pressure gauges include steam pressure, process pressure, draft, and vacuum gauges.

## PRESSURE GAUGES

**Figure 11-1.** Boiler plants use pressure gauges to measure pressure at the various points within a boiler system.

### Pressure Measurements

Amount of pressure can be expressed in many units including pounds per square inch (psi), inches of water column (in. WC), and inches of mercury (in. Hg). Pressure measurement is always measured relative to another pressure. The pressure measured relative to

atmospheric pressure is known as gauge pressure and is typically given in pounds per square inch gauge (psig). This measurement is also given simply as pounds per square inch (psi). Absolute pressure (psia) is pressure measured relative to absolute vacuum. Absolute pressure is equal to gauge pressure plus atmospheric pressure.

Pressure measurements expressed in inches of water column are used for small pressure differences, normally under 1 psi of pressure. A pressure of 1 psi supports a water column 27.7 in. high. A pressure of 1″ WC is equal to 0.0361 psi. A pressure of 12″ WC (1′ WC) is equal to 0.433 psi.

When pressure is less than atmospheric pressure, it is called a vacuum. A vacuum pressure measurement is typically expressed in inches of mercury. Standard atmospheric pressure will support a column of mercury (Hg) 29.92″ high. A perfect vacuum has a pressure of 0 psia, or 29.92″ Hg. A pressure of 1 psia is equal to 2.036″ Hg, and a pressure of 1″ Hg is equal to 0.4912 psia.

Different gauge faces are used to indicate standard ranges of pressure. **See Figure 11-2.** The graduations of pressure on the faces will vary, depending on the pressure gauge range. For example, on a pressure gauge with a range of 0 psi to 15 psi, each graduation can be in ½ psi increments. On a pressure gauge range of 0 psi to 400 psi, the graduations can be in 5 psi increments. The specific pressure gauge range used is determined by the maximum pressure to be measured and how easy the gauge is to read. The pressure gauge range selected should allow the pointer to be in the 12 o'clock position when it is at its normal operating pressure.

## PRESSURE GAUGE RANGES

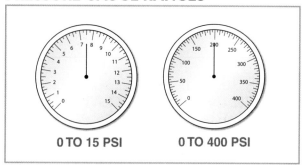

0 TO 15 PSI          0 TO 400 PSI

**Figure 11-2.** Pressure gauges are available in various ranges.

### Steam Pressure Gauges

A steam pressure gauge should have a range 1½ times to 2 times the safety valve pressure setting. When ordering a new steam pressure gauge, the safety valve pressure setting must be known. A steam pressure gauge should be connected to the highest part of the steam side of a boiler and must be easily visible from the operating level. A steam pressure gauge must be mounted in a location that is free from vibration. The face of the gauge must be clean and well lit.

The gauge must be protected by a pigtail siphon. The siphon forms a water leg, preventing live steam from entering the Bourdon tube. **See Figure 11-3.** If steam is allowed to enter the Bourdon tube, the gauge must be tested and/or recalibrated. If the steam pressure gauge is located below the steam drum, an allowance must be made for any existing hydrostatic pressure. The steam pressure gauge is tested by comparing readings with a test gauge or by using a deadweight tester.

## STEAM SIPHONS

**Figure 11-3.** A siphon protects the Bourdon tube in a pressure gauge from live steam.

## Process Pressure Gauges

Pressure gauges are located at critical process points within process systems. Pressures that need to be measured in process systems include feedwater line pressure, atomizing steam line pressure for fuel oil, fuel oil line pressure, and injection air line pressure for pulverized coal. In each case, the pressures at critical locations are monitored and the system flow is adjusted as necessary. The pressure gauge selected must be able to measure the appropriate pressure range.

## Draft Gauges

A *draft gauge* is a pressure gauge used to measure the difference in draft pressure between the atmosphere and the furnace, breeching, or stack. It is typically calibrated in tenths of an inch of water column. Either a manometer or a diaphragm draft gauge can be used to measure draft pressures. **See Figure 11-4.**

The boiler operator must be familiar with the draft readings required at various locations within the boiler system. An improper draft pressure in the boiler furnace can result in gases of combustion leaking into the boiler room through the boiler setting or unneeded air leaking into the boiler.

**Manometers.** A *manometer* is a draft gauge that measures pressure with a liquid-filled tube. A manometer can be a U-tube or inclined-tube gauge. Both types use liquid to indicate pressure. When both columns of the gauge are open to the atmosphere, the liquid in the columns is at the same level. To measure draft, one column of the manometer should be open to the atmosphere, and the other column connected to the point of measurement.

**Diaphragm Draft Gauges.** A *diaphragm draft gauge* is a draft gauge that measures draft using a flexible diaphragm connected with a linkage to a pointer on a scale that is calibrated in inches or fractions of an inch of water column. The top and bottom of the diaphragm are permanently connected to the two points to be measured, one of which may be connected to the atmosphere. Any difference in pressure between the two points causes the diaphragm, and thus the linkage to the pointer, to move.

### TECH TIP

*Steam pressure gauges can be damaged if live steam contacts the internal components. A pigtail or U-tube siphon filled with water forces steam to condense and cool. The pressure is transmitted to the gauge without the damaging effects of the live steam.*

## DRAFT GAUGES

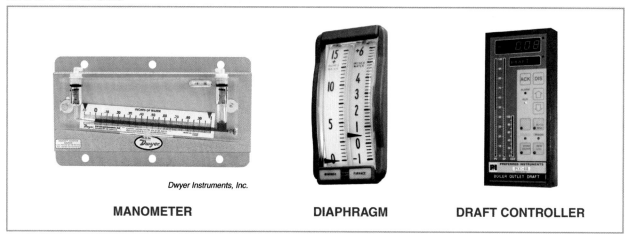

*Dwyer Instruments, Inc.*

**MANOMETER**        **DIAPHRAGM**        **DRAFT CONTROLLER**

**Figure 11-4.** Draft gauges are used to indicate draft pressure. In some cases, a draft gauge is integrated with a draft controller.

The readings are usually transmitted to a control room where they can be monitored. The scales used on diaphragm gauges vary from hundredths of an inch to several inches of water column, depending on which pressure is being measured.

The diaphragm draft gauge can be permanently mounted on a control panel or a signal can be transmitted to the boiler control room. The reading on the gauge will vary depending on where the pressure is measured. Typical locations to measure draft are at the outlet of a forced draft fan, within the furnace, and at various places in the breeching and stack.

### Vacuum Gauges

A *vacuum gauge* is a pressure gauge used to measure pressure less than atmospheric pressure. A vacuum gauge with a Bourdon tube is a common type used in steam plants. Vacuum gauges are typically calibrated to read inches of mercury from 0″ Hg to 30″ Hg. A vacuum gauge is often combined with a pressure gauge into a compound gauge. **See Figure 11-5.**

A vacuum gauge is sometimes called a suction gauge when used where the suction side of a pump or strainer is in a vacuum. The vacuum reading can be used for troubleshooting. For example, an unusually high vacuum (suction) reading indicates a clogged suction line, a dirty strainer, and/or a closed suction valve.

A U-tube manometer using mercury as the liquid can be used to measure vacuum. However, the use of mercury is being discouraged for environmental and safety reasons, and instruments containing mercury are rarely seen.

## VACUUM AND COMPOUND GAUGES

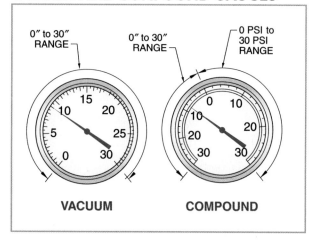

**Figure 11-5.** A compound gauge combines the functions of a pressure gauge and a vacuum gauge.

## TEMPERATURE GAUGES

A temperature gauge is used to measure the intensity of the heat of an object. *Thermometer* is the general term for an instrument used to measure temperature. Thermometers are typically calibrated in degrees Fahrenheit (°F) or degrees Celsius (°C). **See Figure 11-6.** The main types of thermometers are thermocouples, resistance temperature detectors, bimetallic thermometers, and liquid-in-glass thermometers. High-temperature thermometers are sometimes called pyrometers. Pyrometers can be used to measure furnace temperatures in the range of 100°F to 2500°F.

## TEMPERATURE SCALES

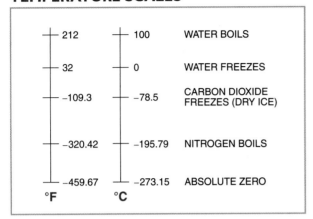

| °F | °C | |
|---|---|---|
| 212 | 100 | WATER BOILS |
| 32 | 0 | WATER FREEZES |
| –109.3 | –78.5 | CARBON DIOXIDE FREEZES (DRY ICE) |
| –320.42 | –195.79 | NITROGEN BOILS |
| –459.67 | –273.15 | ABSOLUTE ZERO |

**Figure 11-6.** Temperature is usually measured using the Fahrenheit scale or Celsius scale.

Temperature gauges provide critical information that allows the operator to assess plant operation. An operator must know the operating temperatures of all the boiler equipment in the plant. For example, a sudden rise in the feedwater heater temperature may indicate live steam blowing through the steam trap. This may cause the feedwater pump to become steambound and stop pumping, resulting in a low-water condition in the boiler.

Tracking the temperatures over time can give valuable troubleshooting information. For example, a decrease in feedwater temperature could indicate a loss of condensate return, causing more makeup water to be used. Too much cold makeup water can result in increased fuel consumption, additional chemical usage, and thermal shock to the boiler. Or, if the temperature of the fuel oil is too low, the combustion process will be affected, resulting in incomplete combustion and the formation of soot and smoke.

### Thermocouples

A *thermocouple* is a thermometer consisting of two dissimilar metals that are joined together to form a circuit that generates a voltage proportional to the difference in temperature between the hot and cold ends of the wires. Thermocouples permit rapid, accurate temperature readings over a wide range. Thermocouples can be used to take readings from several remote locations in a plant. They are usually connected to a transmitter that continuously sends a signal to a controller. The temperature readings are then used to determine the control actions needed to keep the boiler running at optimal conditions.

Thermocouples used in high-temperature applications are typically welded together at one end and sealed in a tube called a thermowell. The amount of electric current induced in the thermocouple depends on the materials used and the temperature difference between the two ends of the thermocouple wires. A thermocouple is commonly used to measure the furnace temperature or the temperatures of the gases of combustion. **See Figure 11-7.**

## THERMOCOUPLES

**Figure 11-7.** Thermocouples can be used to measure furnace temperature.

The hot ends of the wires are exposed to the heat of the gases of combustion. The cold ends of the wires are exposed to ambient temperature. As the heat is sensed, a very small electric voltage is induced. This voltage is proportional to the difference in temperature between the hot end and the cold end. A transmitter is then used to send this signal to the controller.

Thermocouples can be used for many applications in the operation of a boiler. They are used to measure temperatures of superheated steam, desuperheated steam, feedwater entering the boiler, condensate returns, gases of combustion entering the stack, inlet and outlet condenser water, and combustion air.

### Resistance Temperature Detectors

A *resistance temperature detector (RTD)* is an electronic thermometer consisting of a high-precision resistor with resistance that varies with temperature. **See Figure 11-8.** RTDs are accurate and reliable temperature sensors, especially for low temperatures and small ranges. They are generally more expensive than thermocouples and are not used for high temperatures or in corrosive environments.

**RESISTANCE TEMPERATURE
DETECTORS (RTDs)**

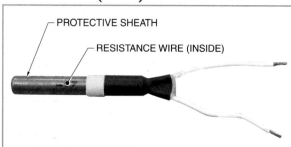

**Figure 11-8.** An RTD contains a resistor that varies with temperature.

An RTD typically increases its resistance when it is exposed to heat. This gives the RTD a positive temperature coefficient. A protective sheath material covers the RTD wires, which are coiled around an insulator that serves as a support. Unlike a thermocouple, an RTD does not generate its own voltage. An external source of voltage or current must be incorporated into the circuit.

RTDs can be used for many applications in the operation of a boiler. Typical applications include monitoring bearing temperatures, monitoring fan motors for overheating, and as a substitute for thermocouples where precise and stable temperature measurements are required.

### Bimetallic Thermometers

A *bimetallic thermometer* is a thermometer that uses a strip of two metal alloys with different coefficients of thermal expansion that are fused together at the ends. Bimetallic thermometers are typically dial-type thermometers used for local temperature indications. **See Figure 11-9.** Bimetallic thermometers are used to measure temperatures from 300°F to 800°F.

## BIMETALLIC THERMOMETERS

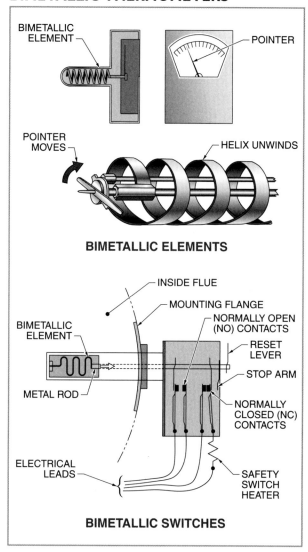

**Figure 11-9.** Bimetallic thermometers work through differential thermal expansion and can be used as switches to activate circuits.

Temperature change causes expansion and contraction of the bimetallic element. Therefore, the bimetallic element can easily be used as a temperature switch. For example, a bimetallic element can be mounted in a boiler or furnace flue and used with a timer as a flame surveillance control, or stack switch. The burner control circuit attempts to start a flame. If the temperature in the stack increases, the bimetallic element expands and forces two electrical contacts together, indicating the presence of a flame. If the temperature in the flue does not increase, the electrical contacts do not touch, the timer runs out, and the burner control circuit shuts the furnace down.

## Liquid-in-Glass Thermometers

In a liquid-in-glass thermometer, the liquid inside the thermometer expands when heated and contracts when cooled to indicate temperature. This type of thermometer comes in a variety of sizes and ranges. Liquid-in-glass thermometers are typically filled with a colored alcohol solution. **See Figure 11-10.** Liquid-in-glass thermometers are normally used as local temperature indicators and are not part of a control circuit.

## LIQUID-IN-GLASS THERMOMETERS

**Figure 11-10.** Liquid-in-glass thermometers are typically used as local temperature indicators.

## LEVEL GAUGES

The simplest way to determine the level of a material in a container or tank is to visually inspect it, such as through a gauge glass. The ASME Code allows for remote monitoring of water level. Therefore, other methods that send a signal to a control room may be used.

Level can be measured in linear units, such as inches of depth of fuel oil. A level measurement may be converted into units of volume or weight. Level-measuring and control instruments are essential for maintaining adequate quantities of materials for a process.

## Gauge Glasses

The ASME Code requires that high pressure steam boilers be equipped with at least one gauge glass. A *gauge glass* is a water level indicator that consists of glass that allows a direct view of the water level. **See Figure 11-11.** A gauge glass is normally attached to a water column by a screw or flanged fitting. Most are equipped with quick-closing stop valves that can be closed in the event of a glass failure.

Quick-closing stop valves are frequently chain-operated so that they can be closed from a location where the boiler operator is out of danger. Some gauge glasses have ball check valves incorporated into the body of quick-closing stop valves that close automatically in the event of a gauge glass failure. Steam boilers that operate at 400 psi and up have either two gauge glasses or one gauge glass and an electrical level indicator. Common types of gauge glasses include tubular, flat glass, reflex, and ported.

**Tubular.** Tubular gauge glasses consist of tubular glass mounted between boiler connections that let the steam and water enter the tube. Tubular gauge glasses are made in many lengths and diameters to ensure they are available in many pressure ranges. Many tubular gauges glasses have a red painted stripe. This helps make the water visible in the glass. Tubular gauge glasses are used in boilers that operate at up to 300 psi. The glass is likely to burst above that pressure.

**Flat Glass.** Flat-glass gauge glasses require a mica shield to protect the glass from the high-alkaline boiler water and steam. Flat-glass gauge glasses are used in boilers that operate at up to about 2000 psi. The glass is usually about ⅜″ to ¾″ thick because of the high pressure. Flat-glass gauge glasses are sometimes called armored gauge glasses.

**Reflex.** Reflex gauge glasses consist of a flat glass with a special vertical sawtooth surface that acts as a prism to improve readability. The refractive index of water is different than that of steam. Light entering the portion of the prism in contact with the liquid is refracted into the boiler and the glass appears dark. The light entering the portion of the prism above the water is reflected back out and the glass appears light. This makes it easy to read the gauge. Reflex gauge glasses are used in boilers that operate at up to about 350 psi.

## GAUGE GLASSES

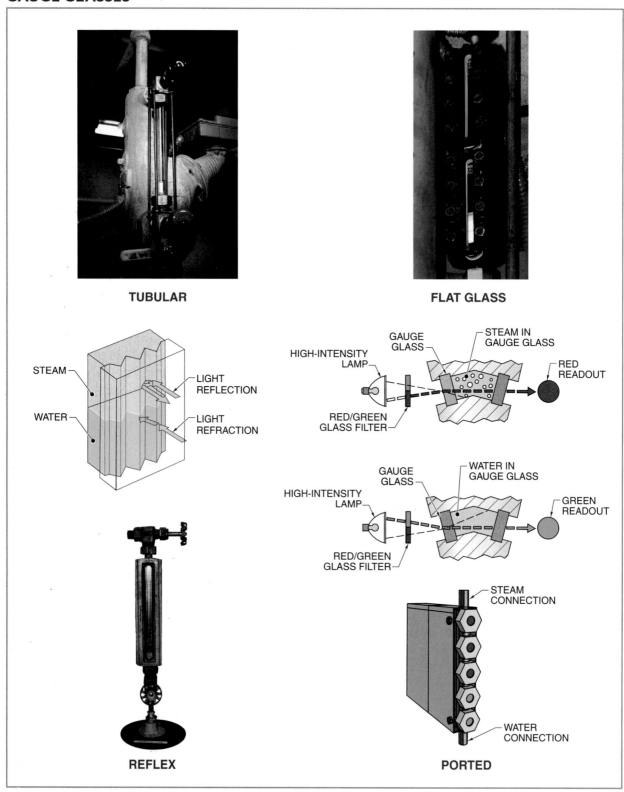

**Figure 11-11.** Gauge glasses are available as tubular, flat glass, reflex, and ported. The different designs are intended for use at different pressure ranges.

**Ported.** Ported gauge glasses are used in boilers that operate at up to about 3000 psi. Ported gauge glasses require a mica shield to protect the glass from the high-alkaline boiler water and steam. Ported gauge glasses display the ports below the boiler water level in green and the ports above the water level in red.

## Conductivity Probes

A *conductivity probe* is a water level indicator that senses the boiler water level by detecting a change in current flow as water covers or uncovers the end of the probe. A conductivity probe is typically wired to display the level remotely at a panel. Depending on the length of the probe, it can be used as a level indicator, a high water or low water alarm switch, or a feedwater pump ON or OFF switch. **See Figure 11-12.**

## CONDUCTIVITY PROBES

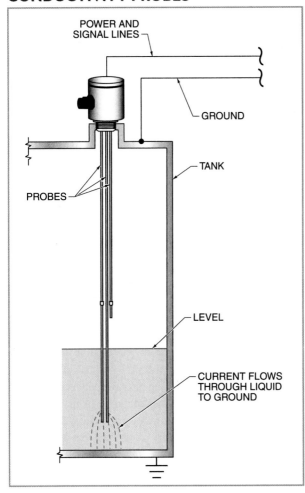

**Figure 11-12.** Conductivity probes use the conductivity of water to measure the water level.

The newest systems have the capability to automatically monitor the condition of the probe. The results can be displayed in a control room to indicate that a blowdown of the probe column is necessary to maintain accuracy. Additionally, system intelligence can identify probes that have become unstable and need to be replaced.

## Magnetic Float Indicators

A *magnetic float indicator* is a water level indicator that consists of a tube chamber and an internal float. The float is nonmagnetic but contains a ring of magnets. The magnets within the float operate a magnetic indicator located on the outside of the pressure chamber. The indicator is either a single follower or a series of flags that rotate and change color as the float magnets pass by. **See Figure 11-13.**

## Floats

A *float* is a level-measuring instrument with a hollow ball attached to it that floats on top of a liquid in a tank. **See Figure 11-14.** Floats are specifically designed and calibrated for each tank according to the size and shape of the tank and the liquid that is held within. Floats are commonly used in level alarms used to measure water level in a boiler. Floats can also be used to indicate tank levels, actuate alarms or shutdown switches, or even mechanically control valves. For example, a switch can start a pump when the float is at one position and stop the pump when the float is at another position.

Special floats are available for use in condensate return tanks. A float is attached to an arm long enough for the float to reach the top and bottom of the tank. The arm is coupled through a seal to a level indicator calibrated for the volume in the tank. Special float control systems are mounted in a housing and connected to a control valve. The housing is mounted at the desired level. An increase in the liquid level in the housing opens the valve, allowing the liquid to be discharged to the liquid return line. As the liquid level falls, the float falls and closes the valve.

## MAGNETIC FLOAT INDICATORS

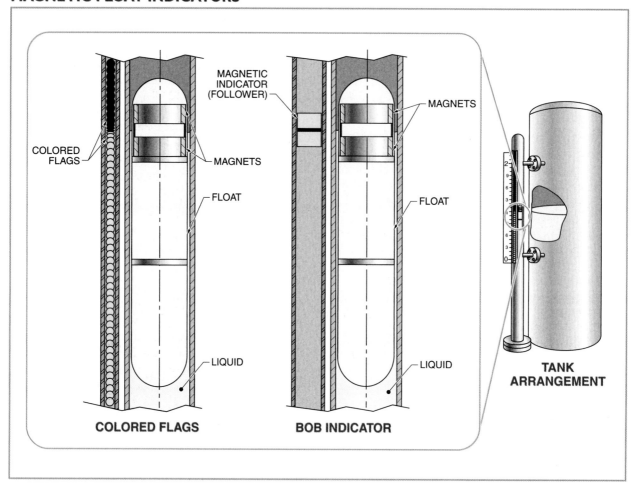

**Figure 11-13.** Magnetic float indicators use a float with magnets to flip flags or move a follower to show the level.

## FLOATS

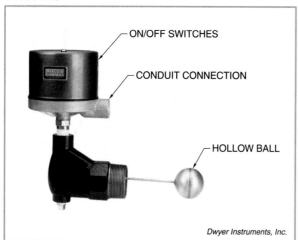

Dwyer Instruments, Inc.

**Figure 11-14.** A float activates an ON/OFF switch when the level in a tank exceeds the alarm level.

### Displacers

A *displacer,* or pneumatic level control, is a level-measuring instrument consisting of a cylinder, heavier than the liquid in which it is immersed, connected to a spring or torsion device that measures its weight. **See Figure 11-15.** The displacer is usually mounted in a housing that eliminates fluid turbulence. When the displacer is supported by an arm connected to the sealed end of a torque tube, the sensing system becomes separated from the output device. A rod running down the center of the torque tube, fastened at the sealed end, transfers the amount of rotation at the sealed end of the torque tube to the various output devices. Displacers are often used to measure the water level in a boiler.

### Pressure Measurements

The pressure variations at the base of a liquid column provide a means of determining liquid level in a storage tank. As long as the liquid in the tank has a constant density, variations in the pressure are caused only by variations in the level. For an open vessel, the simplest way to measure pressure is to connect a pressure gauge, switch, or transmitter to the lowest practical level so that any rise in level creates an increase in pressure.

## DISPLACERS

**Figure 11-15.** A displacer measures the buoyancy of a float in a cylinder to determine the level of a liquid.

For example, an open vertical cylindrical water tank 20′ high is filled with water and has a pressure connection at the bottom. **See Figure 11-16.** To measure the actual tank level, the measurement instrument is calibrated to measure a range of 0′ to 20′.

### Ultrasonic Sensors

An *ultrasonic sensor* is a level-measuring instrument that uses an ultrasonic signal to measure level. **See Figure 11-17.** A transmitter generates a high-frequency sound directed at the surface of the material in the vessel. The transit time is the time it takes for the sound to travel to the surface of the material and back to the receiver. The electronic circuitry in the receiver measures the transit time and calculates the distance.

## OPEN-TANK PRESSURE MEASUREMENT

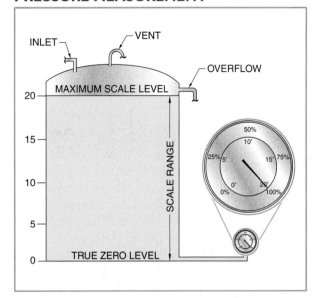

**Figure 11-16.** Pressure measurement can be used to determine liquid level when the density is known.

## ULTRASONIC SENSORS

**Figure 11-17.** Ultrasonic sensors measure the transit time from a transmitter to the surface of a material to determine level.

As the level in the vessel changes, the transit time changes. The distance from the receiver down to the surface is calculated from the transit time. The level is then calculated by subtracting the distance from the height of the vessel. An ultrasonic sensor can be used to measure the level of a liquid in a tank or a solid fuel in a storage bin.

### Bubblers

A *bubbler* is a level-measuring instrument consisting of an air tube extending to the bottom of a tank, a pressure gauge and transmitter, a flow-control meter, and an air-pressure regulator. A bubbler may also be called a pneumercator. Bubblers are specifically designed and calibrated for each tank according to the size and shape of the tank and the liquid that is held in it. **See Figure 11-18.** An air tube is placed vertically in the tank with its open end placed close to the bottom. The top of the tube is connected to an air pressure regulator and a flow-control meter.

## BUBBLERS

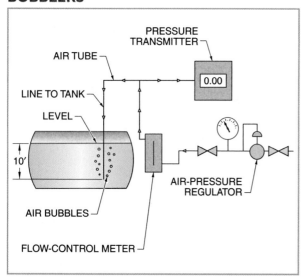

**Figure 11-18.** A bubbler can be used to indicate the level of fuel in a fuel oil tank. The pressure at the transmitter depends on the fuel oil level.

When the pressure of the air in the air tube supports the column of liquid in the tank, the air pressure remains constant. As the level of the liquid in the tank drops, there is less resistance. Therefore, less air pressure is required to balance the column of liquid in the tank and the pressure decreases.

Although bubblers can be used to measure all types of liquids, in the boiler room they are used primarily to measure fuel oil in fuel oil tanks. Routine service on bubblers involves keeping the air tube clean, adding gauge glass fluid when needed, and purging the line to the tank to ensure it is clear.

### Magnetostrictive Sensors

A *magnetostrictive sensor* is part of a level-measuring system for boiler water consisting of an electronics module, a waveguide, and a float containing a magnet that is free to move up and down a brass rod pipe inserted into a vessel from the top. **See Figure 11-19.** Within the rod is a waveguide constructed of magnetostrictive material. Magnetostrictive materials can convert magnetic energy into kinetic energy and back.

## MAGNETOSTRICTIVE SENSORS

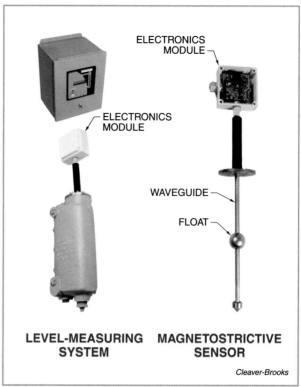

*Cleaver-Brooks*

**Figure 11-19.** The electronics module in a magnetostrictive sensor uses a waveguide and float to determine level.

The electronics module at the top end of the waveguide generates a current pulse that creates a magnetic field in the waveguide. The interaction of the magnetic field with the magnets in the float results in the generation of a second pulse in the waveguide that reflects back to the top. The time between the generated pulse from the electronics module and the return pulse is a function of the distance between the magnets within the float and the waveguide.

The electronics module allows flexible operation of the sensor. It can be used as part of an ON/OFF feedwater pump operation or used with modulating control. It can provide blowdown reminders and messages indicating whether the blowdown has been performed properly. A stuck float can be used to signal an alarm or to shut down a boiler. It can also provide high-water and low-water alarms.

### Guided-Wave Radar

A *guided-wave radar* is a water level indicator that uses a radar pulse that is sent down a waveguide that extends into the water to calculate the level based on the radar transit time. **See Figure 11-20.** Radar is only affected by materials that reflect energy, which means that temperature variations, dust, pressure, and viscosity do not affect accuracy. When the pulse hits a surface, a significant portion of the energy of the pulse is reflected back up the probe to the electronic controller. The level is directly proportional to the time it takes to send and receive the pulse. A portion of the emitted pulse will continue down to the end of the probe.

### GUIDED-WAVE RADAR

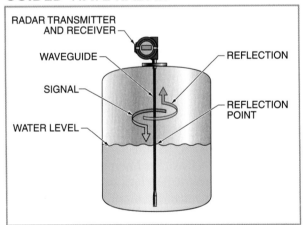

**Figure 11-20.** With a guided-wave radar, the level is calculated based on the radar transit time.

## FLOW METERS

A flow meter is used to measure the flow rate of liquid, gas, or steam in a pipe or duct. Flow meters are used for air, water, steam, fuel oil, and gases. For many types of flow meters, the meter must be calibrated for the specific

liquid or gas to be measured. If this is not taken into account, readings may be in error. Flow meter readings can be expressed in gallons, pounds of steam per hour, or cubic feet per minute. Readings can be taken at the flow meter or sent to a totalizer that compiles the flow information. Flow meters that measure the rate of flow include differential-pressure flow meters, variable-area flow meters, positive-displacement flow meters, and Coriolis flow meters.

### Differential-Pressure Flow Meters

A differential-pressure flow meter can be used to measure the flow of air, steam, or other gases. A pipeline flow restriction is used to cause a pressure drop as fluid flows through the restriction. A differential-pressure flow meter measures the pressure drop across the restriction, which varies with changes in the flow rate. **See Figure 11-21.**

### DIFFERENTIAL-PRESSURE FLOW METERS

**ORIFICE PLATE**

**TRANSMITTER**

**Figure 11-21.** A differential-pressure flow meter measures the pressure drop across a flow restriction to calculate flow.

The differential pressure is measured with a pressure gauge and sent to a transmitter. The transmitter then converts the pressure differential into a rate of flow and transmits it to a controller. Differential-pressure flow meters are easy to use and very reliable. However, the pressure drop they cause wastes energy and increases operating costs.

The most common types of restriction devices are an orifice plate and a venturi tube. An *orifice plate* is a flow restriction device consisting of a thin circular metal plate with a sharp-edged hole in it and a tab that protrudes from a flange. A *venturi tube* is a flow restriction device consisting of a fabricated pipe section with a converging inlet section, a straight throat, and a diverging outlet section.

### Variable-Area Flow Meters

A *variable-area flow meter* is a meter that maintains a constant differential pressure and allows the flow area to change with the flow rate. **See Figure 11-22.** A variable-area flow meter measures flow using the amount of resistance created by a float or piston as it changes the area (size) of the flow path. As fluid flows through the tube, the float rises, changing the flow area. The movement of the float or piston varies depending on the resistance against the flow. The most common type of variable-area flow meter is a rotameter. A *rotameter* is a variable-area flow meter consisting of a tapered tube and a float with a fixed diameter.

## VARIABLE-AREA FLOW METERS

**Figure 11-22.** A variable-area flow meter measures flow by the resistance caused by a float or piston.

### Positive-Displacement Flow Meters

A *positive-displacement flow meter* is a flow meter that admits fluid into a chamber of known volume and then discharges it. This flow meter measures and discharges the volume of a fluid in a manner similar to a positive-displacement pump. With the positive-displacement flow meter, the reading is taken off the meter. The positive-displacement flow meter is commonly used to measure the flow of fuel oil and water.

One type of positive-displacement flow meter uses a wobble plate that rotates as fluid enters the feed chamber and is discharged out of the discharge chamber. **See Figure 11-23.** The wobble plate rotates with the flow and indicates rate of flow by the number of revolutions it makes in a given period. Other types of positive-displacement flow meters use rotating chambers of constant size where the number of rotations is counted to measure the flow.

## POSITIVE-DISPLACEMENT FLOW METERS

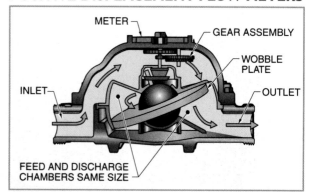

**Figure 11-23.** A positive-displacement flow meter measures flow by counting the number of times a chamber of known size is filled and emptied.

### Coriolis Flow Meters

A *Coriolis flow meter* is a meter that measures the mass flow rate of a fluid. **See Figure 11-24.** It does not measure the volumetric flow rate, such as cubic feet per second, passing through the meter, but the mass flow rate, such as pounds per second. The mass flow rate is then divided by the fluid density to equal the volumetric flow rate. If the density of the measured fluid is constant, Coriolis flow meters provide accurate volumetric flow measurements. If the density of the measured fluid changes, Coriolis flow meters have difficulty measuring the volumetric flow accurately. Coriolis flow meters may also have difficulty handling the condensation in steam. The presence of the condensate interferes with the accuracy of the flow measurement.

## CORIOLIS FLOW METERS

**Figure 11-24.** A Coriolis flow meter measures the mass flow of a fluid.

## ANALYZERS

Water analyzers are necessary for monitoring the level of impurities in boiler water, feedwater, and condensate. Water analyzer results are used to decide on a water treatment process and to adjust the frequency and length of the continuous boiler blowdown. An *analyzer sampling system* is a system of piping, valves, and other equipment used to extract a sample from a process, condition it if necessary, and convey it to an analyzer. **See Figure 11-25.** An in-line analyzer must be directly connected to a process stream in order to obtain a sample. Some analyzers continuously analyze the process samples, while others require time to analyze a sample and thus a sample is only fed to them intermittently.

## ANALYZER SAMPLING SYSTEMS

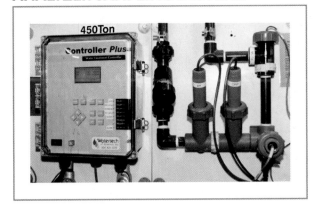

**Figure 11-25.** An analyzer sampling system consists of the components needed to condition a sample and bring it to an analyzer.

## Conductivity Meters

A conductivity meter consists of two electrodes immersed in a solution that measure the conductivity of the solution. **See Figure 11-26.** Conductivity meters are used to monitor the total dissolved solids in the water. If the conductivity of boiler water is higher than a specific limit, the boiler must be blown down.

## CONDUCTIVITY METERS

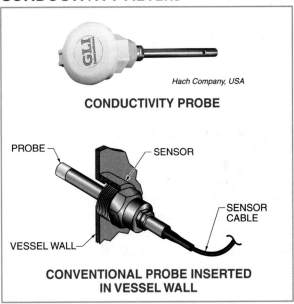

**Figure 11-26.** Conductivity meters measure the conductivity of boiler water to determine the total dissolved solids in the water.

Electrical conductivity in a solution is proportional to the concentration of ions in the solution. Conductivity changes significantly with temperature, so the temperature is measured and used to adjust the measured conductivity. Conductivity measurement works best at lower concentrations. At higher concentrations, it is more complex because the response may be nonlinear.

Since the measurement of conductivity is an indication of the total ions present in a solution the scale of the instrument can be calibrated in ppm or percent concentration of the conductive material in a solution. This is frequently an important consideration in determining the purity of feedwater, boiler water, or condensate.

Conductivity measurement is also used to detect the presence of liquid carryover in a steam line. A common use for a conductivity meter is checking the quality of the recovered steam condensate being used for feedwater.

## pH Analyzers

A *pH meter* is a water analyzer used to monitor the acidity or alkalinity of a boiler water. **See Figure 11-27.** The pH is the measurement of the acidity or alkalinity of a solution caused by the dissociation of chemical compounds in the water. As these compounds dissociate, electrically charged ions are formed in the solution. Some ions are positively charged and some are negatively charged.

Water itself dissociates into hydrogen ions (H+) and hydroxyl ions (OH–). An acidic solution has a greater concentration of hydrogen ions. An alkaline solution has a greater concentration of hydroxyl ions. The pH is usually measured on a scale of 0 to 14, with 7 being neutral, less than 7 being acidic, and greater than 7 being alkaline. The pH level is an important indicator of the corrosive impact of the water.

The pH of the boiler water and condensate should be routinely measured. Boiler water must be maintained in an alkaline condition to prevent corrosion in the boiler. Wastewater must be kept within a certain range before being discharged. **See Figure 11-28.**

## IN-LINE WASTEWATER SAMPLING

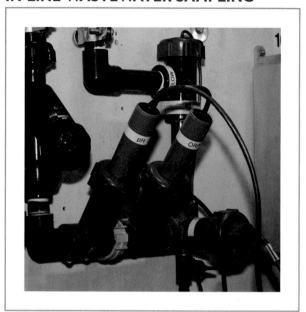

**Figure 11-28.** Wastewater must be kept within a certain pH range before being discharged.

## BOILER WATER pH MEASUREMENT

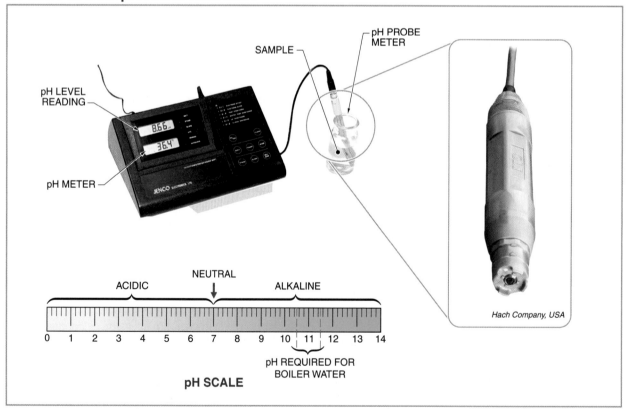

**Figure 11-27.** A narrow pH range must be maintained for boiler water.

## Stack Analyzers

Stack analyzers are needed to meet environmental monitoring requirements. Stack analyzers can be installed in the stack to measure the gases of combustion directly. **See Figure 11-29.**

## STACK ANALYZERS

**Figure 11-29.** A stack analyzer can be installed in a breeching or stack.

Typical values measured include the oxygen level and carbon monoxide concentration. A stack analyzer also provides information on the operation of the boiler, allowing adjustments to be made to increase operating efficiency. The stack analyzer used varies by the location, fuel type, and system capacity. The most common type of gas analyzers are combustion gas analyzers, oxygen analyzers, and opacity analyzers.

**Combustion Gas Analyzers.** Gases of combustion can be analyzed using a gas analyzer and an oxygen analyzer. A gas analyzer uses the principle that different gases absorb different, very specific, wavelengths of electromagnetic radiation in the infrared (IR) or ultraviolet (UV) regions of the electromagnetic spectrum. The drop in intensity of a specific wavelength after it has passed through a sample is directly related to the concentration of the gas in the sample.

Gas analyzers may measure ranges of up to 100% and down to about 1 ppm. Individual analyzers can be selected for carbon monoxide, carbon dioxide, sulfur oxides, nitrogen oxides, and methane. An excessive level of carbon monoxide indicates incomplete combustion and insufficient secondary air. Methane is measured during the purge cycle. Sulfur oxides and nitrogen oxides are monitored because of environmental concerns. Gas analyzers need to be carefully calibrated. **See Figure 11-30.**

## COMBUSTION GAS CALIBRATION

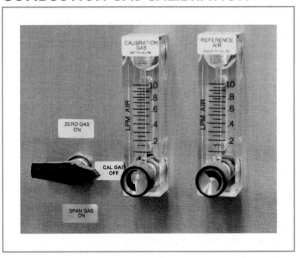

**Figure 11-30.** Gas analyzers need to be carefully calibrated.

**Oxygen Analyzers.** Oxygen is necessary to support the combustion process. Combustion processes require a precise air-fuel ratio to ensure the efficient use of the fuel and the air. Because of the high temperature, a zirconium oxide oxygen analyzer is often used to measure the excess oxygen content. This is a measure of burner efficiency. **See Figure 11-31.**

## OXYGEN ANALYZERS

**Figure 11-31.** Zirconium oxide oxygen analyzers measure the oxygen content of a high-temperature gas stream. In a boiler application, the analyzer output can be used to open or close dampers to control combustion.

**Opacity Analyzers.** An opacity analyzer is a stack analyzer consisting of a focused light source and an analyzer that measures the received light intensity. **See Figure 11-32.** Opacity analyzers are sometimes called smoke indicators.

Stack gas opacity is measured to determine the amount of particulate matter in the gases of combustion. Particulate matter absorbs and scatters the transmitted light, and an increase in particulate matter reduces the amount of light received at the detector. The decrease in light intensity is proportional to the amount of particulate matter. The received light is measured by a photodiode and converted to an output signal.

Coal and heavy oil-fired combustion processes generally require an opacity analyzer. Burning coal produces flyash, which is usually removed from the gases of combustion with an electrostatic precipitator. The opacity analyzer monitors for any particulate breakthroughs. Natural gas and light oil-fired combustion processes rarely produce particulate matter in their gases of combustion.

## OPACITY ANALYZERS

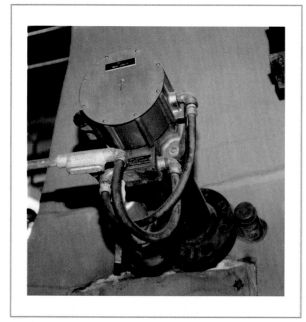

**Figure 11-32.** Opacity analyzers measure the attenuation of a light source as the light passes through a sample.

An opacity smoke indicator alarm can be used to indicate an unexpected amount of particulate matter in the gases of combustion. An alarm delay is set for the desired time. An inadequate alarm delay setting could cause excessive sensitivity of the smoke indicator. This can result in frequent alarms caused when lighting the burner or during other temporary periods of high smoke levels. When the smoke indicator senses smoke density exceeding the alarm point and alarm delay, the alarm light is energized and latched. The smoke indicator must then be reset to cancel the alarm.

---

### SECTION 11.1–CHECKPOINT

1. What is the typical range of a steam pressure gauge compared to the safety valve setting?
2. Which gauges are typically used to measure draft?
3. How does a thermocouple generate voltage?
4. How can a bimetallic thermometer be used as a temperature switch?
5. What are the four main types of gauge glasses?
6. How is a float designed to work in a condensate return tank?
7. How is a differential-pressure flow meter, such as an orifice plate, used to measure flow?
8. How is the mass flow rate through a Coriolis flow meter converted to the volumetric flow rate?
9. Why do conductivity measurements work best at lower concentrations?
10. What types of combustion processes typically require an opacity analyzer?

---

## SECTION II.2
# BOILER CONTROL SYSTEMS

### CONTROL LOOPS

Boiler control systems are used to monitor and automatically control boiler operation. A boiler control system can be mounted on a boiler or operated from a control room. **See Figure 11-33.** A boiler control system uses control loops to operate. A *control loop* is a system that consists of a primary element and transmitter, a control element, and a final element. A local control loop controls one variable, with the controller typically located close to the point of measurement.

### TECH TIP

*A boiler operator should understand the function of each component in a control loop and be able to determine that each component is working properly. This knowledge helps instrumentation maintenance personnel minimize the time needed to return malfunctioning instrumentation to normal service.*

## BOILER CONTROL SYSTEMS

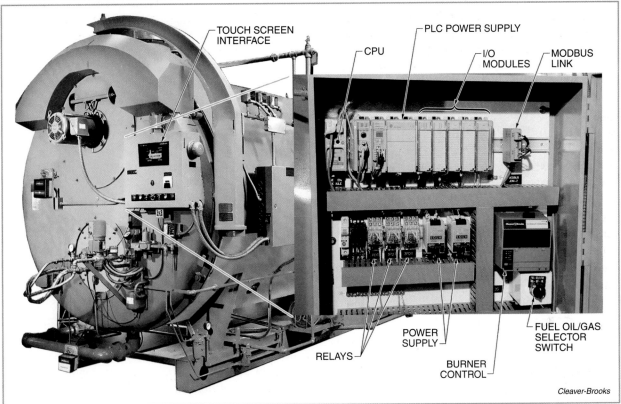

**Figure 11-33.** Boiler control systems are designed to monitor and control boiler operation automatically.

Boiler control loops are used for pressure, temperature, liquid level, and flow control. Pressure control loops control the pressure of the boiler or fuel oil. Temperature control loops control the temperature in the boiler or fuel oil heaters. Level control loops control the boiler water level or the fluid levels within tanks. Flow control loops control the flow of feedwater into a boiler, steam flow out of a boiler, fuel flow to the boiler burners, or burner draft airflow. For example, a typical local control loop controls boiler water level by measuring the level and transmitting the level to a controller. The controller then sends a signal to a control valve. The control valve opens or closes to control the level. **See Figure 11-34.** However, actual implementation of level control uses more variables.

A system control loop controls multiple variables. It can automatically control several functions of a system, such as providing control of the steam pressure while maintaining the proper air-fuel ratio over the entire boiler operating range. A system control loop provides supervisory control over multiple local control loops.

## LOCAL CONTROL LOOPS

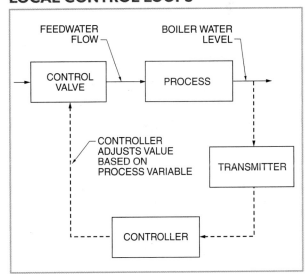

**Figure 11-34.** A typical local control loop controls boiler water level by measuring the level and transmitting the level to a controller. The controller then sends a signal to a control valve. The control valve opens or closes to control the level.

### Primary Elements

A *primary element,* also known as a sensor, is an instrument that measures a process variable and produces a usable output in the form of mechanical movement, electrical output, or instrument air-pressure output. Sensors are located in a process to measure variables such as temperature, pressure, and level. A *transmitter* is a device that conditions a low-energy signal from a sensor and produces a suitable signal for transmission to other components or devices.

### Control Elements

A *control element,* also known as a controller, is a device that takes an input signal, compares the signal to a setpoint, performs a computation, and sends an output signal to a final element. Controllers can be connected to alarms and/or automatic shutdown devices that are activated during abnormal conditions. Controllers can also record information for tracking boiler conditions.

### Final Elements

A *final element,* also known as a control device, is a device that controls the flow of liquid, gas or air, or electrical current. Final elements are used to implement the strategy of an automatic control system. Typical final elements include control valves, actuators, regulators, and dampers. A process control system is only as good as the components that actually regulate the flow of material or energy into or out of a process. Control valves are the most common type of final element. Regulators are self-operating control valves.

**Control Valves.** A *control valve* is a final element that is used to modulate fluid flow in response to signals from a control element. **See Figure 11-35.** The ability of control valves to precisely adjust the flow of a fluid is vital for accurate control. Common types of control valves include quick-opening valves and equal-percentage flow valves. Caution is necessary when quickly changing the flow in high-temperature and high-pressure systems. A *solenoid valve* is an electrically operated control valve that quickly adjusts a valve to an open or closed position.

## CONTROL VALVES

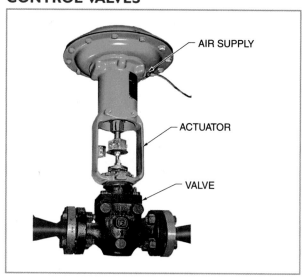

**Figure 11-35.** Control valves are used to regulate the flow of materials and energy into and out of a process.

**Actuators.** An *actuator* is a device that provides the power and motion needed to manipulate the moving parts of a valve or damper used to control fluid flow through a final element. **See Figure 11-36.** The requirements of an actuator are speed, power, precision, and resolution. An actuator must respond quickly to a change in a control signal and have enough power to overcome the process pressure and mechanical friction of the moving parts. Positioning of the final element needs to be repeatable for the same control signal. In addition, the actuator must be able to work with minimal maintenance for long periods.

Actuators can be spring-loaded diaphragm actuators that will fail-safe. The fail-safe condition is a position that a valve defaults to in the event of a power failure in order to minimize dangerous situations. For example, a valve used to control steam flow to a heating process should normally fail closed so that steam flow stops when power is lost in order to prevent overheating. Dampers used to control the flow of air are typically operated by an actuator.

**Regulators.** A *regulator* is a self-operating control valve used for pressure and temperature control. Regulators require no other source of energy than the process itself. A *pressure regulator* is an adjustable valve that is designed to automatically control the pressure downstream. A *pilot-operated pressure regulator* is a regulator that uses upstream fluid as a pressure source to power the diaphragm of a larger valve. **See Figure 11-37.**

## ACTUATORS

**Figure 11-36.** An actuator provides the power and motion needed to manipulate the moving parts of a valve or damper that are used to control fluid flow through a final element.

## PILOT-OPERATED PRESSURE REGULATORS

**Figure 11-37.** A pilot-operated pressure regulator uses the upstream pressure as a power source.

**Dampers.** A *damper* is an adjustable blade or set of blades used to control the flow of air. **See Figure 11-38.** Dampers are normally constructed from welded steel that is treated to resist corrosion and rust. Dampers may also contain blade seals that seal when the blades contact

each other. Damper blade seals can become brittle and crack when exposed to temperature variations and humidity extremes. Cracked blade seals lead to excessive air leakage and poor control.

## DAMPERS

**Figure 11-38.** Dampers use movable blades to control the amount of airflow.

An actuator moves a damper crank arm. A drive blade transmits the force to the other blades through a set of linkages. The linkage from the actuator must be connected to the drive blade only. Blades should be locked with a bolt or key instead of a set screw, which can vibrate loose. If the damper blades are mounted vertically instead of horizontally, thrust bearings should be used on the bottom to support the weight of the blades.

### Control Panels

Control panels can be located either in a local area on or near the boiler or in the control room. **See Figure 11-39.** The control room is usually cleaner and has better temperature control than other areas. Transmitters are used to relay the readings from the instruments to the controller, which then provides signals to activate local switches and other controllers.

Local control panels provide the operator with system information at the point of operation. They are normally connected directly to the instruments and controllers and are usually found on package boilers. Local control panels may have external connections to a central control panel. The control panel may contain indicators, recorders, and annunciators to alert the operator and record the conditions of the boiler system. Indicators provide instrument readings to an operator. Indicators are typically gauges, but they could be simple lights.

**Recorders.** Recorders are used to record the pressures, temperatures, humidity, electricity levels, liquid levels, and flow rates in the various systems within a steam plant. The recorder may store the information electronically or it may use paper strip or circular charts. **See Figure 11-40.** Paper charts are available in different sizes, depending on the data needed.

Recorders can record data concerning one specific variable or several variables present in the boiler system. This data can later be analyzed to assess plant operation. This allows for accurate evaluation of each system and helps increase plant and production efficiency.

## CONTROL PANELS

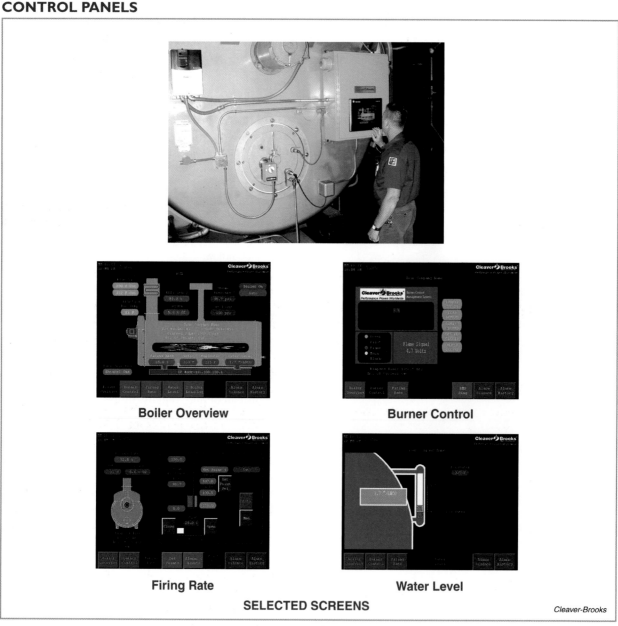

**Boiler Overview**

**Burner Control**

**Firing Rate**

**Water Level**

**SELECTED SCREENS**

*Cleaver-Brooks*

**Figure 11-39.** A touch screen can be integrated into a control panel to allow an operator to switch between different views.

## RECORDERS

**Figure 11-40.** Recorders and gauges are often mounted on control panels near the boiler or in a control room.

A *totalizer* is a recorder that records the total amount measured. Totalizers are used to track the total consumption or discharge of a material. Examples include the consumption of fuel or water, the production of steam, and the discharge of specific gases.

**Annunciators.** An *annunciator* is an audible alarm that is created electrically or electronically. Annunciator systems include common alarms and dedicated alarms. Common alarms indicate general plant conditions that need operator attention. Dedicated alarms provide indications as to the cause of a problem. The simplest alarms turn on an indicator light for further investigation by the operator. Some alarms include safety controls that result in the shutdown of the boiler. Operator attention is normally required to reset the alarm or restart the boiler.

## DIGITAL CONTROL SYSTEMS

A *digital control system* is an integrated system of control loops that use digital control signals to control a process. **See Figure 11-41.** A digital control system can be a direct digital control (DDC) system when the boiler

is being used in a commercial application, such as in an office building, hospital, school, apartment building, or sports stadium. It can also be a programmable logic control (PLC) system when used in an industrial application, such as in a manufacturing plant.

## DIGITAL CONTROL SYSTEMS

**Figure 11-41.** A digital control system includes a digital controller and many types of displays to help the operator manage large, complex processes.

While both DDCs and PLCs are digital control systems, there are specific differences between the two. A DDC system has a fixed program with some adjustable parameters, requires a controlled environment, and is less expensive than a PLC. A PLC is programmable and more flexible, but more costly.

Digital control systems for boilers and related equipment can be a part of a plant automation system used for process applications, as well as a part of heating, ventilating, and air conditioning (HVAC) systems. These systems use a computer to control equipment in a building or production plant. Controls in building or plant automation systems can be networked within a plant and/or connected to remote locations via the Internet.

Each manufacturer has specific control system strategies. However, no control system strategy is designed to replace a safety device. For example, high-temperature and low-temperature limit controls, boiler water high-limit and low-limit controls, and flame safeguard controls are still required when a plant automation system is used. In addition, the installation of an automation system does not replace scheduled testing and preventive maintenance.

## Transducers

Most modern sensors include a measuring device and a solid-state transducer. A *transducer* is a device that receives a signal representing a variable, such as temperature, pressure, or flow, and converts that signal into an electrical or digital signal that is compatible with a digital control system. For example, the signal from the movement of a diaphragm in a pressure gauge can be converted to a digital or electrical signal.

A transducer integrated into a pressure transmitter sends the signal to a controller. **See Figure 11-42.** Electrical signals are processed in a PLC system or other digital controller. The controller can be integrated with an energy management and control system, which provides for more economical plant operation.

## PRESSURE TRANSMITTERS

**Figure 11-42.** A transducer and pressure transmitter are integrated into one device that converts a measurement into useful units and sends a signal to a controller.

## Input and Output Devices

Input and output (I/O) devices are the way in which a computer controller receives and transmits data to and from the field equipment. The four types of I/O modules and devices for digital control systems are analog-in (AI), analog-out (AO), digital-in (DI), and digital-out (DO). Analog devices use continuous measurements or signals, such as 4 mA to 20 mA transmissions. Digital devices use binary signals representing predetermined conditions such as open/closed and ON/OFF. The type of device used determines the required I/O type.

## Programming Devices

A *programming device* is a device used to control the actions of a PLC, DDC system, or computer. Manufacturers have designed programming devices that minimize learning time and programming time. These devices allow plant personnel to solve control problems rather than spending time learning to program the device. The programming languages are distinctively different for PLCs than for DDC systems, but the programming concepts are similar.

---

### SECTION 11.2 – CHECKPOINT

1. What is the role of a primary element?
2. What is the role of a control element?
3. What are four common final elements?
4. Where is a typical location for local control?
5. What is a digital control system?
6. What is a transducer?

## CASE STUDY—BOILER DOWN ON CONTROL AIR COMPRESSOR FAILURE

### Situation

The operating engineer at a correctional facility was responsible for starting, stopping, operating, maintaining, repairing, and logging the boilers, chillers, cooling towers, air compressors, and air handlers throughout the facility. This correctional facility had two boilers. One was a gas-fired scotch marine boiler, and the other was a resistance-element electric boiler. The gas boiler was a four-pass horizontal firetube boiler with a pneumatic feedwater regulator controlling a pneumatic feedwater valve.

### Problem

The pneumatic feedwater regulator and valve arrangement worked well for several years with minimal maintenance. One day, the air pressure was lost when a control problem developed at the air compressors that furnished the control air for the facility. This control air was supplied to the pneumatic feedwater regulator and valve.

The loss of air pressure happened on a Saturday when the only person in the facility from the maintenance department was the operating engineer. The pneumatic feedwater valve was a normally closed valve. The valve closed on loss of air pressure and did not allow any water to be fed into the boiler. The water level dropped, and the low-water fuel cutoff eventually shut the boiler down. The boiler was still hot enough to produce steam and eventually evaporated enough water that the water level fell below the lowest visible level in the gauge glass.

The water level fell to the point where the auxiliary low-water fuel cutoff was tripped and the boiler required a manual reset to start again. The control system sent alarms to the engineer's phone. By the time the engineer responded to the boiler being shut down on low water, the water level was so low that it was out of sight in the gauge glass.

### Solution

Local code required a boiler inspection any time the boiler water level fell to the point where the auxiliary low-water fuel cutoff tripped. Inspection of the boiler found nothing wrong except for the air compressor problem. The boiler inspector required that an alarm system be set up to secure the boiler and sound an alarm any time the control air pressure was lost.

**Name** _____ **Date** _____

_____ **1.** Pressure gauges can be calibrated in pounds per ___.
A. square foot
B. cubic inch
C. linear inch
D. square inch

_____ **2.** When one end of a thermocouple is heated, a small ___ is produced.
A. voltage
B. resistance
C. pressure
D. movement

_____ **3.** A steam pressure gauge should have a range of ___.
A. 1 to 1½ times the safety valve setting
B. 1½ to 2 times the safety valve setting
C. 2 to 2½ times the safety valve setting
D. 110% of the MAWP

_____ **4.** A bubbler may also be called a(n) ___.
A. pneumercator
B. float
C. opacity meter
D. magnetostrictive sensor

_____ **5.** A steam pressure gauge must be protected from live steam by ___.
A. a metal frame
B. steam gauge guards
C. a blower
D. a pigtail siphon

_____ **6.** Opacity smoke indicator alarms can be used to ___ in the gases of combustion.
A. indicate an unexpected amount of particulate matter
B. measure excess oxygen
C. detect an unexpected amount of carbon monoxide
D. measure excess carbon dioxide

_____ **7.** Vacuum gauges are calibrated in inches of ___ atmospheric pressure.
A. mercury below
B. water below
C. mercury above
D. water above

_____ **8.** A control valve is an example of a ___.
   A. primary element
   B. control element
   C. transmitter
   D. final element

_____ **9.** Thermocouples measure ___.
   A. gauge pressure
   B. absolute pressure
   C. temperature
   D. oxygen content

_____ **10.** A measurement of ___ can be made with a thermocouple.
   A. steam pressure
   B. feedwater flow
   C. condensate flow
   D. steam temperature

_____ **11.** Liquid level in a tank may be measured with a ___.
   A. bubbler
   B. thermocouple
   C. pressure control
   D. tank chart

_____ **12.** Flow meters measure the rate of flow of ___.
   A. gases only
   B. liquids only
   C. solids
   D. a variety of fluids

_____ **13.** A differential-pressure flow meter functions by measuring the difference in pressure across a ___.
   A. pipe elbow
   B. restriction
   C. steam trap
   D. thermocouple

_____ **14.** A rotameter consists of a(n) ___.
   A. straight tube and a float
   B. tapered tube and a float
   C. orifice plate
   D. pressure diaphragm

_____ **15.** A variable-area flow meter maintains a constant ___ and allows the flow area to change with the flow rate.
   A. differential pressure
   B. temperature
   C. diameter
   D. flow rate

_____ **16.** A(n) ___ measures a process variable.
   A. primary element
   B. regulator
   C. transmitter
   D. actuator

_____ **17.** Thermometers used in a steam plant typically measure ___.
   A. psig
   B. psia
   C. degrees Fahrenheit
   D. thermocouple voltage

_____ **18.** A ___ is used to measure the difference in draft pressure between the atmosphere and the furnace, breeching, or stack.
   A. steam pressure gauge
   B. vacuum gauge
   C. resistance pressure detector
   D. draft gauge

_____ **19.** A ___ consists of two dissimilar metals that are joined together to form a circuit that generates a voltage proportional to the difference in temperature between the hot and cold ends of the wires.
   A. resistance temperature detector
   B. thermocouple
   C. pyrometer
   D. temperature transmitter

_____ **20.** Absolute pressure is equal to ___.
   A. gauge pressure minus atmospheric pressure
   B. gauge pressure plus atmospheric pressure
   C. atmospheric pressure
   D. gauge pressure plus steam pressure

_____ **21.** A float is a level measuring instrument with a ___ attached to it that floats on top of a liquid in a tank.
   A. displacer
   B. pressure gauge
   C. hollow ball
   D. bubbler

_____ **22.** An ultrasonic sensor measures ___ to determine level.
   A. pressure difference
   B. voltage
   C. differential pressure
   D. transit time

_____ **23.** A boiler steam pressure gauge should be connected to the ___ side of the boiler.
   A. lowest part of the steam
   B. lowest part of the water
   C. highest part of the steam
   D. highest part of the water

_____ **24.** A(n) ___ is a flow restriction device consisting of a fabricated pipe section with a converging inlet section, a straight throat, and a diverging outlet section.

     A. venturi tube

     B. wobble plate

     C. orifice plate

     D. Bourdon tube

_____ **25.** A ___ pressure regulator is a regulator that uses upstream fluid as a pressure source to power the diaphragm of a larger valve.

     A. pilot-operated

     B. Coriolis

     C. damper

     D. digital

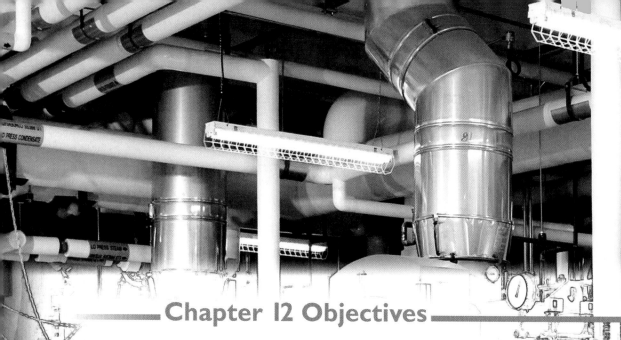

# Chapter 12 Objectives

## SECTION 12.1—OPERATOR DUTIES AND RESPONSIBILITIES

- Describe routine operator duties and responsibilities when taking over a shift.
- Explain operator duties and responsibilities during boiler startup and shutdown.

## SECTION 12.2—BOILER SAFETY CHECKS

- Describe boiler safety checks performed while a boiler is on-line.
- Describe boiler safety checks performed while a boiler is off-line.

## SECTION 12.3—BOILER MAINTENANCE AND INSPECTIONS

- Describe common boiler preventative maintenance duties.
- Explain common practices for boiler inspections and hydrostatic tests after repairs.
- Describe the reason for boiler lay-up and the types of lay-up.

## SECTION 12.4—BOILER ROOM SAFETY

- Describe emergency plans and procedures.
- Describe elements of boiler operator safety.
- Explain boiler room fire prevention.
- Describe proper use of tools and reporting of accidents.

## SECTION 12.5—BOILER OPTIMIZATION

- Describe resources available for energy savings.
- Explain steam and energy management.

## ARROW SYMBOLS

| AIR | GAS | WATER | STEAM | FUEL OIL | CONDENSATE | AIR TO ATMOSPHERE | GASES OF COMBUSTION |
|---|---|---|---|---|---|---|---|
| ⇨ | ⇨ | ⇨ | ➡ | ⇨ | ➡ | ➡ | ➡ |

**Digital Resources**
ATPeResources.com/QuickLinks
Access Code: 472658

# CHAPTER 12

# STEAM BOILER OPERATION

## INTRODUCTION

A boiler operator is responsible for the safe and efficient operation of a boiler plant. Although the duties may vary from plant to plant, certain duties and responsibilities are generally the same. Operators must know how to take over a shift and start up and shut down a boiler. There are many boiler safety checks that must be performed to ensure safe operation.

In addition, boiler operators are usually expected to perform routine maintenance and prepare a boiler for inspections. Boiler operators are also expected to understand boiler room safety requirements. Boiler optimization may also be included in boiler operator responsibilities. In many cases, a large amount of money can be saved by understanding and funding boiler improvements.

## SECTION 12.1
## OPERATOR DUTIES AND RESPONSIBILITIES

### ROUTINE OPERATOR DUTIES

Although the duties of boiler operators may vary from plant to plant, certain duties and responsibilities are generally the same. There are many normal duties that are performed regularly, and many safety checks must be performed to ensure that boiler safety equipment is working properly. In addition, boilers may need to be started or shut down at any time. An effective boiler operator training program will ensure the boiler plant is operated safely and efficiently.

A boiler operator must understand and follow established procedures for operating a boiler. Typical duties that are usually performed daily when taking over a shift include bottom blowdown, testing and inspecting equipment, monitoring, attendance, and communication and documentation. A boiler operator must also understand various types of boiler startup and shutdown.

### Taking Over a Shift

Boilers often operate 24 hr a day. A boiler operator must be prepared to take over a shift from another operator and follow established procedures to ensure that the boiler plant operates safely and efficiently. The normal operator duties for taking over a shift typically include the following:

- Report to work early enough to check the boiler room log and all operating equipment before relieving the operator from the previous shift.
- Check the water level in boilers that are operating by blowing down the water column and gauge glass. **See Figure 12-1.**
- Test the low-water fuel cutoff control by blowing it down.
- Inspect all running auxiliaries and accessories for proper temperature, pressure, lubrication, and lack of excessive vibration. **See Figure 12-2.**
- Listen for any unusual noises.
- Check the water level in the deaerator or feedwater tank.
- Check the burner for the correct flame and the correct fuel oil temperatures and pressures.
- Follow the chief engineer's instructions for correct feedwater treatment and boiler blowdown.
- Use soot blowers as scheduled. Soot acts as an insulator and must be removed from boiler heating surfaces in order to maintain high heat transfer rates and ensure efficient boiler operation.
- Determine that the fuel supply is adequate.
- Review the chief engineer's log for any instructions that are different from routine procedures. **See Figure 12-3.**
- Perform bottom blowdown as required.

## BLOWING DOWN WATER COLUMNS

**Figure 12-1.** Blowing down the water column is one of the duties of a boiler operator.

## CHECKING RUNNING AUXILIARIES

**Figure 12-2.** Running auxiliaries should be routinely checked for proper lubrication.

## CHIEF ENGINEER'S LOG

**Figure 12-3.** The chief engineer's log should be read every shift.

**Bottom Blowdown.** Bottom blowdown is a normal operator duty that is performed to maintain water quality in the boiler. Most boilers have two blowdown valves in series. The valve located farthest from the boiler is always a slow-opening valve. This valve is the blowing valve and takes the wear and tear of the blowdown. The valve located closest to the boiler can be a slow-opening valve or a quick-opening valve. This valve is a sealing valve. **See Figure 12-4.**

## BLOWDOWN VALVES

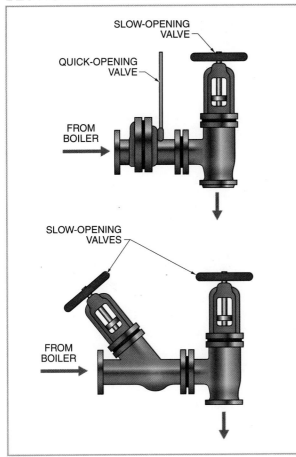

**Figure 12-4.** The valve located closest to the boiler can be a slow-opening valve or a quick-opening valve.

When seatless valves are used on blowdown lines, the manufacturer-recommended operating procedure should be followed. The manufacturer should always be consulted if there is a question regarding the valve-opening sequence. The best time to blow down a boiler is at its lightest load. The procedure for performing a bottom blowdown is as follows:

1. Check the water level to ensure it is at or above the NOWL.
2. Open the valve closest to the boiler first, then open the valve located farthest from the boiler. Open the valves slowly but fully. Never leave the blowdown valves unattended while open. Always keep hands on the valves until they are completely closed.
3. Monitor or have someone else monitor the water level in the gauge glass while blowing down.
4. Close the valve farthest from the boiler first. Close the valve closest to the boiler last.

**Testing and Inspecting Equipment.** Jurisdictions that require a boiler operator's license usually have specific requirements for testing and inspecting equipment. Requirements for testing and inspecting equipment often include the following:

• Periodically inspect the feedwater pump, fuel oil pump, and draft fans. **See Figure 12-5.** This helps reduce maintenance costs and downtime, as minor problems that are corrected immediately can prevent larger problems from developing.

• Periodically inspect the fire in the furnace. This inspection can provide an indication of the quality of combustion. **See Figure 12-6.**

• Identify and document any unusual occurrences, such as steam load changes, steam leaks, low water in the boiler, machinery failures, and any other extraordinary event.

## INSPECTING FEEDWATER PUMPS

**Figure 12-5.** Feedwater pumps are inspected for leaks, unusual noises, and overheating.

## INSPECTING FIRES IN FURNACES

**Figure 12-6.** The boiler operator inspects the fire in a furnace for combustion quality.

**Monitoring.** Jurisdictions that require a boiler operator's license usually have specific boiler monitoring requirements. These monitoring requirements may include the following:

- Maintain the proper water level in the boiler at all times.
- Maintain the correct operating steam pressure. Variations in steam pressure can have an adverse effect on process plant operation.
- Keep the burners in good operating condition. **See Figure 12-7.**
- Maintain the correct fuel pressure and temperature. **See Figure 12-8.**
- Keep a record of the fuel on hand to determine the current supply of fuel and the daily fuel consumption.
- Maintain a record of the amount and temperature of the feedwater being used. Any unusual changes in feedwater makeup indicate a loss of condensate returns.
- Maintain the draft in the boiler necessary to ensure complete combustion of the fuel and to remove the gases of combustion.

## CLEANING FUEL OIL BURNERS

**Figure 12-7.** Fuel oil burners must be cleaned to prevent a carbon deposit buildup.

## MAINTAINING CORRECT FUEL OIL PRESSURE AND TEMPERATURE

**Figure 12-8.** Fuel pressure and temperature are regulated for combustion efficiency.

**Attendance.** Jurisdictions that require a boiler operator's license usually have specific minimum attendance requirements. A jurisdiction will have very specific definitions for "non-automatic," "automatic," "monitored," and "boiler supervisor" types of boilers, along with several grades of licenses and classifications of boilers. For example, the city of Seattle has attendance requirements as follows:

- For a non-automatic steam boiler rated at 20,000,000 Btu/hr input or more, there is a requirement for constant attendance by an appropriately licensed steam engineer. The operator may, on rare occasions, be away from the boiler for up to 20 min.

- For an automatic steam boiler rated at 20,000,000 Btu/hr input or more, there is a requirement that the boiler be checked a minimum of every 2 hr by an appropriately licensed steam engineer.

- For an automatic steam boiler rated at 20,000,000 Btu/hr input or more that is monitored, there is a requirement that the boiler be checked twice daily by an appropriately licensed steam engineer and monthly by a boiler supervisor.

- For a boiler that is not monitored, there is a requirement that the operator not leave the boiler room for a period longer than is considered safe. The operator is responsible for making this judgment based on the minimum time it takes the water in the gauge glass to go from showing in the middle of the gauge glass to not showing in the gauge glass. The time this process takes varies from 2 min to 8 min, depending on the boiler.

**Communication and Documentation.** A boiler operator must maintain a boiler room log to document the operating conditions of a boiler. **See Figure 12-9.** This record typically includes steam flow, water consumption, fuel consumption, and any auxiliaries in operation. In addition, any maintenance done on a shift or operating problems encountered should be recorded in the log. The readings and data recorded over a day allow the chief engineer to determine how efficiently the boiler has been operated.

Another method of communicating with other personnel is speaking directly with them or speaking via telephone. These methods have inherent problems because of the lack of documentation, potential misunderstandings, lack of clear direction, and lack of available personnel when needed. Although none of these traditional methods will ever be totally replaced, many modern methods of electronic communication are now being used. For example, a computerized preventive maintenance and inventory system can be integrated into a building energy management system. Also, in some plants, boiler operators receive their work orders and post their time via the Internet.

There are systems that use a bar code reader to scan an equipment identification number affixed to the equipment. After the bar code is scanned, the preventive maintenance procedure, a list of tools and materials needed, and the specifications for that equipment can be read or printed. Operators can then log any relevant data into the computer system at the end of the day.

Information in the form of company policies, equipment and system diagrams, specifications and descriptions, and procedures for items not encountered every day may be made available on a computer network. Email and text messaging are convenient ways of communicating with management and other personnel. This type of communication allows the receivers to answer when it is convenient and to prioritize their responses. It also leaves a trail of information for documentation purposes if needed.

**Other Operator Duties.** The following are common operating practices that may be performed as needed to prevent damaging a boiler:

- Verify that the water treatment system is operating properly and producing boiler feedwater of sufficient quality for the temperatures and pressures involved.

- Blowdown all the dead legs of the low water trips, water column, etc., on a regular basis to prevent sludge buildup, which leads to device malfunction. Never under any circumstance disable a low water trip.

- Monitor the quality of condensate coming back from the process to enable the diversion of the condensate in the event of a catastrophic process equipment failure.

- Adjust continuous blowdown to maintain conductivity of the boiler water within required operating limits, and operate the mud drum blowdown on a regular basis. Never blowdown a furnace wall header while the boiler is operating.

- The boiler water side should be inspected on a regular basis. If there are any signs of scaling or building up of solids on the tubes, water treatment adjustments should be made, and the boiler should be mechanically or chemically cleaned.

- Verify that the water leaving the deaerator is free of oxygen, that the deaerator is operated at the proper pressure, and that the storage tank water is at saturation temperature. The deaerator internals should be inspected for corrosion on a regular basis.

## BOILER ROOM LOGS

| POWER HOUSE LOG | | | | | | | | | | | | | | | | | | Dates From _____ To _____ | |
|---|---|---|---|---|---|---|---|---|---|---|---|---|---|---|---|---|---|---|---|
| TIME | NO. BOILER IN OPERATION | STEAM PRESSURE ON BOILER | FLUE TEMP. °F | FURNACE DRAFT IN. H₂0 | NO. BURNERS | BOILER FUEL METER READING | STEAM FLOW INTEGRATOR MAIN PLANT | STEAM FLOW INTEGRATOR BLDG. NO. ___ | WATER SOFTENER INTEGRATOR | WATER SOFTENER DIAL METER | FEEDWATER TEMP °F | GALLONS OF OIL IN TANK | OIL TEMP. FROM TANK AT PUMP °F | RETURN OIL TEMP °F | SMOKE ALARM SIGNAL | OUTSIDE TEMP °F | WEATHER | ENGINEER TO SIGN FOR ACTIVITIES DURING WATCH | ENGINEER'S INITIALS |
| 9 A.M. | | | | | | | | | | | | | | | | | | | |
| 10 | | | | | | | | | | | | | | | | | | | |
| 11 | | | | | | | | | | | | | | | | | | | |
| 12 P.M. | | | | | | | | | | | | | | | | | | | |
| 1 | | | | | | | | | | | | | | | | | | | |
| 2 | | | | | | | | | | | | | | | | | | | |
| 3 | | | | | | | | | | | | | | | | | | | |
| 4 | | | | | | | | | | | | | | | | | | | |
| 5 | | | | | | | | | | | | | | | | | | | |
| 6 | | | | | | | | | | | | | | | | | | | |
| 7 | | | | | | | | | | | | | | | | | | | |
| 8 | | | | | | | | | | | | | | | | | | | |
| 9 | | | | | | | | | | | | | | | | | | | |
| 10 | | | | | | | | | | | | | | | | | | | |
| 11 | | | | | | | | | | | | | | | | | | | |
| 12 A.M. | | | | | | | | | | | | | | | | | | | |
| 1 | | | | | | | | | | | | | | | | | | | |
| 2 | | | | | | | | | | | | | | | | | | | |
| 3 | | | | | | | | | | | | | | | | | | | |
| 4 | | | | | | | | | | | | | | | | | | | |
| 5 | | | | | | | | | | | | | | | | | | | |
| 6 | | | | | | | | | | | | | | | | | | | |
| 7 | | | | | | | | | | | | | | | | | | | |
| 8 | | | | | | | | | | | | | | | | | | | |

| BOILER ROOM DAILY RECORD | | | SHIFT | NO. OF BOILER | % BOILER EFFICIENCY | % MAKEUP WATER | LB OF STEAM/ LB OF OIL |
|---|---|---|---|---|---|---|---|
| Actual lb steam made : | Gallons of fuel burned : | | | | | | |
| Heat added/lb steam : | Heat value of fuel/gal. : | | NO. 1 | | | | |
| Equivalent lb steam made : | Total heat in fuel, millions Btu : | | | | | | |
| Boiler pressure : | Efficiency : | | NO. 2 | | | | |
| Temperature feedwater : | Average B.H.P. rated : | | | | | | |
| Quality or superheat steam : | Average B.H.P developed (@ 800,000 Btu/day) : | | NO. 3 | | | | |
| Total heat added to steam, million Btu : | Average percent rating developed : | | | | | | |

**Figure 12-9.** A boiler operator must maintain a boiler room log to document the operating conditions of a boiler.

### Boiler Startup and Shutdown

Start-up and shutdown procedures for boilers vary from plant to plant. The procedures depend on conditions such as the number of boilers, the size of the plant, the type of fuel burned, and whether the plant is operated automatically or manually. Although procedures and methods may vary, they must be performed in a safe and efficient manner. A boiler may be started when it is already hot (live) and ready to go, or it may be started when it is cold.

**Live Plant Startup.** When starting up a boiler with other boilers on-line, the operator must perform all safety checks before lighting the burner. The procedure for live plant startup is as follows:

1. Initiate the firing sequence. A thorough investigation must be performed if the safety systems trip and shut down the boiler. After the main flame has been proven, test the flame scanner for proper operation.

2. Make sure both the boiler vent and the drain between the main steam stop valves and the automatic non-return valve are open.

3. Open the equalizing valve around the main steam stop valves.

4. Open the main steam stop valves.

5. Close the boiler vent when steam pressure on the incoming boiler reaches about 10 psi to 15 psi. **See Figure 12-10.**

6. Test the low-water fuel cutoff by blowing it down. Blow down the water column and gauge glass to test for proper operation.

## CLOSING BOILER VENTS

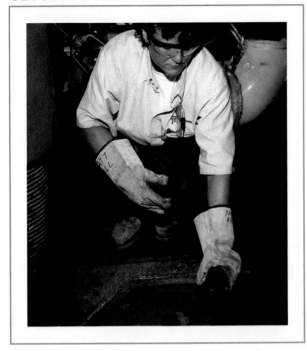

**Figure 12-10.** When steam pressure on the incoming boiler reaches 10 psi to 15 psi, the boiler vent must be closed.

7. Open the automatic nonreturn valve when the steam pressure reaches approximately 75% to 85% of the line pressure.

8. Test the safety valve by hand at this time. The steam pressure must be at a minimum of 75% of the safety valve setting.

9. Bring the boiler pressure up slowly and let the automatic nonreturn valve cut the boiler in on-line.

10. Close the drain between the automatic nonreturn valve and the main steam stop valve.

11. Make sure all automatic controls are functioning. **See Figure 12-11.**

A boiler with two hand-operated main steam stop valves instead of a main steam stop valve and an automatic nonreturn valve is started with a slightly different procedure. The procedure for live plant startup with two hand-operated main steam stop valves is modified as follows:

1. Open the equalizing line around the main steam stop valve when approximately 85% of the line pressure is on the boiler to warm the main steam line and valve.

2. Open the main steam stop valve that is farthest away from the boiler.

3. Slightly open the steam valve closest to the shell of the boiler and open it slowly when the boiler steam pressure is slightly below line pressure. This allows steam to flow from the header back to the boiler coming on-line. Any condensate in the line is forced back into the boiler. This prevents carryover in the event that the operator opens the main steam stop valve too rapidly.

## SWITCHING CONTROLS TO AUTOMATIC OPERATION

**Figure 12-11.** All automatic controls are checked for proper function.

When a boiler is equipped with a superheater, certain precautions must be taken during the warm-up period. The superheater outlet drain valve must be opened and remain open until the boiler is cut in on-line. Steam must flow through the superheater at all times to prevent overheating the superheater tubes. Different types of superheaters require different handling procedures during warm-up. Manufacturer specifications for the recommended start-up and shutdown procedures should be followed in order to prevent damage to the superheater tubes.

### TECH TIP

*The superheater steam flow should never be less than 25% of the capacity of the boiler.*

**Boiler Shutdown.** Boiler shutdown requires a standard procedure to ensure safe shutdown. The procedure for boiler shutdown is as follows:

1. Secure the fuel to the burner. **See Figure 12-12.**

2. Open the superheater outlet drain as soon as the burner is shut down.

3. Close the boiler main steam stop valves and open all steam drains. **See Figure 12-13.**

4. Open the boiler vent when the steam pressure is down to about 10 psi to 15 psi to prevent a vacuum from forming in the boiler.

5. Maintain the water at the NOWL while the boiler is cooling down.

6. Shut down the feedwater pump and the feedwater system when the boiler has cooled down sufficiently.

## SECURING FUEL TO BURNERS

**Figure 12-12.** Fuel to the burner must be secured when shutting down the boiler.

**Cold Plant Startup.** Cold plant startup is performed when boilers have been out of service and the boiler water has cooled. The procedures for startup of a cold plant differ from procedures followed when starting up a live plant.

During cold plant startups, all packing and flange gaskets in the plant and boiler room must be inspected for leaks. Also, fuel must be properly prepared before lighting the burners. The type of fuel used determines how the fuel is prepared. Boilers using heavy fuel oils must have fuel oil circulating at the proper temperature before it can be burned. This requires that fuel oil heaters be used, or that a lighter grade, such as No. 2 or No. 4 fuel oil, be used when first lighting the burner. All fuel oil strainers and the fuel oil burner assembly should be cleaned prior to ignition.

## CLOSING MAIN STEAM STOP VALVES

**Figure 12-13.** The boiler main steam stop valves must be closed on boiler shutdown.

When gas is used, little or no fuel preparation is necessary. The operator only needs to determine if the pressure at the burner is correct. When burning coal, an adequate supply must be available at the stoker or the pulverizing mill. The boiler temperature should be brought up slowly. The boiler warm-up curve should be strictly followed. The standard warm-up curve for a typical boiler does not allow the boiler water temperature to increase by more than 100°F per hour. It is possible for a continuous minimum fire to exceed the maximum warm-up rate. If this is the case, the burner must be intermittently fired to ensure that this rate is not exceeded. The procedure for starting up a cold plant, regardless of the type of fuel burned, is as follows:

1. Check the boiler to verify all inspection opening covers are in place, all stack coverings have been removed, and any hand-operated dampers have been opened. Verify that all valves on the boiler are in the proper position for startup.

2. Make sure the water level in the gauge glass is correct. The boiler should be filled so the water in the gauge glass is slightly below the NOWL. This leaves room in the boiler for the water to expand as it heats up.

3. Check the feedwater system to ensure the correct valves are open. Inspect the feedwater pumps, feedwater regulators, and water level in the feedwater heater and the feedwater tank. Turn on the required feedwater pumps, transfer pumps, and condensate return pumps and verify that they are all operating at the correct pressures.

4. Light the burner following all necessary safety procedures. Test the flame scanner for proper operation.

5. Warm up the boiler slowly to allow uniform expansion of the boiler and all steam lines. Open the automatic nonreturn valve and the main steam stop valve in the main steam line. **See Figure 12-14.** Open all drains in the main steam lines and steam header to make certain that the condensate is removed as the piping warms up.

6. Maintain the recommended NOWL in the boiler. **See Figure 12-15.** Blow down the low-water fuel cutoff and water column gauge glass.

7. Close the boiler vent when the steam pressure reaches 10 psi to 15 psi. By this time, all the air has been expelled from the steam drum.

8. Check the automatic combustion controls, feedwater regulator, and all boiler room auxiliaries to ensure they are operating properly. Switch over all boiler controls to automatic operation as soon as the boiler has reached its normal operating pressure.

## OPENING MAIN STEAM STOP VALVES

**Figure 12-14.** The boiler main steam stop valves and steam line drains must be opened on boiler startup.

At this time, the boiler operator must watch for any sudden and unusual occurrences. The operation of the boiler plant is at its most critical stage. All boiler room auxiliaries and automatic controls must be monitored very closely.

## MAINTAINING NOWLs

**Figure 12-15.** The boiler water level must be maintained at the NOWL during boiler startup.

---

### SECTION 12.1 — CHECKPOINT

1. What are five common duties performed when taking over a shift?

2. What is the procedure for opening and closing valves while performing a bottom blowdown?

3. What is the difference between boilers that receive a live plant startup and boilers that receive a cold plant startup?

## SECTION 12.2
# BOILER SAFETY CHECKS

### SAFETY CHECKS

There are many types of controls and safety interlocks used on boilers today. However, not all boilers have all types of controls and interlocks. The controls used on a boiler depend on the date of manufacture, the type of fuel used, the boiler brand and type, and any relevant state or local codes.

Each boiler plant must have procedures to document the steps needed to perform any required boiler checks. The procedures should include a method to document that the checks were performed, who performed them, and when they were done. The operating manuals for the equipment can provide the specific details used to develop standard operating procedures for the boiler plant.

### Safety Checks with Boiler On-Line

Certain boiler checks are standard and are required on a periodic basis. Boiler operators must review the specific procedures for these checks. They should be conducted when the boiler is being prepared to be placed in service or during a low-use period. When finished, the controls must be reset to their original settings and all covers replaced to protect the controls. The frequency of these checks depends on the boiler and local code requirements.

For safety reasons, these checks should be performed when no other activities are taking place around the boiler. The burner should be in manual mode and at low fire before any actions are taken that could cause the burner to shut off. Typical on-line safety checks include checking the low-water fuel cutoff, auxiliary low-water fuel cutoff, operating pressure control, high-limit pressure control, flame scanner, and safety valves.

**Low-Water Fuel Cutoffs.** The simplest test of a low-water fuel cutoff is to blow down the device and piping. However, blowing down the low-water fuel cutoff does not match actual operating conditions. A better test is an evaporation test, which slowly evaporates the boiler water until the low-water fuel cutoff automatically shuts down the burner. An evaporation test should be performed when the boiler is in

manual low-fire hold. The procedure for performing an evaporation test is as follows:

1. Stop the feedwater flow to the boiler. **See Figure 12-16.**
2. Allow the boiler to steam off naturally.
3. Monitor the water level in the gauge glass.
4. Note that the burner must automatically shut down when the water level reaches the marking on the outside of the low-water fuel cutoff. If the burner does not shut down, the low-water fuel cutoff must be inspected and repaired before restarting the boiler.

**CLOSING FEEDWATER VALVES**

**Figure 12-16.** During an evaporation test, the feedwater valves must be secured.

**Auxiliary Low-Water Fuel Cutoffs.** Many boilers also have an auxiliary low-water fuel cutoff. **See Figure 12-17.** To test the auxiliary low-water fuel cutoff, the low-water fuel cutoff must be bypassed to prevent it from shutting off the burner before the auxiliary low-water fuel cutoff is activated. This test presents significant safety issues and should be done with the utmost care. The wires used to bypass the low-water fuel cutoff should be brightly colored and long enough to be easily visible. The wires must not be left in place after the test has been completed. Considerations must be made with regard to accessing control panels and preventing an arc flash. The procedure for testing an auxiliary low-water fuel cutoff is as follows:

1. Bypass the low-water fuel cutoff.
2. Stop the feedwater flow to the boiler.
3. Allow the boiler to steam off naturally.
4. Monitor the water level in the gauge glass.
5. Note that the burner automatically shuts down when the water level reaches the marking on the outside of the auxiliary low-water fuel cutoff. If the burner does not shut down, the auxiliary low-water fuel cutoff must be inspected and repaired before restarting the boiler.
6. Remove bypass wires and replace access panel.

## TESTING AUXILIARY LOW-WATER FUEL CUTOFFS

**Figure 12-17.** Many boilers also have an auxiliary low-water fuel cutoff.

**Operating Pressure Controls.** The operating pressure control shuts down the burner when the steam pressure increases to the setpoint and restarts the burner when the steam pressure decreases by the differential pressure. **See Figure 12-18.** The procedure for verifying the operating pressure control is as follows:

1. Observe the operating pressure control, pressure gauge, and temperature gauge.
2. Verify that the operating pressure control is correctly starting and stopping the burner at the pressure setpoint and subtractive differential. If not, the operating pressure control must be inspected and repaired.

## TESTING OPERATING CONTROLS

**Figure 12-18.** The boiler operating controls need to be tested regularly.

**High-Limit Pressure Controls.** The high-limit pressure control shuts down the burner when the steam pressure increases above the setpoint of the operating pressure control. The high-limit pressure control is adjusted to make and break the circuit at preselected points. This control needs to be manually reset to start the burner through a restart cycle. The procedure for testing the high-limit pressure control is as follows:

1. Note the current setting for the high-limit pressure control.
2. Reset the control to a pressure below the operating pressure control setting. The burner should automatically shut down.
3. If the burner does not automatically shut down, turn the burner switch off and inspect and repair the high-limit pressure control.
4. Press the reset button on the high-limit pressure control.
5. Return the high-limit pressure control setting to the original pressure and replace any covers.

**Flame Scanners.** The flame scanner is tested to prevent the possibility of a furnace explosion. The flame scanner will not allow the fuel valve to open if the scanner does not prove the pilot or main flame. This prevents fuel from entering the furnace of the boiler and building up when the burner is not lit.

The flame scanner should be tested at least once a week. The procedure for testing the flame scanner is as follows:

1. With the burner on, remove the flame scanner and cover the flame scanner eye with your hand. **See Figure 12-19.** This action simulates a flame failure, causing the programmer to close the fuel valve.

2. If the burner does not shut down, turn the burner switch off and inspect and repair the flame scanner.

3. Reinstall the flame scanner and reset the programmer after it has stopped.

4. Turn the burner switch on and initiate a normal firing cycle for the burner.

## TESTING FLAME SCANNERS

**Figure 12-19.** A flame scanner is tested to verify proper operation.

**Safety Valves.** The safety valve and piping need a simple visual inspection. The drip pan and elbow or flexible connection should be in good condition without evidence of damage. **See Figure 12-20.** The pipe hangars supporting the vent piping should also be inspected to ensure that there is no evidence of damage. The interval between tests will vary from plant to plant because of operating steam pressures, local codes, and plant routines.

## INSPECTING DRIP PANS

**Figure 12-20.** The safety valve drip pan and elbow or flexible connection should be in good condition without evidence of damage.

Typically, with boilers 15 psi to about 400 psi, the safety valves should be manually tested once a month and pop-tested once a year. With boilers above about 400 psi, the safety valves should be manually tested every six months, preferably when the boiler is going to be removed from service. They should also be pop-tested once a year or completely overhauled by the factory. The procedures provided by the manufacturer should always be followed. The procedure for performing a manual test of a safety valve is as follows:

1. Make sure the boiler has a pressure of at least 75% of the safety valve popping pressure.

2. Pull the safety valve test lever by hand, hold it open for a few seconds, and allow the safety valve to snap closed. **See Figure 12-21.** The safety valve should close with no leakage or dripping.

3. If the safety valve leaks after performing this test, open the valve again to allow steam to blow any debris off the valve seat. If the valve still leaks, the boiler must be taken off-line and the valve inspected and repaired by a qualified repair technician.

4. Return the boiler to normal operation.

## TESTING SAFETY VALVES

**Figure 12-21.** The safety valves are tested by hand before the boiler is brought up to line pressure.

A pop test is performed to determine if a safety valve opens at the correct pressure. The procedure for performing a pop test of the safety valves is as follows:
1. Manually increase the firing rate to maximum.
2. Allow the pressure to increase until the setpoint of the safety valve is reached. The safety valve should pop open at the setpoint. If the safety valve does not pop open, the boiler must be taken off-line. The safety valve is not working properly and must be repaired or replaced.
3. Return the boiler to normal operation.

An accumulation test is used to establish the relief capacity of boiler safety valves. This test should not be conducted without a boiler inspector present. Boilers equipped with superheaters should not be subjected to an accumulation test because the superheater tubes could be damaged due to the lack of steam flow, causing them to overheat. The procedure for performing an accumulation test of a safety valve is as follows:

1. Shut off all steam outlets from the boiler and manually increase the firing rate to maximum. The safety valve should relieve all the steam without the pressure increasing more than 6% above the popping pressure. If the pressure rises more than this amount, the boiler must be taken off-line. The safety valve is not adequate for the boiler and a new one must be specified.
2. Open all steam outlets from the boiler and return the boiler to normal operation.

### Safety Checks with Boiler Off-Line

Many safety checks need to be performed when the boiler is off-line. These checks can be performed when the boiler is down for maintenance or inspection. The frequency of performing these checks depends on the boiler and local code requirements. Generally these checks should be done a few times a year. Some of these checks are based on the fuel used and may not be needed for all boilers on site.

These checks are all performed on devices that are part of the burner management system. They are intended to force the boiler or burner to operate outside of the normal operating conditions where the safety equipment is expected to either prevent the burner from igniting or to shut down the burner if it is firing.

Care must be taken to ensure that only one control is checked at a time. If multiple changes are made at a single time, it is impossible to determine exactly which control prevented the burner from igniting. Finally, it is necessary to return the boiler controls to their original operating settings prior to bringing the boiler back on-line.

Typical off-line safety checks include checking the oil drawer switch, combustion air-proving switch, atomizing-media pressure switch, low-gas-pressure switch, high-gas-pressure switch, low-oil-pressure switch, low-oil-temperature switch, and high-oil-temperature switch and testing the gas valve for leaks. Local codes may have specific testing requirements.

**Oil Drawer Proving Switches.** The oil drawer holds the oil gun in place during firing. The internal contacts of the oil drawer proving switch are closed when the oil drawer is in its forward position and latched in place. **See Figure 12-22.** The procedure for performing a test of the oil drawer proving switch is as follows:
1. Turn the gas/oil selector switch to the oil setting.
2. Unlatch the oil gun locking plate from the forward locking pin groove. Pull the gun back and lock the fastener hook in the rear locking pin groove.

3. Initiate the ignition sequence. The ignition sequence should fail.

4. If the ignition sequence does not fail, turn the burner switch off and inspect and repair the oil drawer proving switch.

5. Return the oil gun to its proper location and turn the selector switch to the setting for the fuel that is currently being used.

## OIL DRAWERS

OIL GUN
OIL DRAWER

*Cleaver-Brooks*

**Figure 12-22.** The oil drawer has a proving switch that verifies that the drawer is properly closed before the programmer allows the firing sequence to begin.

**Combustion Air Proving Switches.** The combustion air proving switch verifies that there is sufficient air to ensure proper combustion. The internal contacts of the combustion air proving switch are closed with the presence of sufficient combustion air pressure from the forced draft fan. This switch is actuated by air pressure. The procedure for performing a test of the combustion air proving switch is as follows:

1. Note the current setting for the control.

2. Reset the control to its maximum setting.

3. Initiate the ignition sequence. The ignition sequence should fail.

4. If the ignition sequence does not fail, turn the burner switch off and inspect and repair the combustion air proving switch.

5. Return the control to its original setting and replace any covers.

**Atomizing-Media Pressure Switches.** An oil burner that uses air or steam to atomize the oil must be checked to confirm the atomizing medium is present at the required pressure. The internal contacts of the atomizing-media pressure switch are closed when the atomizing media are present at the required pressure. If the pressure of the atomizing air or steam is low, the atomizing-media pressure switch will open a circuit and shut down the burner. The procedure for performing a test of the atomizing-media pressure switch is as follows:

1. Note the current setting for the control.

2. Reset the control to the maximum setting.

3. Initiate the ignition sequence. The ignition sequence should fail.

4. If the ignition sequence does not fail, turn the burner switch off and inspect and repair the atomizing-media pressure switch.

5. Return the control to its original setting and replace any covers.

**Low-Gas-Pressure Switches.** With gas-fired boilers, the low-gas-pressure switch is normally closed whenever the gas line pressure is above a preselected pressure. **See Figure 12-23.** The low-gas-pressure switch opens when the gas pressure drops below a preselected minimum requirement, deenergizes the main gas valve (or valves), and energizes the vent valve. The procedure for performing a test of the low-gas-pressure switch is as follows:

1. Note the current setting for the control.

2. Reset the control to the maximum setting.

3. Initiate the ignition sequence. The ignition sequence should fail.

4. If the ignition sequence does not fail, turn the burner switch off and inspect and repair the low-gas-pressure switch.

5. Return the control to its original setting and replace any covers.

**High-Gas-Pressure Switches.** The high-gas-pressure switch is normally closed whenever gas line pressure is below a preselected pressure. The switch opens when gas pressure rises above a preselected maximum requirement, deenergizes the main gas valve, and energizes the vent valve. The procedure for performing a test of the high-gas-pressure switch is as follows:

1. Note the current setting for the control.

2. Reset the control to the minimum setting.

3. Initiate the ignition sequence. The ignition sequence should fail.

4. If the ignition sequence does not fail, turn the burner switch off and inspect and repair the high-gas-pressure switch.

5. Return the control to its original setting and replace any covers.

## TESTING LOW-GAS-PRESSURE SWITCHES

**Figure 12-23.** The low-gas-pressure switch verifies that the gas pressure to the burner is at or above the minimum pressure.

**Low-Oil-Pressure Switches.** The low-oil-pressure switch is designed to open the burner limit circuit in the event the oil pressure should drop below the preselected minimum pressure for proper operation. The control may be equipped with a manual reset. The procedure for performing a test of the low-oil-pressure switch is as follows:

1. Note the current setting for the control.

2. Reset the control to the maximum setting.

3. Initiate the ignition sequence. The ignition sequence should fail.

4. If the ignition sequence does not fail, turn the burner switch off and inspect and repair the low-oil-pressure switch.

5. Return the control to its original setting and replace any covers.

**Low-Oil-Temperature Switches.** When heavy oil is being burned, the oil must be heated in order to be atomized properly. The low-oil-temperature switch will prevent the burner from starting and will shut the burner down if the oil temperature drops below the required

temperature for proper operation. The control can be wired into several circuits of the programmer. The procedure for performing a test of the low-oil-temperature switch is as follows:

1. Note the current setting for the control.

2. Reset the control to the maximum setting.

3. Initiate the ignition sequence. The ignition sequence should fail.

4. If the ignition sequence does not fail, turn the burner switch off and inspect and repair the low-oil-temperature switch.

5. Return the control to its original setting and replace any covers.

**High-Oil-Temperature Switches.** The high-oil-temperature switch serves the same function in reverse that the low-oil-temperature switch does. This switch opens the circuits and shuts down the burner when the oil temperature rises above a preset temperature. The procedure for performing a test of the high-oil-temperature switch is as follows:

1. Note the current setting for the control.

2. Reset the control to the minimum setting.

3. Initiate the ignition sequence. The ignition sequence should fail.

4. If the ignition sequence does not fail, turn the burner switch off and inspect and repair the high-oil-temperature switch.

5. Return the control to its original setting and replace any covers.

**Gas Valve Leak Testing.** Gas valves should be tested for leaks at least annually. **See Figure 12-24.** If the bleed pipe does not have a manual test valve installed, one must be installed or be a part of the test apparatus. The vent valve must be closed to perform this test. To ensure an automatic vent valve is closed, it must be bypassed or a deadman switch needs to be installed in the valve solenoid circuit. The procedure for performing a leak test of the gas valves is as follows:

1. Deenergize the control system to ensure that there is no power to the safety shutoff valves.

2. Ensure the manual test valve is closed.

3. Remove the leak test plug and connect the test apparatus to the manual test petcock.

4. Close the downstream manual gas cock.

5. Immerse the ¼″ tubing vertically ½″ into a clear container of water.

6. Slowly open the test valve.

7. When the rate of bubbles coming through the water stabilizes, count the number of bubbles appearing in a 10 sec period. The number of bubbles must not exceed the values established by the manufacturer. The operating manual for each valve model typically shows tables of this information, based on pipe diameter.

8. Close the manual test petcock, remove the test apparatus and replace the leak test plug.

9. Open the downstream manual gas cock.

10. Restore the system to normal operation, remove the jumper wire, if used, and open the vent valve.

## TESTING GAS VALVES

**Figure 12-24.** Gas valves should be tested for leaks at least annually.

---

### SECTION 12.2 – CHECKPOINT

1. What are six common safety checks performed while a boiler is on-line?

2. What is the difference between a manual test and a pop test of a safety valve?

3. What is the purpose of off-line safety checks of burner management system components?

4. What is the purpose of a combustion air proving switch?

## SECTION 12.3
# BOILER MAINTENANCE AND INSPECTIONS

## MAINTENANCE AND INSPECTIONS

Routine boiler plant maintenance can extend the life of the equipment and reduce boiler downtime. The preventive maintenance program used varies with each plant and depends on the type of equipment in use.

### Preventive Maintenance

A preventive maintenance program includes an annual internal and external boiler inspection as recommended by the ASME Code. Preventive maintenance includes general maintenance, valve maintenance, pump maintenance, gauge glass replacement, and cleaning new and retubed boilers.

**General Maintenance.** The general preventive maintenance duties of a boiler operator include the following:

- Drain air compressor tanks of condensate every shift. Change compressor oil according to the manufacturer-recommended procedure.

- Monitor and clean blades on draft fans whenever there is a sign of buildup on the blades or excessive vibration occurs.

- Perform vibration analysis on large rotation equipment. This will establish a baseline vibration signature for the equipment, which can be compared with new readings, for identifying developing problems.

- Perform infrared inspections of all electrical panels and connections. Hot spots detected may indicate lost, bad, or corroded connections. **See Figure 12-25.**

**Valve Maintenance.** Valves are one of the most used components in a steam boiler system. Valves can be automatic or hand-operated. Because of the important role of hand-operated valves in the boiler system, proper functioning is imperative. To ensure the proper operation of valves, undue force should never be used when opening or closing a valve. If the valve is difficult to move, service may be required. A gate valve should never be used for throttling service because it may leak around the gate afterward. Recommended lubricants should be applied to packing per manufacturer specifications.

The packing gland should not be tightened on the valve too tightly. This may damage the valve permanently. Some valves are designed to leak slightly around the packing gland. Routine maintenance for valves includes replacing packing, glands, gaskets, and/or valve stems. In addition, seats, gates, and discs must be reground or replaced as necessary.

## INFRARED INSPECTIONS

*Fluke Corporation*

**Figure 12-25.** Infrared inspections can detect hot spots that may indicate lost, bad, or corroded connections.

Automatic valves have diaphragms and pilot valves, which must also receive preventive maintenance on a regular basis. New gaskets should always be installed before assembling a reconditioned valve. Gaskets for line flanges on a flange-connected valve should be replaced when reinstalling valves.

**Pump Maintenance.** The boiler operator is also in charge of feedwater pumps, vacuum pumps, oil pumps, and sump pumps. In some plants, boiler operators are expected to perform minor repairs and service on boiler room pumps.

Packing glands and mechanical seals require lubrication or flushing with water to ensure proper operation. When shaft pump seals start leaking, the seal must be completely replaced. Shaft packing can be adjusted with the packing gland nuts to ensure minimum leakage from the pump. When there is excessive leakage due to packing or shaft wear, the packing and/or shaft must be replaced.

Bearings, unless sealed, require lubrication. **See Figure 12-26.** The manufacturer specifications will indicate the type of lubricant to use. When bearings are worn, they should be replaced. If worn bearings are not replaced, damage to the shaft or bearing housing can result. The operator should observe the general operating condition of each pump. If a pump starts to drop in efficiency because of mechanical wear, the pump must be removed from service for a general overhaul.

## LUBRICATING BEARINGS

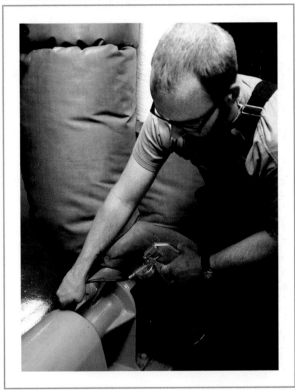

**Figure 12-26.** Bearings are lubricated as part of routine boiler plant maintenance.

Many plants have a preventive maintenance schedule set up for all pump overhauls. Proper maintenance and service to pumps can extend pump life for many years. Normal pump maintenance includes the following:

- Clean and grease pump and motor bearings once or twice a year according to the manufacturer operating procedures.
- Clean and flush oil pumps on turbine auxiliaries according to the manufacturer operating procedures.
- Repack auxiliary pumps when excessive leakage is evident.

**Gauge Glass Replacement.** A gauge glass should be protected by a shield to prevent accidental breakage and to protect the operator. A broken gauge glass can be replaced while a steam boiler is under pressure. The proper eye and hand protection must be worn when replacing a gauge glass. The following procedure should be applied to replace a tubular broken gauge glass:

1. Secure the water and steam gauge glass stop valves. Open the blowdown valve for the gauge glass. Check the water level by a secondary means.

2. Remove the gauge glass guard and nuts. Remove all broken glass and old washers. **See Figure 12-27.**

3. Cut a new glass, leaving a ¼″ clearance and place in position. Place nuts and new washers on the gauge glass. Center the glass so that it does not touch any metal. Tighten nuts by hand and then make a one-quarter turn with an open-end wrench. The use of an adjustable wrench is discouraged because it can slip and cause damage.

4. Crack open the steam gauge glass stop valve and allow the glass to be heated and expand. When the glass has been heated, close the gauge glass blowdown valve and open the water gauge glass stop valve. Open the steam gauge glass valve fully.

5. Check for leaks. If leaks appear, secure the water and steam gauge glass stop valves and open the blowdown valve to take the pressure off the gauge glass. Tighten the leaking nut, one flat of the nut at a time, and retest for leaks.

6. Replace the gauge glass guards.

*A flat gauge glass may also need to be changed.*

## REPLACING GAUGE GLASSES

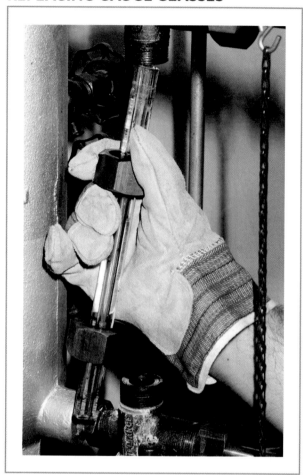

**Figure 12-27.** A gauge glass is removed by loosening the nuts that hold the glass in place.

**Cleaning New or Retubed Boilers.** During the boiler manufacturing process, boiler steam and water sides are contaminated by oil, grease, and other organic matter. These contaminants must be removed to allow optimum heat transfer on the heating surfaces of the boiler tubes. In addition, these contaminants can cause surface impurities in the boiler water, resulting in priming or carryover.

A good cleaning agent, such as an alkaline detergent solution, will remove oil, grease, and other foreign matter from drums, tubes, and headers within the boiler. The manufacturer recommendations should be followed when cleaning a new or retubed boiler.

### Inspections and Hydrostatic Tests

An operator must prepare a boiler for an annual inspection. The boiler inspector should be notified when the boiler will be ready for inspection so an appointment can be arranged. Some states require that a written report be filed with the state department of labor and industry when the inspection of the boiler is complete. A new boiler certificate may then be issued.

The operator should be on hand to assist the boiler inspector and identify any trouble spots. The boiler inspector will examine the boiler on the fire side and the steam and water side, looking for signs of corrosion, overheating, and/or other damage. The chief engineer must address any problems identified by the boiler inspector and make any necessary repairs.

**Inspection Procedures.** Prior to inspection, the boiler must be opened up and thoroughly cleaned. Any repairs to the furnace refractory should be made at this time. Typically, a boiler operator has the following duties unless the boiler is already shut down for maintenance:

• Follow proper lockout/tagout procedures. **See Figure 12-28.**

• Always use approved ladders or scaffolding when cleaning a boiler.

• Follow all confined space permit procedures.

• Cool the boiler slowly to prevent undue stress to the boiler metal and refractory. Never dump a hot boiler because dumping can cause sludge and sediment to bake onto the heating surfaces.

• Verify that the boiler vent is open before opening a manhole cover. This will ensure that there is no vacuum or pressure in the steam drums. A vacuum causes stress on the boiler and is dangerous to operating personnel when opening up the boiler.

• Remove all pipe plugs on cross-connections to the water column, feedwater regulator, gauge glass, and low-water fuel cutoff.

• Open and clean the feedwater regulator and low-water fuel cutoff float chambers. **See Figure 12-29.**

The operator must follow specific procedures when the boiler is ready to be put back into service after inspection. The steam and water side of the boiler must be inspected for any tools, rags, or other equipment left behind. A wrench or screwdriver left inside the boiler could find its way to a bottom blowdown valve, causing it to jam open. Then, the boiler would have to be removed from service, cooled, and dumped so repairs could be made.

All pipe plugs on cross-connections that were removed must be replaced. The feedwater regulator and low-water fuel cutoff must be completely reassembled, using new gaskets when necessary. Chemically treated water must be added to the boiler to bring it close to its NOWL.

## LOCKOUT/TAGOUT

**Figure 12-28.** Lockouts and tagouts are applied during maintenance and repair to prevent injury from energized circuits and equipment operation.

## FLOAT CHAMBERS

**Figure 12-29.** The low-water fuel cutoff float chamber needs to be cleaned during an inspection.

**Hydrostatic Tests.** A hydrostatic test is applied to all new boilers and after any repair work has been done that could compromise the integrity of the shell, drums, or tubes. In addition, a hydrostatic test may need to be done on boilers that have had a low-water condition develop that could have damaged the boiler internals.

When a hydrostatic test is applied on a boiler, the main steam stop valves must be secured. The drains between the two steam stop valves must be left open. This prevents any water buildup if the valve closest to the boiler leaks. The high- and low-water alarm actuators must be removed and the openings plugged.

The safety valves are blanked or gagged, and the boiler vent is closed when the boiler is full of water at a temperature of about 100°F. Gagging a valve means applying a clamp on the valve spindle to keep the valve in full closed position. The water pressure is increased to 1½ times the MAWP by using a hand pump. The pressure is then reduced to the safety valve setting. The boiler is inspected for leaks or bulging. When the hydrostatic test and boiler inspection are completed, gags must be removed and safety valves and alarm actuators replaced.

## Boiler Lay-Up

*Lay-up* is the preparation of the boiler for out-of-service status for an extended period. Idle boilers can corrode badly unless they are stored properly. The extent and rate of corrosion depends on the condition of the boiler. With the need for cleanliness of modern boiler surfaces, more attention must be given to protecting the surfaces from oxygen pitting during downtimes. Even boilers that are idle for only short periods, such as over a weekend, are susceptible to corrosion.

The two ways to lay-up a boiler are the wet method and the dry method. The method of lay-up used depends on the length of lay-up time and on plant conditions. If the boiler may be needed on short notice or if the boiler is going to be down for a short period, the wet method is recommended. If the boiler will not be needed for a long time or if the boiler may freeze, the dry method is recommended. The chief engineer determines which method to use.

**Wet Lay-Up.** The wet lay-up method is preferred if the boiler may be needed on short notice or if the boiler is going to be down for a short period. *Wet lay-up* is the storage of a boiler filled with warm, chemically treated water. The boiler must be thoroughly cleaned on the fire side and the steam and water side to prevent corrosion. After the steam and water side has been checked for

materials that may have been left inside during inspection, the boiler should be closed up with new gaskets installed.

The boiler should then be filled with deaerated feedwater to the NOWL. For each 1000 gal. of water volume in the boiler, 3 lb of caustic soda and 1.5 lb of sodium sulfite should be added. With the vent open, heat should be applied by firing the boiler for an hour. The boiler should be allowed to cool slightly after firing and then completely filled with deaerated feedwater.

If the boiler is equipped with a superheater, the superheater should be filled with condensate or demineralized water treated with 4.8 lb of hydrazine and 0.1 lb of amine per 1000 gal. of water. After filling the superheater, the boiler should be completely filled with deaerated feedwater.

After the boiler is completely filled with treated, deaerated feedwater, a method must be used to prevent oxygen from entering the boiler through the vent line as the water cools and contracts. One method is to connect a drum filled with treated water to the vent line. This ensures that the treated water from the drum will be pulled into the boiler as the boiler water cools and contracts.

Another method used to prevent oxygen from entering a boiler is to fill the space above the water with nitrogen gas. High pressure boilers requiring short outages of only a few days can be protected by maintaining a blanket of nitrogen or another inert gas. A slightly positive pressure of about 5 psi should be maintained in the boiler after it is filled to the NOWL with deaerated feedwater. This can be done by attaching a cylinder of nitrogen to the vent line through a regulator.

**Dry Lay-Up.** The dry lay-up method is preferred if the boiler will not be needed on short notice or if the boiler is going to be down for a longer period. *Dry lay-up* is the storage of a boiler with all water drained. In a multiple boiler plant, individual boilers can be placed in dry lay-up during times of lower demand for periodic boiler inspection. Dry lay-up allows a thorough boiler cleaning or inspection on the fire side and the steam and water side. **See Figure 12-30.**

The boiler must be dried out completely. All horizontal and nondrainable boiler tubes must be blown out with compressed air. Heat should be applied as the best method of drying. Steam and water lines must be secured to ensure moisture does not enter the boiler. Trays of moisture-absorbing chemicals, or desiccants, should be placed in the steam and water side of the boiler. The trays should be placed in each drum of a watertube boiler or on the top flues of a firetube boiler.

## DRY LAY-UP

*Cleaver-Brooks*

**Figure 12-30.** Dry lay-up allows a thorough boiler cleaning or inspection on the fire side and the steam and water side.

The steam and water side of the boiler should then be closed up using new gaskets. The moisture-absorbing chemicals must be inspected periodically, typically about every three or four months, and replaced as necessary. The ASME Code recommends the use of quick lime or silica gel. Manufacturers and chemical suppliers also recommend the use of activated alumina. The amount used depends on the desiccant. For quick lime, about 7 lb/100 cu ft of volume should be used. For silica gel or activated alumina, about 8 lb/100 cu ft of volume should be used. Some manufacturers may recommend a different amount of desiccant.

### SECTION 12.3 — CHECKPOINT

1. What are five common preventive maintenance tasks?
2. What is the purpose of performing infrared inspections of electrical connections?
3. When is a hydrostatic test typically performed?
4. What is the difference between wet and dry lay-up?

## SECTION 12.4
# BOILER ROOM SAFETY

## OPERATOR SAFETY

Boiler room safety must be constantly exercised by all boiler room personnel. Most large industrial plants have safety engineers who identify hazards in the boiler room. Safety engineers develop safe operating procedures to prevent plant accidents.

Accidents can cause injuries and death. In addition, plant downtime caused by accidents is very costly. Key factors in maintaining boiler room safety include boiler operator safety, the reporting of accidents, and fire prevention.

### Emergency Plans and Procedures

An emergency plan documents procedures, exit routes, and assembly areas for facility personnel in the event of an emergency. Employers must have an emergency plan, with a designated official or safety committee responsible for developing and implementing the plan. The boiler operator must be aware of all duties required as part of the emergency plan. Evacuation, if required, must occur in a safe and orderly manner. In any emergency, the proper authorities must be notified immediately.

Emergency boiler procedures specify the exact actions to be taken in the event of an emergency and will vary from plant to plant. The chief engineer is responsible for establishing emergency boiler procedures for handling any situations that may occur in the plant. The chief engineer is also responsible for training operators in the established emergency procedures for a specific steam plant. Emergency boiler procedures should be clearly posted and available to all operating personnel.

The boiler operator must act quickly when correcting the two most common conditions that could lead to an emergency: high- or low-water conditions in a boiler and flame failure. If improper action is taken during an emergency, the boiler or furnace could explode.

**Low-Water Condition.** Steam boilers are equipped with low-water fuel cutoff controls and alarms or low-water alarms to alert the boiler operator so proper action can be taken in the event of a low-water condition. The extent of overheating or damage to the boiler is determined by how quickly the operator reacts to remedy the condition. Tube and furnace distortion or an explosion can result from a low-water condition in a boiler. **See Figure 12-31.**

## TUBE AND FURNACE DAMAGE FROM OVERHEATING

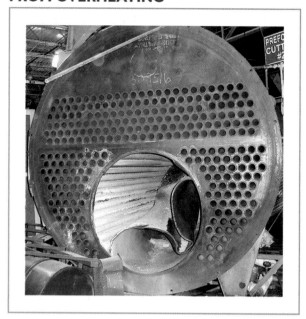

**Figure 12-31.** Tube and furnace distortion or an explosion can result from a low-water condition in a boiler.

If the operator cannot see water in the gauge glass or get a water-level reading via a secondary means, a dangerous low-water condition exists and the water and fuel must be secured. Low-water condition procedures vary depending on the type of boiler and the fuel being used. When in doubt about the water level in a boiler, the fuel should be shut off immediately.

Airflow must be reduced to allow the boiler to cool slowly. The boiler should be taken out of service and thoroughly examined for any indication of overheating. If more than one boiler is on-line and only one boiler is having difficulty maintaining its water level, the following steps should be taken:

1. Secure the feedwater to the boiler with the low-water condition.

2. Secure the fuel to the burner of the boiler with the low-water condition.

3. Maintain the NOWL on all boilers on-line.

Boiler explosions due to a low-water condition typically occur when water is added to the boiler. This water comes in contact with the exposed heating surface and flashes to steam. When water flashes to steam, it expands by a factor of about 1600. This overpressurizes the boiler, causing a rupture and ultimately an explosion. Any time a dangerously low water condition occurs, the boiler must be inspected for possible damage. Some municipalities and insurance companies require that the incident be reported and that an inspector determine if the boiler can safely be put back in service.

With a low-water condition, the steam boiler drum, tubes, and furnace are endangered. With a high-water condition, the superheater tubes, headers, and steam equipment such as pumps and turbines are endangered.

**High-Water Condition.** Steam boilers are equipped with high-water alarms to alert the boiler operator of a high-water condition so that proper action can be taken. Any time the water level in a steam boiler is three-quarters of a gauge glass or higher, water may be carried over with the steam. The water level should be corrected immediately. This may require using the bottom blowdown valves. In most plants today, high-water alarms and annunciators warn the operator of a high-water condition in the boiler.

Both steam and water lines to the gauge glass must be kept clean to prevent false water level readings. If the top line to the gauge glass is closed or clogged, the glass will fill with water. If the bottom line to the gauge glass is closed or clogged, there will be a stationary water level reading (no movement). This will allow the gauge glass to slowly fill with water caused by steam condensing at the top. If the reading on the gauge glass is questionable, the operator should use a secondary means to determine the boiler water level and then clean or replace the clogged lines.

**Flame Failure.** Flame failure occurs when the flame in the furnace has been unintentionally lost. Flame failure is one of the most common causes of a furnace explosion. A furnace explosion happens when explosive or highly flammable gas, vapor, or dust ignite within the furnace. In minor explosions, known as puffs, flarebacks, or blowbacks, flames may blow suddenly from all firing and access doors for a distance of many feet. In major furnace explosions, there can be extensive damage to the doors, air box, smoke box, refractory, and breeching. Furnace explosions can cause injuries and loss of life.

Flame failures in a plant fired by fuel oil are caused by cold fuel oil, water in the fuel oil, air in the fuel oil lines, clogged strainers, a clogged burner tip, or loss of fuel oil pressure. Flame failures in a gas-fired plant are caused by insufficient pressure in gas lines or water in the gas. Flame failures in a plant fired by pulverized coal are caused by wet coal, loss of primary air, or loss of coal to the pulverizing mill because of feeder failure or blockage.

## Boiler Operator Safety

A boiler operator is the person responsible for the safe and efficient operation of the boiler. The boiler operator must develop safety habits to prevent personal injury, injury to others, and damage to equipment. Safety rules vary depending on the type and size of the plant. However, the basic safety rules listed are common to all boiler rooms. Typical general safety rules include the following:

- Use proper personal protective equipment (PPE) at all times.
- Follow proper lockout/tagout procedures.
- Follow confined space procedures.
- Follow proper procedures for handling hazardous materials.
- Follow proper procedures for fire prevention.
- Follow proper procedures for using tools.

**PPE.** *Personal protective equipment (PPE)* is any device worn by a boiler operator to prevent injury. **See Figure 12-32.** All PPE must meet the requirements specified in Occupational Safety and Health Administration (OSHA) 29 CFR 1910 — Occupational Safety and Health Standards and applicable ANSI and other safety mandates.

## PERSONAL PROTECTIVE EQUIPMENT (PPE)

**Figure 12-32.** PPE is any device worn by a boiler operator to prevent injury.

For example, proper eye protection must be worn to prevent injury from flying particles, radiant energy, liquid chemicals, or other hazards. Protective helmets should be worn in areas where overhead hazards are possible. Safety shoes with steel toes should be worn to provide protection from falling objects. Coveralls and clothing made of durable material like denim should be worn to provide protection from contact with hot equipment or sharp objects. Long hair should be secured. Loose jewelry should be removed or secured. Clothing made of flammable synthetic materials should not be worn. Some tasks require fire-resistant coveralls for additional protection.

**Lockout and Tagout.** Lockout and tagout procedures are applied to equipment to prevent equipment operation during maintenance and repair. *Lockout* is the use of locks, chains, or other physical restraints to positively prevent the operation of specific equipment. **See Figure 12-33.** *Tagout* is the process of attaching a danger tag to the source of power to indicate that the equipment may not be operated until the tag is removed. A tagout does not prevent the startup of equipment but serves as a warning to operating and service personnel. Steam stop valves, bottom blowdown valves, and feedwater valves need to be locked out or tagged out when a boiler is removed from service for cleaning and inspection.

The space on the tag may be used to specify lockout and tagout information. Written lockout and tagout procedures must be established for each piece of equipment in the facility. Lockouts and tagouts should only be removed by authorized personnel. A multiple lockout or tagout requires installation and removal of more than one lock on a multiple lockout hasp or tag.

**Confined Spaces.** A *confined space* is a space large enough and so configured that an employee can physically enter and perform assigned work, has limited or restricted means for entry and exit, and is not designed for continuous employee occupancy. Confined spaces have a limited means of egress and are subject to the accumulation of toxic or flammable contaminants or an oxygen-deficient atmosphere.

## LOCKOUT/TAGOUT PROCEDURES

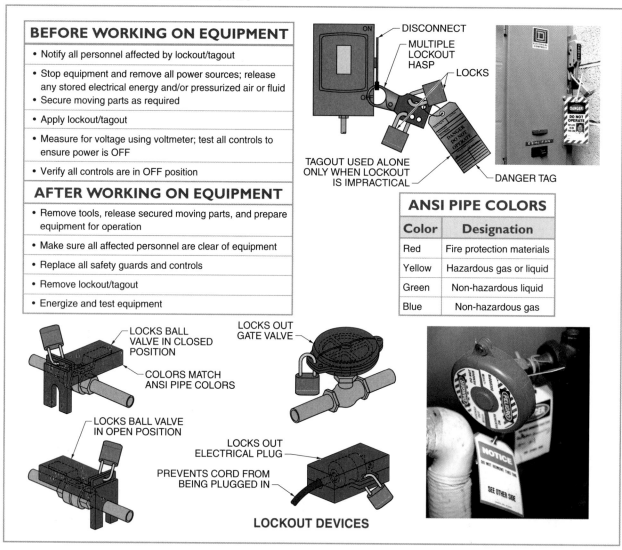

### BEFORE WORKING ON EQUIPMENT

- Notify all personnel affected by lockout/tagout

- Stop equipment and remove all power sources; release any stored electrical energy and/or pressurized air or fluid
- Secure moving parts as required

- Apply lockout/tagout

- Measure for voltage using voltmeter; test all controls to ensure power is OFF

- Verify all controls are in OFF position

### AFTER WORKING ON EQUIPMENT

- Remove tools, release secured moving parts, and prepare equipment for operation

- Make sure all affected personnel are clear of equipment

- Replace all safety guards and controls

- Remove lockout/tagout

- Energize and test equipment

DISCONNECT

MULTIPLE LOCKOUT HASP

LOCKS

TAGOUT USED ALONE ONLY WHEN LOCKOUT IS IMPRACTICAL

DANGER TAG

#### ANSI PIPE COLORS

| Color | Designation |
|-------|-------------|
| Red | Fire protection materials |
| Yellow | Hazardous gas or liquid |
| Green | Non-hazardous liquid |
| Blue | Non-hazardous gas |

LOCKS BALL VALVE IN CLOSED POSITION

COLORS MATCH ANSI PIPE COLORS

LOCKS BALL VALVE IN OPEN POSITION

LOCKS OUT GATE VALVE

LOCKS OUT ELECTRICAL PLUG

PREVENTS CORD FROM BEING PLUGGED IN

**LOCKOUT DEVICES**

**Figure 12-33.** Lockouts and tagouts should only be removed by authorized personnel.

Confined spaces in and around the boiler room may include boilers, fuel oil tanks, feedwater heaters, deaerators, sewers, and underground tunnels. Confined spaces may contain entrapment hazards and life-threatening atmospheres caused by oxygen deficiency and the presence of combustible and/or toxic gases. Oxygen deficiency is caused by the displacement of oxygen by leaking gases or vapors, the combustion or oxidation process, oxygen being absorbed by the vessel or product stored, and/or oxygen being consumed by bacterial action. Oxygen-deficient air can suffocate a person and may result in injury or death.

A confined space permit is required for work in confined spaces based on safety considerations for workers. **See Figure 12-34.** A permit-required confined space

is a confined space that has specific health and safety hazards associated with it. OSHA 29 CFR 1910.146—Permit-Required Confined Spaces covers the requirements for practices and procedures needed to protect workers from the hazards of entry into permit-required confined spaces. Permit-required confined spaces are grouped into categories containing or having a potential to contain a hazardous atmosphere, containing a material that has the potential for engulfing an entrant, having an internal configuration such that an entrant could be trapped or asphyxiated by inwardly converging walls or a floor that slopes downward and tapers into a smaller cross-section, or containing any other recognized safety or health hazard.

## CONFINED SPACE ENTRY PERMITS

---

**WASHINGTON HEATING PLANT**

√ CONFINED SPACE    √ HAZARDOUS AREA

PERMIT VALID FOR 8 HOURS ONLY. ALL COPIES OF PERMIT WILL REMAIN AT JOB SITE UNTIL JOB IS COMPLETED

SITE LOCATION and DESCRIPTION ___Fuel Oil Tank #4___

PURPOSE OF ENTRY ___Routine Inspection___

SUPERVISOR(S) in charge or crews.    Type of Crew                Phone #

___Marty Smith___    ___Maintenance Shift II___ · x3002

\* BOLD DENOTES MINIMUM REQUIREMENTS TO BE COMPLETED AND REVIEWED PRIOR TO ENTRY \*

| REQUIREMENTS COMPLETED | DATE | TIME | REQUIREMENTS COMPLETED | DATE | TIME |
|---|---|---|---|---|---|
| **Lock Out/Deenergized/Try-out** | 3/2 | 09:00 | **Full Body Harness w/"D" ring** | 3/4 | 08:00 |
| **Line(s) Broken-Capped-Blanked** | 3/2 | 11:00 | **Emergency Escape Retrieval Equip** | 3/4 | 08:00 |
| **Purge-Flush and Vent** | 3/3 | 09:00 | **Lifelines** | 3/4 | 08:00 |
| **Ventilation** | 3/3 | 10:00 | Fire Extinguishers | 3/4 | 08:00 |
| **Secure Area (Post and Flag)** | 3/2 | 08:00 | Lighting (Explosionproof) | 34 | 08:00 |
| **Breathing Apparatus** | 3/4 | 08:00 | Protective Clothing | 3/4 | 08:00 |
| **Resuscitator-Inhalator** | 3/4 | 08:00 | Respirator(s) (Air Purifying) | 3/4 | 08:00 |
| **Standby Safety Personnel** | 3/4 | 08:00 | Burning and Welding Permit | N/A | N/A |

Note: Items that do not apply enter N/A in the blank.

\*\* RECORD CONTINUOUS MONITORING RESULTS EVERY 2 HOURS

| CONTINUOUS MONITORING\*\* | Permissible Entry Level | | 3/4 | | | | | | | |
|---|---|---|---|---|---|---|---|---|---|---|
| TEST(S) TO BE TAKEN | | | | | | | | | | |
| **PERCENT OF OXYGEN** | 19.5% to 23.5% | | 20.5 | 20.6 | 20.7 | 20.5 | 20.5 | | | |
| **LOWER FLAMMABLE LIMIT** | Under 10% | | 5 | 5 | 5 | 5 | 6 | | | |
| **CARBON MONOXIDE** | +35 PPM | | 0 | 0 | 0 | 0 | 0 | | | |
| Aromatic Hydrocarbon | +1 PPM | \* 5 PPM | 2 | 1 | 2 | 1 | 1 | | | |
| Hydrogen Cyanide | (Skin) | \* 4 PPM | N/A | | | | | | | |
| Hydrogen Sulfide | +10 PPM | \* 15 PPM | N/A | | | | | | | |
| Sulfur Dioxide | +2 PPM | \* 5 PPM | 3 | 2 | 2 | 2 | 2 | | | |
| Ammonia | | \* 35 PPM | N/A | | | | | | | |

\*Short-term exposure limit: Employee can work in the area up to 15 min.

+8 hr time-weighted avg.: Employee can work in the area 8 hr (longer with appropriate respiratory protection).

REMARKS:

GAS TESTER NAME & CHECK #    INSTRUMENT(S) USED MODEL & / OR TYPE    SERIAL & / OR UNIT #

___Terry Green___    ___Combination Gas Meter___  ___Industrial Scientific___    ___15 A___

SAFETY STANDBY PERSON IS REQUIRED FOR ALL CONFINED SPACE WORK

SAFETY STANDBY PERSON(S)    CHECK #    NAME OF SAFETY STANDBY PERSON(S)    CHECK #

___Kate Washington___    3312

___Tony Linder___    3318

SUPERVISOR AUTHORIZING ENTRY    AMBULANCE 2800    FIRE 2900

ALL ABOVE CONDITIONS SATISFIED ___Marty Smith___    Safety 4901    Gas Coordinator 4529/5387

---

**Figure 12-34.** A confined space entry permit documents and verifies that pre-entry preparations have been completed prior to entry.

Permit-required confined spaces require assessment of procedures in compliance with OSHA standards prior to entry. A non-permit-required confined space is a confined space that does not contain or, with respect to atmospheric hazards, have the potential to contain any hazards capable of causing death or serious physical harm. These conditions can change with tasks such as cleaning with solvents, welding, or painting in the confined space.

An entry permit must be posted at confined space entrances or otherwise made available to entrants before entering a permit-required confined space. The permit is signed by the entry supervisor and verifies that pre-entry preparations have been completed and that the space is safe to enter.

A permit-required confined space must be isolated before entry. This prevents hazardous energy or materials from entering the space. Plant procedures for lockout and tagout of permit-required confined spaces must be followed. Training is required for all boiler operators who are required to work in or around permit-required confined spaces.

**Hazardous Materials.** A *hazardous material* is a substance that can cause injury to personnel or damage to the environment. Proper handling procedures minimize potential injury or damage caused by hazardous materials. Worker education provides the best defense against accidents involving hazardous materials.

Employers must develop, implement, and maintain a written, comprehensive hazard communication program. The program must include provisions for container labeling, chemical inventory, data collection, availability of safety data sheets, and an employee training program. It must also contain a list of hazardous chemicals in each work area. Information must be provided in a language or manner that employees understand.

Containers that contain hazardous materials must be labeled, tagged, or marked with the identity of the hazardous material and appropriate hazard warnings per OSHA 29 CFR 1910.1200(f)—Labels and Other Forms of Warning. Container labeling differs from manufacturer to manufacturer. However, all container labels must include basic right-to-know (RTK) information to convey hazards of the chemical according to federal and state standards. The National Fire Protection Association (NFPA) Hazard Signal System may be used to provide information at a glance. The NFPA Hazard Signal System uses a four-color diamond-shaped sign to display basic information about hazardous materials. **See Figure 12-35.**

A *safety data sheet (SDS)* is printed material used to relay chemical hazard information from the manufacturer, importer, or distributor to the employer. This information is used to inform and train employees on the safe use of hazardous materials. All chemical products used in a facility must be inventoried and have an SDS. This includes chemicals used by boiler operators, custodial staff, building occupants, and contracted maintenance and pest control workers.

## Boiler Room Fire Prevention

Boiler room fire prevention procedures are necessary because of the combustible nature of the materials used in a boiler room. A *combustible material* is any material that burns when it is exposed to oxygen and heat. Combustible materials burn readily and require special handling by the boiler operator.

Insurance statistics reveal that there are approximately 400 industrial fires in the United States every day. The boiler operator must know what is necessary to start and sustain a fire in order to know how to prevent a fire and how to put out a fire if one occurs. Fuel, heat, and oxygen are required to start and sustain a fire. The fire will go out when any one of these is removed. **See Figure 12-36.** Fuel may be fuel oil, wood, paper, textiles, or any other material that burns readily. The fuel must be heated to its ignition temperature. Oxygen is required to support the combustion process.

## HAZARDOUS MATERIAL CONTAINER LABELING

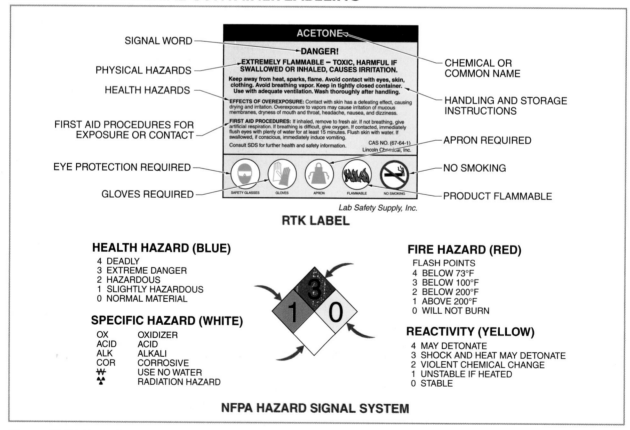

**Figure 12-35.** Hazardous material container labeling identifies appropriate hazard warnings.

## COMBUSTION REQUIREMENTS

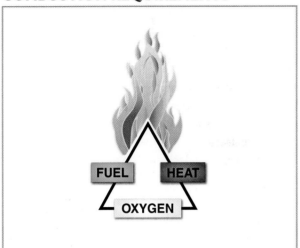

**Figure 12-36.** Fuel, heat, and oxygen are required to support combustion. A fire will go out when any one of these is removed.

Since the main ingredient is the combustible material, waste or oily rags must be stored in safety containers and volatile liquids in safety cans. By maintaining careful control of the combustible materials in a boiler room, the fire hazards are reduced.

**Classes of Fire.** The class of fire is a method of determining the type of fire based on the combustible material burned. Fires are classified as Class A, Class B, Class C, Class D, and Class K.

Class A fires include burning wood, paper, textiles, and other ordinary combustible materials containing carbon. Class B fires include burning oil, gas, grease, paint, and other liquids that convert to a gas when heated. Class C fires include burning electrical devices, motors, and transformers. Class D is a specialized class of fires including burning metals such as zirconium, titanium, magnesium, sodium, and potassium. Class K fires include burning grease in commercial cooking equipment. Fire extinguishers are selected for the class of fire based on the combustible material. **See Figure 12-37.**

Fire extinguishing equipment does not take the place of plant fire protection personnel or the local fire department. Proper authorities must be notified whenever there is a fire in a plant. Boiler operators must know the locations of all fire extinguishing equipment in the facility. Fire extinguishing equipment, such as fire extinguishers, water hoses, and sand buckets, must be routinely checked according to plant procedures.

## FIRE EXTINGUISHER CLASSES

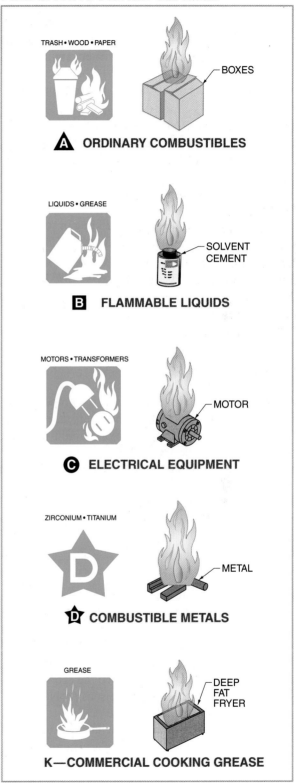

**Figure 12-37.** The fire extinguisher required is determined by the combustible material to be extinguished.

**Fire Hazard Areas.** The number and type of fire extinguishers required is determined by the authority having jurisdiction and is based on how fast a fire may spread, potential heat intensity, and accessibility to the fire. Additional fire extinguishers are installed in hazardous areas. The NFPA lists these hazardous areas as light hazard, ordinary hazard, and extra hazard areas.

A *light hazard area* is a building or room that is used as a church, office, classroom, or assembly hall. The contents in buildings and rooms of this nature are either noncombustible or are not anticipated to be arranged in a manner that would be conducive to the rapid spread of fires.

An *ordinary hazard area* is a building or room that is used as a shop and related storage facility, light manufacturing plant, automobile showroom, or parking garage. An *extra hazard area* is a location that includes woodworking shops, manufacturing plants using painting or dipping, and automotive repair shops. In extra hazard areas, Class A and Class B flammables exceed those expected in ordinary hazard areas.

### Proper Use of Tools

A boiler operator must always use the proper tools. Using the wrong tools can cause injury or property damage. A low-voltage drop light should be used when working inside a boiler to prevent potential injury from electrical shock. Ladders that meet OSHA requirements must always be used. Loose tools should never be left on ladders, catwalks, scaffolds, or any other elevated surfaces. The tools could fall and cause injury.

### Reporting Accidents

Even when safety precautions are followed, the possibility of an accident still exists. All accidents should be reported regardless of their nature. Most plants have a specific procedure for reporting accidents. The boiler operator must become familiar with this procedure. If complications arise as the result of an accident that was not reported or put on file, serious problems can occur regarding insurance claims.

Accident reports commonly include the following information: date, time, and place of accident; immediate superior; name of injured person; nature of injury; what injured person was doing at time of accident; and cause of accident. **See Figure 12-38.** Accident reports are also used to document plant safety records.

---

**SECTION 12.4–CHECKPOINT**

1. What is the purpose of emergency procedures?
2. What are possible consequences of a low-water condition?
3. What is lockout?
4. What are five classes of fires?
5. What is the difference between ordinary hazard areas and extra hazard areas?
6. Why is it important to use the proper tools?
7. Why should all accidents be reported?

---

## SECTION 12.5
# BOILER OPTIMIZATION

### BOILER OPTIMIZATION

Operating boilers efficiently reduces the use of fuel and lowers the operating costs. It also reduces the discharge of gases of combustion into the atmosphere. Proper maintenance extends the life of the equipment, reducing the net equipment operating costs while improving steam availability.

### Resources for Energy Savings

Many resources for energy savings are available from the government, energy suppliers, and equipment manufacturers. Most programs provide technical support and advice to improve energy efficiency, and some programs provide partial funding to implement energy-saving improvements.

**Department of Energy.** The Department of Energy (DOE) has created several offices to encourage energy efficiency in commercial and industrial facilities. One is the Office of Energy Efficiency and Renewable Energy (EERE). The Database of State Incentives for Renewables & Efficiency (DSIRE) provides information about all state, local, utility, and federal programs for grants, loans, and rebates sponsored by the DOE. This includes financial incentives, rules, regulations and policies, and any related programs and initiatives for each state.

**Environmental Protection Agency.** The Environmental Protection Agency (EPA) addresses a number of environmental issues that include emissions and wasted energy from both commercial and industrial boilers. For example, the EPA's *Climate Leaders Greenhouse Gas Inventory Protocol* is a guide to reducing greenhouse gas emissions.

## ACCIDENT REPORT FORMS

| Company Name | Plant Protection Department | AMBULANCE, INJURY, AND SICK REPORT |
|---|---|---|
| Date | Time | |

Type of Call:  ☐ Ambulance  ☐ Doctor    ☐ Injury  ☐ Sickness

| Requested By | Phone Extension |
|---|---|

Name of Sick or Injured

| Dept. No. | Index No. | Division |
|---|---|---|

Street Address

City, State, and Zip Code

Plant Location of Sick or Injured Person

| Taken to *(Name of Hospital or Doctor)* | Location *(City of Hospital or Doctor)* |
|---|---|

Taken By:  ☐ Ambulance   ☐ Taxi    ☐ Private Auto

Describe Injury or Illness

### WITNESSES

| Full Name | Dept. No. | Index No. | Division |
|---|---|---|---|
| Full Name | Dept. No. | Index No. | Division |
| Full Name | Dept. No. | Index No. | Division |

### GUARDS RESPONDING

| Name | Name | Name |
|---|---|---|

### BRIEF SUMMARY

| Name of Guard Submitting Report | Date Report Submitted |
|---|---|

**Figure 12-38.** All accidents must be reported and recorded using an accident report form.

**Energy Monitoring Software.** Energy monitoring software (EMS) includes a variety of energy-related software applications including utility bill tracking, real-time metering, HVAC and lighting control systems, building simulation and modeling, carbon and sustainability reporting, IT equipment management, demand response, and/or energy audits. EMS uses collected data for reporting, monitoring, and engagement purposes. Reporting includes the verification of energy data, benchmarking, and setting energy-use reduction targets. Monitoring includes energy trend analysis and tracking energy consumption. Engagement includes real-time responses or data collection to promote energy conservation.

**Energy Optimization.** Many states provide specific programs for large energy users. For example, the Large-Customer Energy Analysis Program (LEAP) in the state of Illinois works with large energy users to help manage and reduce energy costs. The state claims that participants in the LEAP program are typically able to reduce energy use by 10% to 30%.

In addition, many equipment manufacturers provide help in optimizing energy use. For example, Cleaver-Brooks has a program called BOOST that helps customers with reports covering the costs and financial returns of boiler upgrades that improve efficiency or environmental performance. The reports can be used to help justify making capital improvements to boilers. Typical improvements could include the addition of economizers and blowdown heat recovery systems.

## Steam and Energy Management

The primary objectives in the design and operation of boiler plants are continuity of service and optimum operating efficiency. A boiler is designed to transfer the heat released in the process of combustion to the boiler water, then produce steam from the boiler water. In the boiler furnace, the chemical energy in the fuel is converted into heat, and this heat is transferred to the boiler water. The boiler's thermal efficiency is the percentage of the heat liberated that is transferred into the boiler water. The boiler's combustion efficiency is the percentage of the Btu content of the fuel that is released as heat by the fuel-burning equipment of the boiler.

Total plant efficiency is also an important consideration. Examples of items to consider are steam piping insulation that has been damaged or missing, the removal of steam piping where equipment has been decommissioned or removed, the repair or replacement of defective or inoperable steam traps, the minimization of condensate loss, and the minimization of heat losses due to venting at the deaerator. **See Figure 12-39.**

## STEAM SYSTEM INSULATION

**Figure 12-39.** The integrity of insulation can be checked during an audit.

**Thermal Efficiency.** *Thermal efficiency* of a boiler is the ratio of the heat absorbed by the boiler (output) to the heat available in the fuel (input). Thermal efficiency can be used to determine the amount of fuel required to generate a given quantity of steam. The purpose of calculating the thermal efficiency is to track the efficiency over time. If the thermal efficiency changes unexpectedly, it often indicates a problem with the fuel, furnace, boiler, draft, or feedwater systems.

The heat from combustion is absorbed by the water in the boiler. However, not all heat released from the fuel is used to heat the water. Some heat is wasted in the process. The heat losses that occur in a boiler can result from the gases of combustion released to the atmosphere, radiation, water vapor produced from burning hydrogen, incomplete combustion, moisture in the fuel, and moisture in the combustion air. Determining thermal efficiency of a boiler consists of accounting for all the heat units in the fuel, used and wasted.

Thermal efficiency calculations can be complex. In some plants, it is convenient to determine a daily evaporation rate rather than determining thermal efficiency. In any given plant, the feedwater temperature is usually constant, and the heat content of the steam and the fuel remain relatively constant. The only variables are the rate of steam generation and the amount of fuel consumption. The flow rate of steam can be determined using a steam flow meter, and the fuel consumption data is found from fuel flow meters or coal scales. The evaporation rate is calculated using the following equation:

$$ER = \frac{W_s}{W_f}$$

where

$ER$ = evaporation rate (in lb steam/unit of fuel)

$W_s$ = steam flow (in lb/hr)

$W_f$ = fuel flow (in lb/hr or therms/hr)

For example, a boiler generates 100,500 lb of steam/hr while using 1200 therms of natural gas. The evaporation rate is calculated as follows:

$$ER = \frac{W_s}{W_f}$$

$$ER = \frac{100,500}{1200}$$

**$ER$ = 83.75 lb steam/therm**

For plants where the feedwater temperature, heat content of the steam, and heat content of the fuel vary, the thermal efficiency can be calculated. The calculations can be automated with a combustion control system or with a simple spreadsheet. Thermal efficiency is calculated by using the following equation:

$$TE = \frac{W_s\left[H_s - \left(T_{fw} - 32\right)\right]}{W_f \times C}$$

where

$TE$ = thermal efficiency

$W_s$ = steam flow (in lb/hr)

$H_s$ = enthalpy of steam (in Btu/lb)

$T_{fw}$ = feedwater temperature (in °F)

$32$ = base temperature for determining enthalpy (in °F)

$W_f$ = fuel flow (in lb/hr or therms/hr)

$C$ = fuel heat content (in Btu/unit of fuel)

For example, a boiler uses 1200 therms/hr of natural gas while generating 100,500 lb/hr of steam. A therm is equal to 100,000 Btu. The enthalpy (energy) of the steam is 1300 Btu/lb. The enthalpy of steam can be found in steam tables. The feedwater temperature is 300°F. The thermal efficiency is calculated as follows:

$$TE = \frac{W_s\left[H_s - \left(T_{fw} - 32\right)\right]}{W_f \times C}$$

$$TE = \frac{100,500\left[1300 - \left(300 - 32\right)\right]}{1200 \times 100,000}$$

$$TE = \frac{100,500\left(1300 - 268\right)}{120,000,000}$$

$$TE = \frac{100,500\left(1032\right)}{120,000,000}$$

$$TE = \textbf{0.86} \text{ or } \textbf{86\%}$$

**Combustion Efficiency.** *Combustion efficiency* is a measure of the ability of the burners to burn fuel efficiently. The most efficient combustion occurs at the stoichiometric ratio, where each molecule of fuel reacts with the appropriate amount of oxygen in the air for 100% combustion efficiency. In practice, 100% combustion efficiency is not possible because excess air is required due to limitations in the mixing of the air and fuel.

**Energy Audits.** An *energy audit* is an audit that identifies how energy is used in a facility and recommends ways to improve energy efficiency and reduce energy costs. **See Figure 12-40.** ASHRAE, formerly the American Society of Heating, Refrigerating and Air-Conditioning Engineers, identifies three levels of energy audits: a walk-through-assessment, an energy survey and analysis, and a detailed analysis of capital-intensive modifications.

A walk-through-assessment is an energy audit in which energy bills are analyzed and a visual survey of the facility is conducted. A energy analysis report identifies no-cost and low-cost opportunities for improvement. An energy survey and analysis is a more detailed audit and includes how energy is used within the facility. With this type of audit, potential capital-intensive opportunities are identified in a final report for further research and analysis. A detailed analysis of capital-intensive modifications is an audit that focuses on capital-intensive opportunities and provides a higher degree of monitoring, data collection, and engineering analysis.

## AUDIT REPORT ANALYSIS

| Problem | Cost of Energy Loss | |
| --- | --- | --- |
| Compressed air distribution system has seven leaks | $560/month | |

| Possible Solutions | Estimated Project Costs | Payback |
| --- | --- | --- |
| Repair leaks in existing distribution system | $800 | 1.5 months |
| Replace all piping to eliminate unnecessary connections and reroute around areas where piping is prone to physical damage | $3,600 | 6.5 months |

**Figure 12-40.** An energy audit identifies how energy is used in a facility and then recommends ways to improve energy efficiency and reduce costs.

Local utilities can arrange for one of their engineers to assist with an energy audit or make recommendations of third-party providers. Most utilities have testing equipment, such as portable meters and diagnostic software, that in-house staff can use to conduct an energy audit.

The energy audit reveals all of the major factors affecting energy consumption. The audit should include utility bill analysis based on historical data provided by the client, on-site survey of the property, and diagnostic testing.

**Consultants.** Consultants are often brought into a boiler plant to help with projects intended to improve efficiency or environmental performance. Consultants are used when the scope of the work is too large to be handled by maintenance staff, when the existing maintenance staff does not have the skills or tools to accomplish the work, when the owner needs the objectivity of a third party, or for other reasons.

When the scope of the work of a proposed project is too large to be handled by in-house maintenance staff, additional help may be needed. In many cases, the maintenance staff has been reduced, making it difficult for the staff to complete additional duties. When projects are added to the work load, it is not practical to hire more staff and train them for the short-term only. By hiring a consultant, the company can have the extra manpower for as long as it is needed. The consultant is paid by the hour, and the hours can be spread out over any period.

Most maintenance staff are hired to maintain, repair, and operate the equipment and may not have the skills required for certain projects. Consultants are more expensive per hour than maintenance staff employees, but a consultant can be expected to know what needs to be done instead of having to be trained.

Consultants can then be used to train the staff on the new equipment or systems.

A typical project that a consultant may be hired to perform is to monitor energy use throughout the boiler plant. **See Figure 12-41.** Accurate information about actual energy use is seldom available. A temporary monitoring program can provide the information needed to justify making changes to operations or equipment to optimize energy use.

## ENERGY USE DATA LOGGERS

*Fluke Corporation*

**Figure 12-41.** A typical task that a consultant may be hired to perform is monitoring energy use throughout a boiler plant.

The company may need the objectivity of a third party, even though the maintenance staff may have the time and expertise to handle the proposed project. It is common for maintenance staff to work for the same employer for years and in the same building. Some maintenance staff members do not keep up on the newest technology, materials, and methods needed for the proposed projects. A third party can provide another point of view and help justify changes. In some cases, consultants are used to add credibility to a project.

**ENERGY STAR® Rating.** The ENERGY STAR® program was created in the 1990s by the EPA and the DOE. The program was designed to reduce energy consumption and greenhouse gas emissions by power plants. It was developed as a voluntary labeling program designed to identify and promote energy-efficient products. **See Figure 12-42.** The program was expanded in 1995 to include labels for residential heating and cooling systems and new homes. The ENERGY STAR label can also be found on new homes and in commercial and industrial buildings.

## ENERGY STAR LABEL

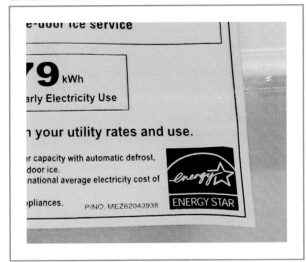

**Figure 12-42.** The ENERGY STAR® program was designed to reduce energy consumption and greenhouse gas emissions by power plants. It was developed as a voluntary labeling program designed to identify and promote energy-efficient products.

*The Leadership in Energy and Environmental Design (LEED®) standard is a rating system that indicates the energy efficiency of a building.*

Commercial and industrial buildings include hospitals, hotels and motels, K–12 schools, offices, retail stores, supermarkets, warehouses, data centers, automobile assembly plants, refineries, pharmaceutical facilities, and wastewater facilities. By entering data into a program provided by ENERGY STAR, a rating can be generated for each building. This rating process compares the energy consumption to that of other buildings of the same type.

A building must earn a rating of 75 out of 100 or above to qualify for the ENERGY STAR rating for a building. The ENERGY STAR energy performance ratings have been incorporated into some green buildings standards, such as LEED® for Existing Buildings.

---

### SECTION 12.5—CHECKPOINT

1. How can the DOE EERE be used to encourage energy-efficiency improvements?
2. What is the purpose of energy monitoring software?
3. What is the difference between thermal efficiency and combustion efficiency?
4. What is the purpose of an energy audit?

## CASE STUDY—TESTING SAFETY VALVES

### Situation

The operating engineer at a correctional facility was responsible for starting, stopping, operating, maintaining, and repairing the boilers, chillers, cooling towers, air compressors, and air handlers throughout the facility. This correctional facility had a gas-fired scotch marine boiler. The boiler operating control was set at 24 psi. A high-limit pressure control with manual reset was set at 26 psi. The boiler had two safety valves. The lower pressure safety valve was set at 50 psi and the higher pressure safety valve was set at 53 psi. The feedwater pumps had been replaced during the last year.

The preventive maintenance required for the boiler included testing the safety valves before the annual cleaning and inspection. The standard testing procedure was to secure the main steam stop valve, set the pressure controls above the safety valve setpoint, allow the steam pressure to increase until the safety valves popped open, and record the pressure at which the valves opened. Then, the boiler would be shut down, allowing the steam pressure to decrease, and the pressure at which the safety valves closed would be recorded.

As the steam pressure continued to decrease below the setpoint, the pressure controls would be reset to the original pressure setting. The pressure controls would then be tested by starting the boiler and increasing the steam pressure to the setpoint to verify the controls were working correctly. This was done for each of the pressure controls (operating and high-limit). All of the test results and setpoints would be recorded in the boiler room log.

### Problem

The main steam stop valve was secured and the setpoints on the pressure controls were set above the safety valve setpoint. The boiler was started in low fire and the steam pressure was allowed to increase until the first safety valve popped open. It opened at 50 psi as expected.

However, the boiler was producing steam that was escaping from the safety valve. The firing rate had to be increased to allow the pressure to continue to increase to the second safety valve setpoint. The escaping steam also required water to be fed into the boiler to replace the water that was evaporating. There was water in the surge tank and the feedwater pumps were operating, but the water level in the gauge glass continued to fall. The test was stopped before the water level was low enough for the boiler to shut down on low water.

### Solution

An investigation determined that the feedwater pumps were selected to operate at a pressure of 60 psi of total dynamic head. With line pressure losses along the pipe and valves before entering the boiler, the pump could only overcome a pressure of 51 psi. When the boiler pressure was increased to allow for testing of the 53 psi safety valve, the feedwater pumps could not pump water into the boiler at the higher pressure. Since the safety valves could not be tested properly, the safety valves were removed from the boiler and tested and reset at an authorized facility for repairing safety valves.

---

## CASE STUDY—MANHOLE GASKET FAILURE

### Situation

The operating engineer in a sports stadium was responsible for starting, stopping, operating, maintaining, and repairing the boilers, chillers, cooling towers, air compressors, and air handlers throughout the stadium. The boilers were classified as power hot water boilers. A power hot water boiler is a boiler used for heating water or other liquids to a pressure exceeding 160 psi or to a temperature exceeding 250°F. This boiler plant operated at 180 psi at 260°F.

During annual maintenance and inspection, the boilers would be removed one at a time from service. The shift engineers would work on the boilers as time allowed around the various scheduled stadium events. It would normally take about a week for each boiler.

### Problem

During stadium events, most of the shift engineer's time was spent in the stadium monitoring temperatures, taking readings, and answering calls about controlling room temperatures. Local code required that the boilers be checked and logged every two hours during operation. Returning to the boiler room after a service call, the shift engineer noticed steam coming from two large air grilles in each end of the boiler room. These grilles were used to supply air to the boilers for combustion. Opening the boiler room door, the shift engineer found hot water on the floor and a room full of steam.

There was a large rollup door next to the entrance door. The engineer opened the door to help dissipate the steam. The steam dissipated quickly and the engineer could see the hot water was coming from the boiler.

### Solution

The circulating pump for the boiler was secured along with the isolation valves. Another boiler was started, brought up to temperature, and put on-line. Further investigation found that the manhole gasket had failed and was leaking. The manhole gasket installed during the previous annual inspection was not the correct gasket. It was intended for the large horizontal hot water tanks that were installed in the stadium and for a low water temperature of 125°F. The high boiler temperature caused the gasket to deform and eventually fail. The proper gasket was installed and tested. The boiler was put back into service. To ensure the correct gasket was used in the future, a note on the repair was placed in the boiler room log and maintenance schedule.

**Name** _____ **Date** _____

_____ 1. When taking over a shift, the boiler operator must first check the ___.
   A. steam pressure
   B. fuel oil supply
   C. water level on all boilers
   D. bottom blowdown valves

_____ 2. The low-water fuel cutoff can be tested by blowing down the low-water fuel cutoff or by ___.
   A. securing the feedwater and allowing the water level in the boiler to drop
   B. closing the blowdown valves
   C. modulating the firing rate
   D. a hydrostatic test

_____ 3. In order to prevent uneven expansion of the boiler on cold startup, the boiler must be ___.
   A. warmed up as quickly as possible
   B. thoroughly vented
   C. heated using steam from outside sources
   D. warmed up slowly

_____ 4. During startup, the boiler vent should remain open until a pressure of approximately ___ is reached.
   A. 0.1 psi
   B. 15 psi
   C. 50% of the MAWP
   D. 75% of the MAWP

_____ 5. A safety valve can be manually tested when the ___.
   A. pressure is at least 25% of MAWP
   B. pressure is at least 75% of MAWP
   C. burner is going through the firing sequence
   D. boiler is down for inspection

_____ 6. Boilers equipped with a superheater must be protected during warm-up by ___.
   A. keeping the superheater drain valve closed until the boiler is cut in on-line
   B. keeping the superheater drain valve open until the boiler is cut in on-line
   C. passing the feedwater through the superheater tubes
   D. bypassing the superheater tubes

_____ **7.** Superheaters are prevented from overheating by the circulation of ___.
A. air
B. steam
C. water
D. gases of combustion

_____ **8.** To ensure high heat transfer rates, ___ should be removed from boiler heating surfaces.
A. flue gas
B. condensate
C. carbon dioxide
D. soot

_____ **9.** During boiler shutdown, the ___ should be opened when the steam pressure in the boiler has dropped to 10 psi to 15 psi.
A. safety valve
B. free-blowing drain
C. boiler vent
D. bottom blowdown valve

_____ **10.** As a boiler is cooling down, the boiler operator must maintain ___.
A. the NOWL
B. the normal draft condition
C. the normal feedwater temperature
D. a 10% $CO_2$ reading

_____ **11.** After a boiler has had major repair work done that could affect the boiler internals or has developed a low-water condition, it should be subjected to ___.
A. an accumulation test
B. a hydrostatic test
C. an evaporation test
D. both an accumulation and a hydrostatic test

_____ **12.** A ___ is used to verify the flame in a burner.
A. thermodynamic trap
B. flame scanner
C. high-limit control
D. combustion air proving switch

_____ **13.** Boilers that are out of service for an extended period can be laid up with ___.
A. the water level above the NOWL
B. oxygen present to prevent corrosion
C. carbon dioxide present to prevent corrosion
D. dry lay-up

_____ **14.** Plant emergency procedures should be established by the ___.
A. operator on duty
B. insurance inspector
C. ASME Code
D. chief engineer

_____ **15.** A low-water fuel cutoff ___.
    A. shuts off the water supply
    B. shuts off the fuel supply
    C. shuts off the condensate return
    D. starts the water supply

_____ **16.** A boiler that has experienced a low-water condition should be ___.
    A. thoroughly examined for signs of overheating
    B. thoroughly examined for scale buildup
    C. brought up to full steam pressure to test for leaks
    D. brought up to the NOWL

_____ **17.** A high-water condition in the boiler can be corrected by ___.
    A. closing the main steam stop valve
    B. increasing the firing rate
    C. performing a bottom blowdown
    D. decreasing the continuous blowdown

_____ **18.** ___ is a measure of the ability of burners to burn fuel efficiently.
    A. Combustion efficiency
    B. Thermal efficiency
    C. An energy audit
    D. A consultant report

_____ **19.** If there is a danger of the boiler freezing, the boiler should be laid up ___.
    A. with a light fire
    B. by using steam from the header to keep the boiler warm
    C. completely dry
    D. with antifreeze

_____ **20.** A furnace explosion can be caused by ___.
    A. excess draft
    B. a low-water condition
    C. loss of feedwater
    D. an accumulation of combustible gases

_____ **21.** The ASME Code recommends that boilers be inspected internally and externally every ___.
    A. six months
    B. year
    C. two years
    D. three years

_____ **22.** Maintenance work performed inside a boiler requires a(n) ___, which contains procedures based on safety considerations for workers.
    A. accident report
    B. boiler log
    C. inspector notice
    D. confined space permit

_____ **23.** New or retubed boilers should be cleaned with ___.
    A. alkaline detergent
    B. muriatic acid
    C. softened water
    D. compressed air

_____ **24.** The three ingredients needed to start a fire are fuel, heat, and ___.
    A. carbon dioxide
    B. combustible material
    C. nitrogen
    D. oxygen

_____ **25.** A fire caused by oil, gas, grease, or paint would be classified as a Class ___ fire.
    A. A
    B. B
    C. C
    D. D

# Chapter 13 Objectives

## SECTION 13.1—LICENSING

- Describe the role of training for boiler operators.
- Identify the information resources available to boiler operators.
- Describe different licensing jurisdictions and organizations.
- Identify typical license examination question formats.
- Practice taking sample examinations.

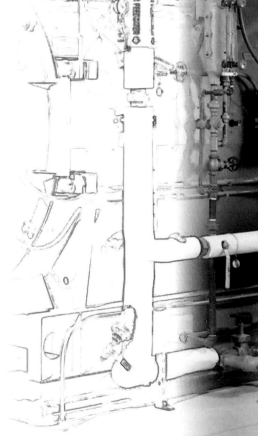

## ARROW SYMBOLS

| AIR | GAS | WATER | STEAM | FUEL OIL | CONDENSATE | AIR TO ATMOSPHERE | GASES OF COMBUSTION |
|-----|-----|-------|-------|----------|------------|-------------------|---------------------|
| ⇨ | ⇨ | ➡ | ➡ | ⇨ | ➡ | ⇨ | ➡ |

**Digital Resources**
ATPeResources.com/QuickLinks
Access Code: **472658**

# LICENSING

## INTRODUCTION

Licensing examinations are a part of the process of licensing boiler operators and stationary engineers. The purpose of the licensing examination is to ensure that the individual is qualified to operate and maintain steam boilers and related equipment. Licensing laws vary from state to state. In some cases, a local municipality may have specific licensing laws. Insurance companies may also have licensing requirements.

An individual preparing to take a licensing examination should be aware of all requirements set by the licensing agency. Licensing agencies may have specific licensing examination requirements governing age, experience, and expected knowledge of equipment operation. These requirements may also include registration fees and certain types of identification. Practicing with sample licensing examination questions will help prepare for taking a licensing examination.

## SECTION 13.1

# LICENSING

### TRAINING

To pass a licensing examination, boiler operation knowledge and skills must be acquired. Once basic skills are mastered, higher levels of skills are necessary. In addition, new boiler operation equipment is continually being developed. New skills are needed to remain current with advancing technology and to grow in the boiler operation profession. The ability to adapt as the industry changes is crucial for continued success.

Boiler operators often upgrade their skills in the profession with time and experience. They use tasks they perform on a daily basis to refine their knowledge and skills. In some instances, boiler operators may start in maintenance positions and move into boiler operator positions after successfully passing licensing examinations. Others become boiler operators through continuing education or through an apprenticeship program.

### International Union of Operating Engineers

The International Union of Operating Engineers (IUOE) offers apprenticeship programs for boiler operators. The IUOE represents stationary engineers who work in operations and maintenance in building and industrial complexes. The IUOE also represents operating engineers who work as heavy equipment operators, mechanics, and surveyors in the construction industry.

Their boiler operation apprenticeship program is a program that combines work experience and training. During the apprenticeship, the apprentice prepares to take the appropriate boiler operator licensing examination. Apprenticeship availability is contingent on the local authorities.

### National Association of Power Engineers

The National Association of Power Engineers (NAPE) is an organization of boiler operators and power engineers whose primary interest is in promoting the craft of power engineering. The organization is dedicated to the education of power engineers. For over a century, NAPE has educated its members in chapters across the country. Members can benefit from job opportunities, numerous conventions, and associating with fellow professionals. Classes in all areas of boiler operation are routinely conducted by NAPE.

### Federal Buildings Personnel Training Act

The Federal Buildings Personnel Training Act requires personnel responsible for the operation and maintenance, energy management, safety, and design functions of federally owned facilities to be properly trained. This includes workers employed by the federal government and its contractors.

According to the act, the administrator for General Services Administration (GSA) shall identify, annually, the core competencies necessary to comply with requirements of this act. The identified competencies shall include those relating to building operations and maintenance, energy management, sustainability, water efficiency, safety (including electrical safety), and building performance measures. A strong focus is to be on the facility management and the operation of high-performance buildings (green buildings).

The administrator shall identify a course, certification, degree, license, or registration to demonstrate each of these competencies. There are provisions in the act for ongoing training with respect to each competency. An individual employed by the federal government and its contractors shall demonstrate each core competency identified no later than one year after their date of hire.

Continuing education courses shall be developed or identified to ensure the operation of federal buildings in accordance with industry best practices and standards. A high percentage of federally owned buildings use private contractors to maintain and manage these facilities. The results of this act should demonstrate what can be accomplished in building and maintaining energy-efficient buildings along with good indoor environmental quality.

### Boiler Operation Information Resources

The boiler operator, as a professional, must be familiar with information pertaining to boilers and related plant equipment. Sources of information include magazines, journals, operator's manuals, technical bulletins, government documents, and industry standards. In addition, attending classes, technical update seminars, and factory schools and participating in professional organizations are activities that provide valuable information on current topics and trends.

Industry standards are periodically revised to reflect changes in the field. The American Society of Mechanical Engineers (ASME) publishes the ASME Code. The sections of primary interest to high pressure boiler operators are Section I, Rules for Construction of Power Boilers; Section IV, Rules for Construction of Heating Boilers; Section VI, Recommended Rules for the Care and Operation of Heating Boilers; and Section VII, Recommended Guidelines for the Care of Power Boilers. In addition, ASME publishes *Controls and Safety Devices for Automatically Fired Boilers* (CSD-1). Other sections of the ASME Code are available for specific topics.

## BOILER OPERATOR LICENSES

Boiler operation is an occupation requiring properly trained operators. Licensed boiler operators have passed a licensing examination to demonstrate a minimum level of competency. A boiler operator license documents that the holder is qualified to safely operate and maintain steam boilers and related equipment. Securing a boiler operator license often offers additional opportunities for career advancement.

### Licensing

Boiler operator licensing requirements vary. Not every state has licensing requirements. States that do not issue state licenses may have jurisdictions within them that establish licensing requirements. For example, Pennsylvania does not have a state licensing requirement, but the city of Philadelphia does. Each licensing jurisdiction may have its own requirements. Efforts to standardize licensing requirements nationally are being promoted by the American Society of Power Engineers, Inc. (ASOPE) and the National Institute for the Uniform Licensing of Power Engineers, Inc. (NIULPE).

**Licensing Jurisdictions.** A licensing jurisdiction will often have specific grades of licensing examinations for different sizes of boilers and related equipment. License classification varies with the licensing agency. For example, in New Jersey, the licenses range in order from the Black Seal License–Low Pressure Boiler Operator to the Gold Seal License–First Grade Engineer. To advance, a prerequisite licensing examination for each type or level of license must be passed. Information about boiler operator licensing can be obtained from the appropriate city, county, or state authorities.

Licensing requirements should be researched well in advance of the testing date. Licensing jurisdictions have specific requirements governing applicant age, work experience, and boiler operation knowledge. The licensing examination site may require registration fees, identification, proof of work experience, proof of age and residence, and specific writing utensils.

Some jurisdictions require experience in the field prior to taking a licensing examination. For example, the state of New Jersey requires that, "to be eligible for a boiler operator's black seal license, the applicant shall have had at least three months experience as a helper, apprentice or assistant to a licensed operator of equipment requiring such license. The Office of Boiler and Pressure Vessel Compliance may request a copy of the boiler operator's log for the Examiner's review during the approval of the application."

**American Society of Power Engineers, Inc.** The American Society of Power Engineers, Inc. (ASOPE) is an independent third-party licensing agency whose objective is to establish national standards and a formal, structured level of competence for national recognition of the power engineer.

The three ASOPE boiler license program categories are Main Power Engineer Competency, Supplementary Power Engineer Competency, and Hobbyist and Specialty Power Engineer Competency. The Main Power Engineer Competency license program consists of facility, power plant, and combustion turbine cycle classifications. The Power Engineer Supplementary Competency license program consists of refrigeration, diesel, and hydro classifications. The Hobbyist and Specialty Power Engineer Competency license program is designed for the power engineer operating and maintaining traction, vintage, and locomotive engines.

**National Institute for the Uniform Licensing of Power Engineers, Inc.** The National Institute for the Uniform Licensing of Power Engineers, Inc. (NIULPE) is a third-party licensing agency that acts on a national level for power engineers and those associated with the profession. NIULPE defines a power engineer as a person skilled in the management of energy conversion such as operating a steam boiler and auxiliaries. Registration with NIULPE is voluntary; however, it offers a formal structure through which a person in power engineering may establish a level of national competence and national professional recognition.

---

**TECH TIP**

*The National Association of Power Engineers (NAPE) is a national organization with local chapters throughout the country. Chapter meetings provide an opportunity for professionals to exchange knowledge, experiences, and industry news with other professionals in a particular region. This interaction can enhance skills and growth within the field.*

---

**Canadian Licensing.** In Canada, the title Stationary Engineer is typically used. In most provinces there are four classes of stationary engineers. Some of the provinces also have a fifth class. The highest license is the First Class Stationary Engineer. Stationary engineers are licensed under the Technical Standards and Safety Authority (TSSA).

Stationary engineers are responsible for the operation, maintenance, renovation, and repair of boilers and all other mechanical systems within a facility. All Canadian territories and provinces allow only certified stationary engineers to operate and maintain this equipment. Stationary engineers in Canada are employed in a large variety of facilities, such as office buildings, hospitals, and power plants.

There are two ways of attaining a certificate of qualification for a Fourth Class Operating Engineer in Canada. The practical experience method requires a person to gain twelve months or 1920 hours practical operating experience in a registered power plant or steam plant. The academic and practical method requires a person to attend six months in an approved full-time training course (in doing so a person receives an additional three-month "incentive credit") plus three months of practical training in a registered plant.

TSSA has compiled a list of approved training providers. All of the providers offer full-time programs, approved for the qualifying experience time. A person must complete the practical experience requirement, as well as pass the standardized fourth class examinations before becoming certified.

## Licensing Examination Preparation

Licensing examination preparation is determined by type of license desired. Licensing agencies also have specific requirements governing experience, registration fees, and other licensing examination criteria. Licensing examinations vary in cost, question type, and question format. Some questions may require a sketch and identification of boiler components.

A boiler operator licensing examination typically consists of a variety of true-false, multiple choice, completion, and essay questions. Some licensing examinations may require the demonstration of practical boiler operation skills, such as blowing down a gauge glass. The licensing agency may provide a list of recommended study materials for preparing for the licensing examination. General suggestions for licensing examination preparation include the following:

- Learn as much as possible about the examination. Obtain information from the licensing agency about the examination. Make sure that the examination is the right one for your qualifications. Talk with individuals who have successfully passed the examination.
- Pace your preparation for the examination. Research the topics covered on the examination. Develop a study schedule and focus on specific topics one at a time over an extended period of time.
- Review the examination study material over several days prior to the examination. Any review the night before the examination should be very limited.
- If unfamiliar with the examination location, drive to the site a few days before the examination date.
- Schedule a normal amount of sleep the night before the examination.
- Allow ample time for travel to the examination site.

*The purpose of the boiler operator licensing process is to ensure that an individual is qualified to operate and maintain boilers and related equipment.*

## Licensing Examination Completion

There are a few principles that should be followed when taking any examination. General suggestions for completing the licensing examination include the following:

- Remain calm. The preparation necessary for the examination has been completed, and now, that preparation will ensure the expected results.
- After receiving the examination, assess the amount of time available to complete it. Scan the first few questions of each section to obtain a feel for the examination. Pace your work according to the time allotted for each section.
- Read each question and all answers carefully. Know what is being asked before attempting to answer. Take time to verify that the proper answer blank has been filled in completely.
- On answer sheets to be scanned by a machine, correctly fill in the space on the answer sheet as instructed. Erase any changes completely.
- If calculations are required, be sure to neatly show all work. Some examinations may allow partial credit for an incorrect final answer if the proper formula and procedures are shown.
- If you are unsure of an answer, lightly mark the question number on the answer sheet. When finished with the other questions, return to the unanswered question. If you are still unsure after checking the question again, eliminate the obvious incorrect answers. Make an educated guess using the remaining answers.
- Review the entire answer sheet for any missing information or answers, misplaced answers, or stray marks.

## SAMPLE LICENSING EXAMINATION QUESTIONS

Licensing examinations are generally developed and administered by a specific agency. In some cases, licensing examinations may be developed and administered by a third party. For example, ASOPE administers examinations for the city of Milwaukee. The number and type of licensing examination questions vary. The examination may include true-false, multiple choice, matching, and/or essay questions. Generally, higher levels of examination have greater emphasis on essay questions. The licensing examination may be given on scheduled dates or as the need arises in specific locations. Contact the licensing examination agency for specific dates and times.

The questions in this chapter can be used to practice taking a licensing examination. The following agencies, municipalities, provinces, and states have provided sample test questions that are included in this chapter.

Alaska Department of Labor and Workforce Development
Mechanical Inspection Section
Anchorage, AK

Province of Alberta
Municipal Affairs, Safety Services
Boilers and Pressure Vessels
Edmonton, AB

Arkansas Department of Labor
Boiler Inspection Division
Little Rock, AR

Block and Associates
Gainesville, FL

City of Dearborn
Department of Public Works
Building & Safety Division
Dearborn, MI

City of Elgin
Elgin, IL

The Commonwealth of Massachusetts
Department of Public Safety
Engineering Section, Division of Inspection
Boston, MA

City of Milwaukee
Department of Neighborhood Services
Boiler and Mechanical Safety
Milwaukee, WI

National Institute for the Uniform Licensing of Power
  Engineers, Inc. (NIULPE)

Ohio Department of Commerce
Division of Industrial Compliance
Bureau of Operations and Maintenance
Reynoldsburg, OH

City of Philadelphia
Department of Licenses and Inspections
Philadelphia, PA

Salt Lake City Corporation
Department of Building and Housing Services
Salt Lake City, UT

City of Sioux City
Board of Examiners of Mechanical Stationary Engineers
Inspection Services Division
Sioux City, IA

City of Terre Haute
Office of the Board of Examining Engineers
Terre Haute, IN

**Name** _____ **Date** _____

_____ 1. The bottom blowdown on a boiler ___.
   A. removes sludge and sediment from the mud drum
   B. reduces boiler steam pressure
   C. adds makeup water to the boiler
   D. increases boiler priming

_____ 2. In a(n) ___ boiler, the water circulates inside tubes that are surrounded by flue gases and the products of combustion.
   A. cast iron
   B. electric
   C. firetube
   D. watertube

_____ 3. A ___ is a system programmed for automatic burner sequencing and flame supervision.
   A. high-limit control
   B. burner management system
   C. modulating control
   D. flame sensor

_____ 4. Spalling in a boiler refers to ___.
   A. hairline cracks in the steam drum
   B. hairline cracks in the refractory
   C. slugs of water in the steam
   D. water in the fuel oil

_____ 5. Boiler horsepower is the evaporation of ___ lb/hr of water from and at a feedwater temperature of 212°F.
   A. 34.5
   B. 180
   C. 345
   D. 970.3

_____ 6. A siphon installed between the boiler and the pressure gauge protects the Bourdon tube from ___.
   A. water
   B. the gases of combustion
   C. steam temperature
   D. steam pressure

_____ 7. The pH of the boiler water should have a(n) ___ pH.
  A. acid
  B. neutral
  C. alkaline
  D. low

_____ 8. The boiler superheater raises the temperature of the ___.
  A. feedwater
  B. gases of combustion
  C. air for combustion
  D. steam leaving the boiler

_____ 9. An economizer is used to preheat ___.
  A. air for combustion
  B. fuel oil
  C. feedwater
  D. coal

_____ 10. Fires caused by spontaneous combustion would probably start in the ___.
  A. fuel oil tank
  B. oily waste rag storage
  C. boiler firebox
  D. boiler ash pit

_____ 11. A high suction reading on the fuel oil pressure gauge indicates ___.
  A. dirty strainers
  B. a high tank temperature
  C. dirty burners
  D. leaky pump packing

_____ 12. The automatic nonreturn valve should be opened during startup when the steam pressure on a boiler reaches ___.
  A. the line pressure
  B. 75% to 85% of the line pressure
  C. 5 lb to 10 lb above line pressure
  D. 5 lb to 10 lb below line pressure

_____ 13. A(n) ___ is used to measure furnace temperature.
  A. mercury thermometer
  B. pyrometer
  C. oxygen analyzer
  D. pressure gauge

_____ 14. The range of a boiler pressure gauge must not be less than ___ times the safety valve setting.
  A. 1½
  B. 2
  C. 2½
  D. 3

_____ **15.** Boiler feedwater is chemically treated to ___.
    A. increase circulation in the boiler
    B. increase oxygen concentration
    C. prevent formation of scale
    D. increase boiler makeup water

_____ **16.** The ___ opens when the gas pressure drops below a preselected minimum requirement, deenergizes the main gas valve, and energizes the vent valve.
    A. safety relief valve
    B. main gas valve
    C. high-gas-pressure switch
    D. low-gas-pressure switch

_____ **17.** In a firetube boiler, soot settles on the ___.
    A. inside tube surface
    B. outside tube surface
    C. waterwall surface
    D. lowest part of the water side

_____ **18.** Carbon dioxide in the flue gas is a sign of ___ combustion.
    A. imperfect
    B. complete
    C. incomplete
    D. half-complete

_____ **19.** A manometer measures ___.
    A. flue gas temperature
    B. volume of airflow
    C. atmospheric pressure
    D. difference in pressure between two points

_____ **20.** ___ lay-up is the storage of a boiler with all water drained.
    A. Dry
    B. Evaporative
    C. Hydrostatic
    D. Pressurized

_____ **21.** The main purpose of conducting a(n) ___ test is to verify the strength of the boiler shell and any parts under pressure.
    A. condensation
    B. evaporation
    C. hydrostatic
    D. try lever

_____ **22.** The pressure applied on the boiler during a hydrostatic test is ___ times the MAWP.
    A. 1½
    B. 2
    C. 2½
    D. 3

_____ **23.** Boiler safety valve capacity must be such that the steam pressure can never go higher than ___% above the MAWP.
    A. 2
    B. 4
    C. 6
    D. 10

_____ **24.** As fuel oil is heated, its viscosity ___.
    A. stays the same
    B. increases
    C. decreases
    D. ignites

_____ **25.** No. ___ fuel oil has the highest Btu content per gallon.
    A. 1
    B. 2
    C. 4
    D. 6

T    F    **26.** All steam boilers must have one or more safety valves.

T    F    **27.** An open feedwater heater vents gases to the atmosphere.

T    F    **28.** A hydrostatic test will determine whether a steam boiler has sufficient relieving capacity.

T    F    **29.** When boilers are set in battery, the main steam stop valves, bottom blowdown valves, and feedwater valves should be locked closed before anyone enters the boiler drums.

T    F    **30.** A programmer is a boiler system that modulates the flow of air to and from a burner.

T    F    **31.** In an open feedwater heater, the water and steam come into direct contact.

T    F    **32.** An economizer is a fuel-saving device that reclaims heat from the gases of combustion.

T    F    **33.** Soot accumulation on the tubes of a watertube boiler does not affect the efficiency of the boiler.

T    F    **34.** Scale protects the boiler heating surfaces.

T    F    **35.** Boiler corrosion cannot be prevented as long as water is in contact with metal.

T    F    **36.** Priming can cause water hammer.

T    F    **37.** Dry saturated steam is in a gaseous state.

T    F    **38.** Pressure on water does not change the boiling point of the water.

T    F    **39.** During cold startup, a boiler must be warmed slowly to prevent uneven expansion of the boiler and piping.

T    F    **40.** The primary function of a gas solenoid valve is to shut down the burner if the water level drops below a safe operating level.

T    F    **41.** The interior of the Bourdon tube in a steam pressure gauge is always filled with live steam when it is operating.

T    F    **42.** Safety valves on superheaters should be set to open before main safety valves on boilers.

T    F    **43.** An operating pressure control shuts down the burner when the steam pressure increases to the setpoint.

T    F    **44.** Waterwalls permit greater heat absorption per cubic foot of furnace volume than refractory walls.

T    F    **45.** A steam trap is a device that removes air and condensate without loss of steam.

T    F    **46.** Steam traps are only used on high-temperature hot water systems.

T    F    **47.** Steam traps are used wherever condensate collects.

T    F    **48.** A boiler feedwater pump cannot become steambound.

T    F    **49.** In both firetube and watertube boilers, the hot gases of combustion are inside the tubes.

T    F    **50.** Foaming can lead to carryover.

**Name** _____ **Date** _____

_____ 1. ___ makes a pop-type safety valve pop open.
A. Steam entering the huddling chamber exposing a larger area for the steam pressure to act on
B. Weakening of the spring by the steam entering the valve
C. The momentum of the steam through the valve into the huddling chamber
D. Increasing steam temperature

_____ 2. Manual testing of the safety valve is accomplished by ___.
A. removing and sending it out to a repair shop to have it tested
B. closing the main steam stop valve and letting the pressure build to its setpoint
C. pulling the test lever by hand and then releasing it to allow it to snap closed
D. securing the feedwater to allow the water level to fall below the NOWL

_____ 3. A(n) ___ prevents fuel from entering the furnace of the boiler and building up when the burner is not lit.
A. low-water fuel cutoff
B. operating control
C. safety valve
D. flame scanner

_____ 4. To ensure good heat transfer in a watertube boiler, the outside of the tubes are kept clean by ___.
A. soot blowers
B. cold lime water softeners
C. ion-exchange water softeners
D. bottom and continuous blowdown

_____ 5. When beginning a bottom blowdown, it is important to ___.
A. measure the total dissolved solids
B. maintain the water at the NOWL
C. measure the feedwater temperature
D. set the burner on high fire

_____ 6. When economizers are used to heat boiler feedwater, every 10°F rise in feedwater temperature is accompanied by a ___% savings in fuel.
A. 1
B. 10
C. 25
D. 50

_____ 7. Proper safety procedure requires that ___ before anyone is allowed to enter the steam drum.
     A. safety valves be tagged
     B. the inspector be notified
     C. a safety light be made available
     D. the boiler be properly locked out

_____ 8. Impulse steam turbines use ___ as a force acting in a forward direction on a blade or on a bucket mounted on a wheel.
     A. saturated steam
     B. low-quality steam
     C. deaerated feedwater
     D. steam velocity

_____ 9. The ___ must be tested when the boiler is on-line.
     A. combustion air proving switch
     B. oil drawer proving switch
     C. safety valve
     D. low-gas-pressure switch

_____ 10. The induced draft fan is located between the ___ and ___.
     A. feedwater pump; heater
     B. first pass; heater
     C. boiler; economizer
     D. boiler; stack

_____ 11. The flash point of fuel oil is the minimum temperature at which the fuel oil will ___.
     A. support combustion
     B. flash and no longer flow
     C. flash when exposed to an open flame
     D. have its highest Btu content

_____ 12. The fire point of fuel oil is the minimum temperature at which the fuel oil will ___.
     A. burn continually
     B. no longer flow
     C. flash when exposed to an open flame
     D. have its highest Btu content

_____ 13. The pour point of fuel oil is the ___ temperature at which fuel oil will ___.
     A. lowest; burn
     B. lowest; flow
     C. highest; burn
     D. highest; flow

_____ 14. Fuel oil with a low flash point is ___.
     A. used with high pressure boilers
     B. dangerous to handle
     C. only used in low pressure plants
     D. heated to increase its viscosity

_____ **15.** ___ dioxide is a major source of air pollution and is caused by burning fuel oil.
  A. Nitrous
  B. Vanadium
  C. Hydrogen
  D. Sulfur

_____ **16.** In a closed feedwater heater, the steam and water ___.
  A. do not come into direct contact
  B. only mix on the discharge side
  C. only mix on the feedwater side
  D. are not used at the same time

_____ **17.** The operating range of a steam boiler is controlled by a(n) ___.
  A. operating pressure control
  B. airflow interlock
  C. scanner
  D. program clock

_____ **18.** Water hammer in steam lines is caused by ___.
  A. low steam pressure
  B. high steam pressure
  C. condensation in the line
  D. a sudden drop in plant load

_____ **19.** Carbon monoxide in the flue gas is a sign of ___ combustion.
  A. perfect
  B. complete
  C. incomplete
  D. half-complete

_____ **20.** A(n) ___ area is a location where Class A and Class B flammables exceed those expected in ordinary hazard areas.
  A. extra hazard
  B. combustibles
  C. flammable
  D. fire prevention

_____ **21.** In boiler room work, HRT means ___.
  A. highly regulated therm
  B. high return tubular
  C. hot return tubular
  D. horizontal return tubular

_____ **22.** The safety valves on a boiler drum and superheater safety valves of a steam boiler are set so that ___.
  A. both drum safety valves open together
  B. only one drum safety valve opens
  C. the superheater safety valve opens first
  D. one drum safety valve opens and then the superheater safety valve opens

_____ **23.** The best time to blow down a boiler is when it is ___.
    A. at its peak load
    B. at 75% of its peak load
    C. at its lightest load
    D. being taken out of service

_____ **24.** The frequency of blowing down a boiler can best be determined by ___.
    A. the boiler operator
    B. the boiler inspector
    C. the blowdown flow meter
    D. a boiler water analysis

_____ **25.** In most plants, the suggested operating procedure is to blow down the water column and gauge glass ___.
    A. at least once a week
    B. at least once a shift
    C. only when taking the boiler off-line
    D. except when the boiler is under pressure

T    F    **26.** Sodium sulfite is used in boiler water to prevent pitting of the boiler metal.

T    F    **27.** An airtight boiler room does not allow enough air for complete combustion.

T    F    **28.** If the top line to a boiler gauge glass were closed, the water in the glass would rise to the top.

T    F    **29.** The boiling point of water is not affected by boiler pressure.

T    F    **30.** Priming can be a direct result of maintaining too high a water level in the boiler.

T    F    **31.** In a closed feedwater heater, steam and water come into direct contact with each other.

T    F    **32.** The automatic nonreturn valve is on the main steam line.

T    F    **33.** An atomizing burner can be designed to use steam to atomize fuel oil.

T    F    **34.** A combustible material is any material that burns when it is exposed to oxygen and heat.

T    F    **35.** A high-pressure cutoff control can take the place of a safety valve.

T    F    **36.** Both stop and check valves must be installed on the boiler feedwater line.

T    F    **37.** Try cocks on a water column can be used to determine the water level if the gauge glass is broken.

T    F    **38.** Boiler explosions would not occur if all boilers had an MAWP of 15 psi.

T    F    **39.** The purpose of safety valves is to protect the boiler from exceeding its MAWP.

T    F    **40.** The duty rating of a boiler is the ratio of the heat absorbed by the boiler (output) to the heat available in the fuel (input).

T    F    **41.** An energy audit of a facility identifies how energy is used in a facility and recommends ways to improve energy efficiency and reduce energy costs.

T  F  **42.** The low-water fuel cutoff shuts down the burner when the steam pressure increases above the setpoint of the operating pressure control.

T  F  **43.** In a pop-type safety valve, the huddling chamber assists in opening the valve.

T  F  **44.** If a safety valve leaks after testing, the valve should be opened again using the test lever.

T  F  **45.** The low-water fuel cutoff control shuts off the fuel to a burner when a low-water condition occurs.

T  F  **46.** A low-water fuel cutoff control should be tested by blowing down the float chamber at least once a shift.

T  F  **47.** The water column reduces the turbulence of the water to give a more accurate reading of the water level.

T  F  **48.** Boiler blowdown tanks do not require vents because the water is discharged to the sewer.

T  F  **49.** Blowing down the water column and gauge glass too frequently can cause a false water level reading.

T  F  **50.** A false water level reading can occur if the water column or gauge glass blowdown valves are not fully closed.

**Name** _____ **Date** _____

_____  1. Boilers that are laid up dry have trays of ___ put in the steam drums to absorb moisture.
    A. sodium chloride
    B. hot soda lime
    C. silica gel
    D. potash

_____  2. To obtain complete combustion of a fuel ___ are required.
    A. mixture, atomization, temperature, and time
    B. $CO_2$, $O_2$, CO, and $CO_3$
    C. turbulence, feedwater, air, and flue gas
    D. hot refractory, air, draft, and gases of combustion

_____  3. Oxygen in the boiler causes ___.
    A. scale
    B. pitting
    C. foaming
    D. carryover

_____  4. A ___ valve is a manual valve that is typically used to isolate the feedwater to the boiler or the steam discharge from the boiler.
    A. pressure-reducing
    B. low-water
    C. relief
    D. stop

_____  5. Scale buildup in the tubes of a watertube boiler can be removed by ___.
    A. washing down the boiler with high pressure water
    B. a steam-driven or water-driven turbine
    C. wire brushes
    D. boiling out the boiler with caustic soda

_____  6. When no boiler water analysis is made, the boiler should be blown down at least once ___.
    A. every 8 hours
    B. every 24 hours
    C. a week
    D. a month

_____ **7.** When the boiler low-water alarm is sounding, the operator should ___.
    A. call the chief engineer
    B. increase the firing rate
    C. decrease the draft
    D. secure the fuel

_____ **8.** A ¾″ blowdown line may be used if the boiler has ___ sq ft of heating surface.
    A. more than 100
    B. no more than 100
    C. less than 199
    D. more than 500

_____ **9.** A superheater drain ___ the superheater during startup and shutdown.
    A. removes gases from
    B. removes pressure from
    C. establishes airflow through
    D. establishes steam flow through

_____ **10.** An infrared sensor ___.
    A. can be used as a flame scanner
    B. detects smoke
    C. measures pressure
    D. prevents low water

_____ **11.** A furnace explosion can be prevented by ___.
    A. checking the water level once a shift
    B. testing the safety valves once a month
    C. purging after ignition failure
    D. testing the safety valve regularly

_____ **12.** The bottom blowdown valve located farthest from a boiler is always a ___ valve.
    A. quick-opening
    B. slow-opening
    C. check
    D. sealing

_____ **13.** A lead sulfide cell, which can be used for flame detection, is sensitive to ___.
    A. infrared light
    B. light
    C. heat
    D. temperature increases

_____ **14.** An economizer heats ___.
    A. steam
    B. air for combustion
    C. fuel oil
    D. feedwater

_____ **15.** A combustion air proving switch verifies that there is sufficient ___.
        A. gas flow at the pilot
        B. air for combustion
        C. steam flow
        D. fuel oil pressure

_____ **16.** The Btu content of natural gas is ___ Btu per ___.
        A. 100,000; therm
        B. 100,000; cubic foot
        C. 152,000; therm
        D. 152,000; cubic foot

_____ **17.** Chemicals are added to boiler water by ___.
        A. the feedwater pump
        B. the vacuum pump
        C. the proportional chemical pump
        D. mixing in the reservoir

_____ **18.** The boiler feedwater stop valve is between the boiler and the ___.
        A. check valve
        B. feedwater pump
        C. feedwater control valve
        D. main header

_____ **19.** A pop-type safety valve is a ___ type of safety valve.
        A. counterweighted
        B. pin and disc
        C. deadweight
        D. spring-loaded

_____ **20.** An economizer is used to heat feedwater before it enters a boiler by using ___.
        A. a gas fired preheater
        B. exhaust steam from a turbine
        C. exhaust gases from a gas or diesel engine
        D. gases of combustion as they leave the boiler

_____ **21.** Tubes in a firetube boiler are ___.
        A. rolled and beaded
        B. welded
        C. collared and peened
        D. bolted and nutted

_____ **22.** When burning fuel oil with a firetube boiler, soot can settle on the ___.
        A. stays
        B. waterwalls
        C. outside tube surface
        D. inside tube surface

_____ **23.** Pressure gauges are calibrated with a(n) ___.
    A. pyrometer
    B. Orsat analyzer
    C. lightweight tester
    D. electronic calibrator

_____ **24.** To test a low-water fuel cutoff using an evaporation test, the boiler operator must ___.
    A. open the feedwater stop valve
    B. secure all of the feedwater going to the boiler
    C. completely drain the regulator chamber
    D. have an inspector on hand

_____ **25.** A blowdown flash tank ___.
    A. acts as a separator
    B. recovers condensate
    C. controls blowdowns
    D. recovers low-pressure steam

T    F    **26.** A closed boiler room that does not permit the proper flow of outside air to the furnace will experience combustion problems.

T    F    **27.** All firetube boilers are horizontal boilers.

T    F    **28.** Weighted lever-type safety valves can be used on steam boilers that operate at pressures not exceeding 50 psi.

T    F    **29.** Steam boilers that are out of service can be laid up wet or dry.

T    F    **30.** The dry method of boiler lay-up requires the boiler to be opened and air to be circulated to keep the boiler drums and tubes open to the air.

T    F    **31.** Chemical treatment of the boiler water is not necessary when laying a boiler up wet if the water has been deaerated.

T    F    **32.** A flash point is the lowest temperature at which the vapor of a fuel oil ignites when exposed to an open flame.

T    F    **33.** The fire point of fuel oil is the minimum temperature at which the fuel oil burns continuously.

T    F    **34.** No. 6 fuel oil burns with a clean flame when it is not heated.

T    F    **35.** Poor grades of No. 6 fuel oil can be used quite well with a steam atomizing burner.

T    F    **36.** A check valve is a manual valve that is typically used to isolate the feedwater to the boiler or the steam discharge from the boiler.

T    F    **37.** A steam pressure gauge is typically calibrated in pounds per square foot.

T    F    **38.** The Bourdon tube of a steam pressure gauge has to be protected from the pressure.

T    F    **39.** The presence of soot on the heating surfaces of a boiler increases efficiency.

T    F    **40.** Gauge pressure is pressure below atmospheric pressure.

T    F    **41.** A dry pipe separator is in a steam drum.

T    F    **42.** A dry pipe separator increases the quality of steam leaving a steam boiler.

T    F    **43.** Excess air for combustion is indicated by testing for $CO_2$ in the gases of combustion.

T    F    **44.** A change in steam pressure will change the boiling point of water.

T    F    **45.** Boiler tube dimensions are based on the inside diameter of the tube.

T    F    **46.** The safety valve on a superheater header should be set to a lower popping pressure than the safety valve on the steam drum.

T    F    **47.** A globe valve offers less resistance to flow than a gate valve.

T    F    **48.** To control the flow of steam, a gate valve is opened halfway.

T    F    **49.** A large heat loss in a boiler is caused by heat being carried with the gases of combustion to the atmosphere.

T    F    **50.** The bottom blowdown valves can be used to lower the boiler steam pressure.

**Name** _____   **Date** _____

_____  **1.** In a rotary cup burner, atomization is achieved by ___.
   A. fuel oil pressure
   B. air pressure
   C. the pressurized burner
   D. the rotary cup and primary air

_____  **2.** To prevent a vacuum from forming on a boiler that is coming off-line, the ___.
   A. boiler vent should be opened
   B. boiler should be blown down
   C. boiler should be dumped
   D. safety valve should be opened

_____  **3.** The ___ is a law that was passed to allow for monitoring and controlling environmental air emissions.
   A. Ozone Emissions Act (OEA)
   B. Clean Air Act (CAA)
   C. New Source Performance Standards (NSPS)
   D. Particulate Matter Plan (PMP)

_____  **4.** When warming up a boiler during cold plant startup, the ___ should remain open until the steam pressure reaches approximately 10 psi to 25 psi.
   A. safety valve
   B. boiler vent
   C. nonreturn valve
   D. breeching damper

_____  **5.** High pressure boilers equipped with quick-opening and slow-opening valves are blown down by opening the ___ valve first and closing it ___.
   A. quick-opening; last
   B. slow-opening; last
   C. quick-opening; first
   D. slow-opening; first

_____  **6.** ___ are designed to collect condensate from the steam system so that it can be returned to the boiler through the condensate return tank.
   A. Flash tanks
   B. Steam traps
   C. Bottom blowdown lines
   D. Safety valves

_____ 7. To ensure good heat transfer in a firetube boiler, the outside of the tubes are kept clean by ___.
  A. soot blowers
  B. recommended feedwater treatment
  C. daily surface blowdowns
  D. overfire air

_____ 8. The viscosity of fuel oil is the measurement of the ___ in the fuel oil.
  A. Btu content
  B. fire point
  C. flash point
  D. internal resistance to flow

_____ 9. The maximum size of a boiler bottom blowdown line is ___″.
  A. 1½
  B. 2
  C. 2½
  D. 3

_____ 10. A ___ valve is a manual valve that is typically used to isolate the feedwater to the boiler or the steam discharge from the boiler.
  A. pressure-reducing
  B. low-water
  C. relief
  D. stop

_____ 11. Any repairs done on a boiler safety valve must be done by a ___.
  A. boiler inspector
  B. qualified repair technician
  C. watch engineer
  D. shift engineer

_____ 12. Feedwater is treated chemically before it enters the boiler to ___.
  A. prevent foaming
  B. eliminate blowing down the boiler
  C. allow more frequent blowdown
  D. reduce the boiling temperature

_____ 13. Soot blowers are most commonly found ___.
  A. on an HRT boiler
  B. on a watertube boiler
  C. in main breeching
  D. on vertical firetube boilers over 150 HP

_____ 14. In a furnace, the amount of $NO_x$ decreases as the ___.
  A. combustion air flow decreases
  B. amount of FGR decreases
  C. combustion temperature decreases
  D. combustion temperature increases

_____ **15.** Complete combustion is defined as burning all fuel using ___.

      A. the theoretical amount of air

      B. the minimum amount of excess air

      C. no excess air

      D. $CO_2$ and CO

_____ **16.** When testing a safety valve by hand, ___.

      A. there should be no pressure in the boiler

      B. notify the boiler inspector

      C. the boiler should have a pressure of at least 75% of the popping pressure

      D. first secure the fires

_____ **17.** The absolute pressure of a steam boiler at 160 psig would be ___ psia.

      A. 145.3

      B. 174.7

      C. 200

      D. 320

_____ **18.** The ___ analysis test is used to test the density of smoke leaving a boiler.

      A. proximate

      B. opacity

      C. Orsat

      D. Fyrite

_____ **19.** An automatic nonreturn valve is found on the ___.

      A. discharge side of the feedwater pump

      B. inlet line to the open feedwater heater

      C. boiler main steam line between the main stop valve and the header

      D. boiler main steam line between the boiler and the main stop valve

_____ **20.** A boiler steam pressure of 100 psi with a temperature of 500°F indicates ___.

      A. dry steam

      B. supersaturated steam

      C. superheated steam

      D. steam at its corresponding pressure and temperature

_____ **21.** A ___ regulates fuel supply, air supply, air-fuel ratio, and draft in a boiler in order to deliver the required amount of steam to a load.

      A. programmer

      B. firing-sequence controller

      C. flame scanner

      D. combustion control system

_____ **22.** To aid in the delivery of moisture-free steam, the steam leaves the steam drum through the ___.

      A. dry pipe

      B. surface blowdown valve

      C. desuperheater

      D. internal feedwater pipe

_____ **23.** A broken baffle in a watertube boiler causes ___.
        A. smoke to come from the stack
        B. a rise in stack temperature
        C. a drop in stack temperature
        D. steam to come from the stack

_____ **24.** The water column on a high pressure steam boiler ___.
        A. reduces the fluctuation in the water level to prevent carryover
        B. reduces the turbulence of the water in the gauge glass
        C. provides a place to install the safety valves
        D. provides a place to mount the fuel oil pump

_____ **25.** Safety valves on high pressure boilers can be tested ___.
        A. only by hand
        B. only by pressure
        C. by hand or pressure
        D. automatically

T     F     **26.** One of the purposes of a bottom blowdown valve is to remove sludge from a boiler.

T     F     **27.** The main function of a steam trap is to remove gases and condensate without the loss of steam.

T     F     **28.** A therm of natural gas has 10,000 Btu.

T     F     **29.** In a natural draft system, the lowest draft reading is obtained at the furnace.

T     F     **30.** Burning fuel produces a chemical reaction that releases heat.

T     F     **31.** The main function of a continuous blowdown system is to reduce the total concentration of solids in the boiler.

T     F     **32.** An evaporation test is a test in which the boiler operator allows the water level in the boiler to drop in order to test the proper operation of the feedwater pump.

T     F     **33.** A throttling calorimeter measures the excess air being used for combustion.

T     F     **34.** When conducting a hydrostatic test, all safety valves are either gagged or blanked, and the water pressure is increased to 1½ times the MAWP.

T     F     **35.** Boiler drum safety valves must open before the superheater safety valve to ensure a backflow.

T     F     **36.** Standard atmospheric pressure will support a column of mercury approximately 30″ high.

T     F     **37.** A flow meter totalizes the temperature and pressure of the steam.

T     F     **38.** Purging a furnace before firing can prevent a furnace explosion.

T     F     **39.** Try cocks are most effective at a steam pressure below 250 psi.

T     F     **40.** A pyrometer measures the amount of moisture in the steam leaving a boiler.

T  F  **41.** The forced and induced draft fans on a boiler move the same volume of gas.

T  F  **42.** A high pressure steam boiler has a pressure of more than 15 psi.

T  F  **43.** The size of a boiler tube is determined by the inside diameter of the tube.

T  F  **44.** Superheating steam is possible only at high steam pressures.

T  F  **45.** Preheaters heat air for combustion.

T  F  **46.** An economizer is a heat exchanger that absorbs heat from the gases of combustion leaving a boiler.

T  F  **47.** A desuperheater is only used with saturated steam.

T  F  **48.** Safety valves are located on the uppermost part of the steam space and the outlet of the superheater header.

T  F  **49.** Cavitation occurs when water leaves a steam drum along with the steam.

T  F  **50.** An adequate supply of air is the only thing needed for good combustion.

**Name** _____ **Date** _____

_____  **1.** Before laying up a boiler, the boiler operator must ___.
- A. notify the boiler inspector
- B. remove the boiler certificate
- C. thoroughly clean the fire and water sides
- D. clean only the water side because soot acts as an insulator

_____  **2.** Neutral on the pH scale is ___.
- A. 0
- B. 4
- C. 7
- D. 14

_____  **3.** ___ is a type of particulate matter (PM) consisting of noncombustible material found in the gases of combustion.
- A. Combustion air
- B. Pulverized coal
- C. Fly ash
- D. Induced carbon

_____  **4.** A ___ can be used as part of a draft control system to adjust the speed of a fan.
- A. damper linkage
- B. variable-speed drive
- C. modulating motor starter
- D. jackshaft

_____  **5.** According to the ASME Code, the setting or adjusting of a boiler safety valve popping point must be done by a ___.
- A. boiler inspector
- B. plant manager
- C. watch engineer
- D. qualified repair technician

_____  **6.** ___ is produced from incomplete combustion due to poor burner design or firing conditions.
- A. Carbon monoxide
- B. Nitrogen oxide
- C. Sulfur
- D. Mercury

_____   **7.** A scotch marine boiler is a ___ boiler.
  A. watertube
  B. cast iron
  C. cast iron firetube
  D. firetube

_____   **8.** A boiler that has more than ___ sq ft of heating surface must have two or more safety valves.
  A. 100
  B. 300
  C. 500
  D. 1000

_____   **9.** A(n) ___ test is used to test the relieving capacity of a safety valve.
  A. hydrostatic
  B. accumulation
  C. Orsat analysis
  D. opacity

_____   **10.** An open feedwater heater is located ___.
  A. above the feedwater pump on the discharge side of the pump
  B. above the feedwater pump on the suction side of the pump
  C. below the feedwater pump
  D. between the feedwater pump and the boiler

_____   **11.** A draft system regulates the flow of ___ through a boiler.
  A. air
  B. steam
  C. fuel oil
  D. condensate

_____   **12.** To start a turbine pump, the discharge valve must be ___.
  A. closed
  B. opened
  C. primed
  D. hydrostatically closed

_____   **13.** When performing a hydrostatic test on a boiler, the pressure is ___.
  A. 75% of the safety valve setting
  B. 1½ times the safety valve setting
  C. 2 times the safety valve setting
  D. operating pressure plus atmospheric pressure

_____   **14.** The float feedwater regulator element is located ___.
  A. at the NOWL
  B. on the right side of the steam drum
  C. on the left side of the steam drum
  D. on the feedwater heater

_____ **15.** At a water temperature of 212°F, 1 BHP is equivalent to the evaporation of ___ lb/hr of water.
  A. 34.5
  B. 62.5
  C. 180
  D. 970.3

_____ **16.** The valves located between the boiler and the water column are ___ valves and must be locked and sealed open.
  A. flow control
  B. os&y or lever
  C. stop
  D. nonreturn

_____ **17.** To cut a boiler in on a line equipped with hand-operated main steam stop valves, the steam pressure on the boiler coming on-line should be ___.
  A. slightly lower than line pressure
  B. slightly higher than line pressure
  C. equal to line pressure
  D. no different because the pressure will equalize

_____ **18.** The suggested range of a boiler steam pressure gauge is ___.
  A. 1½ to 2 times the safety valve setting
  B. 2 times the MAWP
  C. not more than 6% above the MAWP
  D. gauge pressure plus atmospheric pressure

_____ **19.** An evaporation test is a test in which the boiler operator allows the water level in a boiler to drop in order to test the proper operation of the ___.
  A. low-water fuel cutoff
  B. water column
  C. main steam stop valve
  D. safety valve

_____ **20.** With all of the safety valves popping, the boiler pressure should not go higher than ___.
  A. 1½ times the MAWP
  B. 6% above the MAWP
  C. 8% above the set pressure
  D. 10% above the set pressure

_____ **21.** Boiler water containing foreign material can cause an increase in ___.
  A. foaming
  B. soot
  C. flame failure
  D. safety valve operation

_____ 22. A ___ ensures that a burner will not start until all fuel safety equipment is proven.
  A. furnace purge
  B. flame scanner
  C. burner management system
  D. modulating cam

_____ 23. Pressure at the discharge side of a forced draft fan is ___ atmospheric pressure.
  A. greater than
  B. less than
  C. the same as
  D. balanced at

_____ 24. A safe maximum water concentration of dissolved solids is maintained by ___.
  A. daily blowdowns
  B. bottom blowdown
  C. surface blowdowns
  D. increased makeup water

_____ 25. Pyrometers measure ___.
  A. air duct velocity
  B. draft
  C. superheater flow
  D. temperature

_____ 26. A ___ is a rotary mechanical device used to drive rotating equipment, such as a generator, by extracting thermal energy from pressurized steam.
  A. steam turbine
  B. gear pump
  C. cavitation pump
  D. steam condenser

_____ 27. A pressure control maintains the ___ at or near the setpoint.
  A. pressure
  B. purge time
  C. firing rate
  D. duty cycle

_____ 28. An atomizing burner can use ___ to atomize fuel oil.
  A. water
  B. gas
  C. vacuum
  D. steam

_____ 29. ___ is the difference in pressure between two points of measurement that causes air or gases of combustion to flow.
  A. Draft
  B. Priming
  C. Flow zone
  D. Carryover

_____ **30.** According to the ASME Code, if a boiler has an MAWP that exceeds 100 psi, the boiler must have ___ blowdown valve(s).
    A. zero
    B. one
    C. two
    D. three

_____ **31.** The minimum size pipe connecting a water column to a boiler is ___″.
    A. ½
    B. 1
    C. 1½
    D. 2

_____ **32.** No. 6 fuel oil has ___ viscosity compared to No. 2 fuel oil.
    A. no
    B. a high
    C. a medium
    D. a low

_____ **33.** A(n) ___ steam turbine allows condensate to be reclaimed for reuse.
    A. atmospheric
    B. saturated
    C. condensing
    D. noncondensing

_____ **34.** An evaporation test should be performed ___ to check the low-water fuel cutoff control.
    A. daily
    B. weekly
    C. monthly
    D. annually

_____ **35.** If a boiler has a heating surface of more than 100 sq. ft, the blowdown lines and fittings must be at least ___″ and not more than 2½″.
    A. ½
    B. ¾
    C. 1
    D. 1½

_____ **36.** Boilers taken off-line for cleaning and inspection may be dumped (emptied) when the ___.
    A. boiler pressure is 50 psi or higher
    B. boiler has cooled
    C. boiler water temperature is 175°F or higher
    D. automatic nonreturn valve cuts the boiler off-line

_____ **37.** A lead sulfide cell is sensitive to ___.
    A. pressure
    B. temperature
    C. electricity
    D. infrared light

_____ **38.** Anthracite coal has a ___ content.
   A. high volatile
   B. low hydrogen
   C. high fixed carbon
   D. high ash

_____ **39.** Bituminous coal has a high ___ content.
   A. volatile
   B. hydrogen
   C. fixed carbon
   D. ash

_____ **40.** For the efficient combustion of pulverized coal, it is necessary to maintain ___.
   A. 50% excess air
   B. a high furnace temperature
   C. a low furnace temperature
   D. positive furnace pressure

_____ **41.** The surface blowdown line on a steam boiler is located ___.
   A. at the lowest visible part of the gauge glass
   B. on the mud drum
   C. at the highest part of the steam drum
   D. near the NOWL

_____ **42.** A dry pipe ___.
   A. removes moisture from the steam
   B. connects the dry drum to the steam drum
   C. increases water circulation
   D. acts as a receiver

_____ **43.** The function of the fuel oil burner is to ___.
   A. preheat the fuel oil
   B. spray fuel oil into the furnace
   C. circulate the fuel oil
   D. pump fuel oil back to the fuel oil tank

_____ **44.** When performing a hydrostatic test on a boiler, the safety valves must be ___.
   A. kept free to pop at set pressure
   B. set at 1½ times the MAWP
   C. tested at the same time
   D. gagged or removed, and the openings must be blank-flanged

_____ **45.** In a three-element feedwater regulating system, ___ are measured.
   A. steam flow, fuel flow, and feedwater flow
   B. feedwater flow, water level, and condensate return
   C. makeup water, condensate return, and water level
   D. steam flow, feedwater flow, and water level

_____ **46.** The check valve on the boiler feedwater line ___.

A. prevents backflow from the boiler into the feedwater line

B. allows the stop valve to be repaired without dumping the boiler

C. allows the water level in the boiler to equalize

D. prevents the bottom blowdown from backing into the feedwater lines

_____ **47.** On boilers that have both quick-closing and slow-opening blowdown valves, the quick-closing valve must be located ___.

A. after the slow-opening valve

B. at the NOWL

C. closest to the boiler

D. on the bottom blowdown tank

_____ **48.** An automatic feedwater regulator ___.

A. warns of high or low water

B. maintains a constant water level

C. shuts the burner off in the event of low water

D. adds makeup water automatically

_____ **49.** A common type of steam pressure gauge operates by the movement of a(n) ___.

A. safety valve

B. Bourdon tube

C. os&y valve

D. try lever

_____ **50.** A(n) ___ sensor in a flame scanner senses visible light from the burner.

A. lead sulfide

B. photocell

C. sodium sulfite

D. ionic

**Name** _____ **Date** _____

_____ **1.** A pH of 11 indicates ___ water.
    A. alkaline
    B. neutral
    C. acidic
    D. hard

_____ **2.** The primary function of the ___ is to shut down the burner if the water level in the boiler drops below a safe operating level.
    A. water flow proving switch
    B. low-water fuel cutoff
    C. safety valve
    D. backflow preventer

_____ **3.** Foaming may be caused by ___.
    A. a high water level
    B. a low water level
    C. a leaking surface blowdown valve
    D. impurities on the boiler water surface

_____ **4.** Fuel, heat, and ___ are required to sustain combustion.
    A. oxygen
    B. carbon
    C. gas
    D. nitrogen

_____ **5.** The Bourdon tube in a steam pressure gauge is protected by a ___.
    A. steam trap
    B. siphon
    C. steam strainer
    D. stopcock

_____ **6.** A dry pipe on a boiler ___.
    A. removes moisture from steam
    B. supplies adequate makeup water
    C. maintains an NOWL
    D. is used in the event of a fire

_____ 7. A ___ valve is a valve that permits fluid flow in only one direction and closes automatically to prevent backflow.
   A. globe
   B. check
   C. butterfly
   D. gate

_____ 8. ___ blowdown is performed to remove sludge and sediment and control high water.
   A. Continuous
   B. Surface
   C. Bottom
   D. Feedwater

_____ 9. Scale deposits on the heating surfaces in a boiler can cause ___.
   A. a higher rate of heat transfer
   B. overheating of the boiler metal
   C. carryover
   D. purging

_____ 10. A deaerating feedwater heater removes dissolved gases such as oxygen and $CO_2$ from the feedwater that can cause ___.
   A. corrosion
   B. foaming
   C. carryover
   D. pumps to become vaporbound

_____ 11. A steam boiler is blown down to ___.
   A. reduce the amount of dissolved oxygen
   B. test the safety valve
   C. remove sludge and sediment
   D. clean the blowdown lines

_____ 12. An air heater employs a counterflow principle in which the gases of combustion ___.
   A. and air move in opposite directions
   B. and air move in the same direction
   C. bypass the flow meter
   D. and air mix together

_____ 13. Boiler operating data and maintenance records are commonly recorded in a ___.
   A. logbook
   B. maintenance planner
   C. combustion control system
   D. consultant report

_____ 14. Mechanical draft is produced by ___.
   A. the height of the stack
   B. the diameter of the stack
   C. power-driven fans
   D. steam jets

_____ **15.** A scotch marine boiler has ___.
      A. an internal furnace
      B. an external furnace
      C. a brick setting
      D. water tubes

_____ **16.** The tubes in a straight-tube watertube boiler are ___ into drums, headers, or sheets to remain stationary.
      A. screwed
      B. rolled and flanged
      C. expanded and flared
      D. welded and bolted

_____ **17.** A compound pressure gauge indicates ___.
      A. differential pressure
      B. pressure and vacuum
      C. absolute pressure
      D. the sum of the steam pressures on two boilers

_____ **18.** A(n) ___ valve prevents a boiler from exceeding its MAWP.
      A. main steam stop
      B. safety
      C. os&y
      D. globe

_____ **19.** ___ coal has the highest amount of fixed carbon.
      A. Bituminous
      B. Anthracite
      C. Subbituminous
      D. Semianthracite

_____ **20.** ___ lay-up is used when the boiler is expected to be out of service for an extended period.
      A. Wet
      B. Dry
      C. Bottom
      D. Surface

_____ **21.** A ___ test is used to determine if a safety valve and discharge piping are capable of discharging steam.
      A. try lever
      B. steam flow
      C. hydrostatic
      D. pressure-reducing

_____ **22.** Soot on the outside of boiler tubes is removed by a ___.
      A. steam turbine
      B. water turbine
      C. soot blower
      D. wire brush and vacuum cleaner

_____ 23. Overheating of tubes on a watertube boiler can be caused by ___.
    A. excessive soot in the tubes
    B. the water level being too high
    C. the water circulation being too rapid
    D. scale buildup inside the tubes

_____ 24. A high pressure boiler has an MAWP of more than ___ psi.
    A. 1
    B. 15
    C. 100
    D. 212

_____ 25. A(n) ___ boiler comes completely assembled with its own feedwater system, fuel system, combustion control, and draft system.
    A. package
    B. electric
    C. field-erected
    D. cast iron

_____ 26. ___ combustion occurs when all the fuel is burned using the minimum amount of excess air.
    A. Perfect
    B. Imperfect
    C. Incomplete
    D. Complete

_____ 27. The most common type of pump used as a feedwater pump is a(n) ___ pump.
    A. centrifugal
    B. steam
    C. reciprocating
    D. injector

_____ 28. Sludge and sediment from the mud drum of a boiler are discharged into a ___.
    A. condensate tank
    B. settling tank
    C. blowdown tank
    D. river or pond

_____ 29. The atomization of fuel oil in a rotary cup burner is caused by the ___.
    A. steam atomizer
    B. rotating cup and primary air
    C. pressure of the fuel oil
    D. secondary and primary air

_____ 30. A(n) ___ boiler is a boiler in which heat and gases of combustion pass through tubes surrounded by water.
    A. cast iron
    B. electric
    C. watertube
    D. firetube

_____ **31.** ___ air is used to atomize fuel oil or convey pulverized coal and control the rate of combustion, thus determining the amount of fuel that can be burned.
    A. Complete
    B. Incomplete
    C. Secondary
    D. Primary

_____ **32.** Ultimate analysis of coal is used to determine the ___ in the coal.
    A. amount of volatile matter, fixed carbon, moisture, and ash
    B. elements present
    C. heat content
    D. moisture level

_____ **33.** In a closed feedwater heater, the ___.
    A. steam and water come into direct contact
    B. feedwater pressure is higher than the boiler pressure
    C. feedwater pressure is lower than the boiler pressure
    D. noncondensable gases are removed from the top

_____ **34.** The ___ of a fuel oil indicates how well or poorly the oil will flow during pumping and atomization.
    A. viscosity
    B. flash point
    C. fire point
    D. specific gravity

_____ **35.** Economizers are used in large plants so that ___.
    A. the dew point can be reduced
    B. impure feedwater can be used
    C. higher flue gas temperatures can be maintained
    D. heat from the gases of combustion can be reclaimed

_____ **36.** A(n) ___ protects against loss of boiler water and potential catastrophic explosion.
    A. airflow proving switch
    B. safety valve
    C. low-water fuel cutoff
    D. stop valve

_____ **37.** A pyrometer measures ___.
    A. temperature
    B. steam pressure
    C. draft
    D. the quality of steam leaving the boiler

_____ **38.** A steam pressure gauge is typically calibrated in ___.
    A. cubic feet per minute (cfm)
    B. pounds per square foot (psf)
    C. cubic inches per second (cis)
    D. pounds per square inch (psi)

_____ **39.** An Orsat analyzer is used to determine the ___ in a boiler.
    A. required stack gas temperatures
    B. combustion efficiency
    C. steam quality
    D. heating value of the fuel

_____ **40.** The bottom blowdown valve located farthest from a boiler is always a ___ valve.
    A. slow-opening
    B. quick-opening
    C. stop
    D. check

_____ **41.** A ___ automatically removes air and condensate from steam lines.
    A. steam header
    B. pressure-relief valve
    C. gate valve
    D. steam trap

_____ **42.** When using No. 6 fuel oil, the boiler operator must ___.
    A. blend it with No. 2 fuel oil
    B. cool the fuel oil
    C. heat the fuel oil
    D. pressurize the fuel oil tank

_____ **43.** Smoke is the result of ___.
    A. complete combustion
    B. incomplete combustion
    C. too much excess air
    D. too much primary air

_____ **44.** In a plant that uses forced draft only, the pressure in the furnace is ___.
    A. at atmospheric pressure
    B. greater than atmospheric pressure
    C. less than atmospheric pressure
    D. determined by the height of the stack

_____ **45.** A throttling calorimeter determines ___.
    A. heat in the condensate
    B. saturated steam temperature
    C. superheated steam temperature
    D. the amount of moisture in steam

_____ **46.** The try cocks on a water column are used ___.
    A. as a secondary means of determining the water level
    B. to blow down the water column
    C. as sampling cocks for water testing
    D. to remove impurities from the surface of the water

_____ **47.** Boiler water is chemically treated to ___.
        A. increase oxygen concentration
        B. prevent scale formation
        C. increase carbon dioxide concentration
        D. improve water circulation

_____ **48.** A(n) ___ system is a boiler system that regulates the flow of air into the burner and the gases of combustion from the burner to the stack.
        A. gas
        B. combustion
        C. oil
        D. draft

_____ **49.** ___ uses steam pressure as the input signal and outputs a signal to a modulating motor that turns a jackshaft to modulate the air and fuel flow.
        A. Metering control
        B. ON/OFF control
        C. Parallel positioning
        D. Single-point positioning

_____ **50.** A superheater header safety valve is set to pop at ___ the steam drum safety valves.
        A. the same pressure as
        B. the MAWP of
        C. a lower pressure than
        D. a higher pressure than

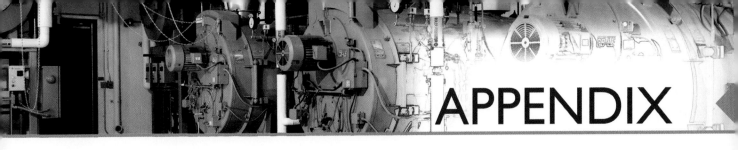

# APPENDIX

## STEAM TABLES (GAUGE PRESSURE)...

| Gauge Pressure* | Absolute Pressure† | Temperature‡ | Heat Content§ | | | Specific Volume Steam $v_g$‖ |
|---|---|---|---|---|---|---|
| | | | Sensible $h_f$ | Latent $h_{fg}$ | Total $h_g$ | |
| 27.96 | 1 | 101.7 | 69.5 | 1032.9 | 1102.4 | 333.0 |
| 25.91 | 2 | 126.1 | 93.9 | 1019.7 | 1113.6 | 173.5 |
| 23.87 | 3 | 141.5 | 109.3 | 1011.3 | 1120.6 | 118.6 |
| 21.83 | 4 | 153.0 | 120.8 | 1004.9 | 1125.7 | 90.52 |
| 19.79 | 5 | 162.3 | 130.1 | 999.7 | 1129.8 | 73.42 |
| 17.75 | 6 | 170.1 | 137.8 | 995.4 | 1133.2 | 61.89 |
| 15.70 | 7 | 176.9 | 144.6 | 991.5 | 1136.1 | 53.57 |
| 13.66 | 8 | 182.9 | 150.7 | 987.9 | 1138.6 | 47.26 |
| 11.62 | 9 | 188.3 | 156.2 | 984.7 | 1140.9 | 42.32 |
| 9.58 | 10 | 193.2 | 161.1 | 981.9 | 1143.0 | 38.37 |
| 7.54 | 11 | 197.8 | 165.7 | 979.2 | 1144.9 | 35.09 |
| 5.49 | 12 | 202.0 | 169.9 | 976.7 | 1146.6 | 32.35 |
| 3.45 | 13 | 205.9 | 173.9 | 974.3 | 1148.2 | 30.01 |
| 1.41 | 14 | 209.6 | 177.6 | 972.2 | 1149.8 | 28.00 |
| **Gauge Pressure#** | | | | | | |
| 0 | 14.7 | 212.0 | 180.2 | 970.6 | 1150.8 | 26.80 |
| 1 | 15.7 | 215.4 | 183.6 | 968.4 | 1152.0 | 25.20 |
| 2 | 16.7 | 218.5 | 186.8 | 966.4 | 1153.2 | 23.80 |
| 3 | 17.7 | 221.5 | 189.8 | 964.5 | 1154.3 | 22.50 |
| 4 | 18.7 | 224.5 | 192.7 | 962.6 | 1155.3 | 21.40 |
| 5 | 19.7 | 227.4 | 195.5 | 960.8 | 1156.3 | 20.40 |
| 6 | 20.7 | 230.0 | 198.1 | 959.2 | 1157.3 | 19.40 |
| 7 | 21.7 | 232.4 | 200.6 | 957.6 | 1158.2 | 18.60 |
| 8 | 22.7 | 234.8 | 203.1 | 956.0 | 1159.1 | 17.90 |
| 9 | 23.7 | 237.1 | 205.5 | 954.5 | 1160.0 | 17.20 |
| 10 | 24.7 | 239.4 | 207.9 | 952.9 | 1160.8 | 16.50 |
| 11 | 25.7 | 241.6 | 210.1 | 951.5 | 1161.6 | 15.90 |
| 12 | 26.7 | 243.7 | 212.3 | 950.0 | 1162.3 | 15.30 |
| 13 | 27.7 | 245.8 | 214.4 | 948.6 | 1163.0 | 14.80 |
| 14 | 28.7 | 247.9 | 216.4 | 947.3 | 1163.7 | 14.30 |
| 15 | 29.7 | 249.8 | 218.4 | 946.0 | 1164.4 | 13.90 |
| 16 | 30.7 | 251.7 | 220.3 | 944.8 | 1165.1 | 13.40 |
| 17 | 31.7 | 253.6 | 222.2 | 943.5 | 1165.7 | 13.00 |
| 18 | 32.7 | 255.4 | 224.0 | 942.4 | 1166.4 | 12.70 |
| 19 | 33.7 | 257.2 | 225.8 | 941.2 | 1167.0 | 12.30 |
| 20 | 34.7 | 258.8 | 227.5 | 940.1 | 1167.6 | 12.00 |
| 22 | 36.7 | 262.3 | 230.9 | 937.8 | 1168.7 | 11.40 |
| 24 | 38.7 | 265.3 | 234.2 | 935.8 | 1170.0 | 10.80 |
| 26 | 40.7 | 268.3 | 237.3 | 933.5 | 1170.8 | 10.30 |
| 28 | 42.7 | 271.4 | 240.2 | 931.6 | 1171.8 | 9.87 |
| 30 | 44.7 | 274.0 | 243.0 | 929.7 | 1172.7 | 9.46 |
| 32 | 46.7 | 276.7 | 245.9 | 927.6 | 1173.5 | 9.08 |
| 34 | 48.7 | 279.4 | 248.5 | 925.8 | 1174.3 | 8.73 |
| 36 | 50.7 | 281.9 | 251.1 | 924.0 | 1175.1 | 8.40 |
| 38 | 52.7 | 284.4 | 253.7 | 922.1 | 1175.8 | 8.11 |
| 40 | 54.7 | 286.7 | 256.1 | 920.4 | 1176.5 | 7.83 |
| 42 | 56.7 | 289.0 | 258.5 | 918.6 | 1177.1 | 7.57 |
| 44 | 58.7 | 291.3 | 260.8 | 917.0 | 1177.8 | 7.33 |
| 46 | 60.7 | 293.5 | 263.0 | 915.4 | 1178.4 | 7.10 |
| 48 | 52.7 | 295.6 | 265.2 | 913.8 | 1179.0 | 6.89 |
| 50 | 64.7 | 297.7 | 267.4 | 912.2 | 1179.6 | 6.68 |
| 52 | 66.7 | 299.7 | 269.4 | 910.7 | 1180.1 | 6.50 |
| 54 | 68.7 | 301.7 | 271.5 | 909.2 | 1180.7 | 6.32 |
| 56 | 70.7 | 303.6 | 273.5 | 907.8 | 1181.3 | 6.16 |
| 58 | 72.7 | 305.5 | 275.3 | 906.5 | 1181.8 | 6.00 |
| 60 | 74.7 | 307.4 | 277.1 | 905.3 | 1182.4 | 5.84 |

## ...STEAM TABLES (GAUGE PRESSURE)...

| Gauge Pressure[#] | Absolute Pressure[†] | Temperature[‡] | Heat Content[§] | | | Specific Volume Steam $v_g$[‖] |
|---|---|---|---|---|---|---|
| | | | Sensible $h_f$ | Latent $h_{fg}$ | Total $h_g$ | |
| 62 | 76.7 | 309.2 | 279.0 | 904.0 | 1183.0 | 5.70 |
| 64 | 78.7 | 310.9 | 280.9 | 902.6 | 1183.5 | 5.56 |
| 66 | 80.7 | 312.7 | 282.8 | 901.2 | 1184.0 | 5.43 |
| 68 | 82.7 | 314.3 | 284.5 | 900.0 | 1184.5 | 5.31 |
| 70 | 84.7 | 316.0 | 286.2 | 898.8 | 1185.0 | 5.19 |
| 72 | 86.7 | 317.7 | 288.0 | 897.5 | 1185.5 | 5.08 |
| 74 | 88.7 | 319.3 | 289.4 | 896.5 | 1185.9 | 4.97 |
| 76 | 90.7 | 320.9 | 291.2 | 895.1 | 1186.3 | 4.87 |
| 78 | 92.7 | 322.4 | 292.9 | 893.9 | 1186.8 | 4.77 |
| 80 | 94.7 | 323.9 | 294.5 | 892.7 | 1187.2 | 4.67 |
| 82 | 96.7 | 325.5 | 296.1 | 891.5 | 1187.6 | 4.58 |
| 84 | 98.7 | 326.9 | 297.6 | 890.3 | 1187.9 | 4.49 |
| 86 | 100.7 | 328.4 | 299.1 | 889.2 | 1188.3 | 4.41 |
| 88 | 102.7 | 329.9 | 300.6 | 888.1 | 1188.7 | 4.33 |
| 90 | 104.7 | 331.2 | 302.1 | 887.0 | 1189.1 | 4.25 |
| 92 | 106.7 | 332.6 | 303.5 | 885.8 | 1189.3 | 4.17 |
| 94 | 108.7 | 333.9 | 304.9 | 884.8 | 1189.7 | 4.10 |
| 96 | 110.7 | 335.3 | 306.3 | 883.7 | 1190.0 | 4.03 |
| 98 | 112.7 | 336.6 | 307.7 | 882.6 | 1190.3 | 3.96 |
| 100 | 114.7 | 337.9 | 309.0 | 881.6 | 1190.6 | 3.90 |
| 102 | 116.7 | 339.2 | 310.3 | 880.6 | 1190.9 | 3.83 |
| 104 | 118.7 | 340.5 | 311.6 | 879.6 | 1191.2 | 3.77 |
| 106 | 120.7 | 341.7 | 313.0 | 878.5 | 1191.5 | 3.71 |
| 108 | 122.7 | 343.0 | 314.3 | 877.5 | 1191.8 | 3.65 |
| 110 | 124.7 | 344.2 | 315.5 | 876.5 | 1192.0 | 3.60 |
| 112 | 126.7 | 345.4 | 316.8 | 875.5 | 1192.3 | 3.54 |
| 114 | 128.7 | 346.5 | 318.0 | 874.5 | 1192.5 | 3.49 |
| 116 | 130.7 | 347.7 | 319.3 | 873.5 | 1192.8 | 3.44 |
| 118 | 132.7 | 348.9 | 320.5 | 872.5 | 1193.0 | 3.39 |
| 120 | 134.7 | 350.1 | 321.8 | 871.5 | 1193.3 | 3.34 |
| 125 | 139.7 | 352.8 | 324.7 | 869.3 | 1194.0 | 3.23 |
| 130 | 144.7 | 355.6 | 327.6 | 866.9 | 1194.5 | 3.12 |
| 135 | 149.7 | 358.3 | 330.6 | 864.5 | 1195.1 | 3.02 |
| 140 | 154.7 | 360.9 | 333.2 | 862.5 | 1195.7 | 2.93 |
| 145 | 159.7 | 363.5 | 335.9 | 860.3 | 1196.2 | 2.84 |
| 150 | 164.7 | 365.9 | 338.6 | 858.0 | 1196.6 | 2.76 |
| 155 | 169.7 | 368.3 | 341.1 | 856.0 | 1197.1 | 2.68 |
| 160 | 174.7 | 370.7 | 343.6 | 853.9 | 1197.5 | 2.61 |
| 165 | 179.7 | 372.9 | 346.1 | 851.8 | 1197.9 | 2.54 |
| 170 | 184.7 | 375.2 | 348.5 | 849.8 | 1198.3 | 2.48 |
| 175 | 189.7 | 377.5 | 350.9 | 847.9 | 1198.8 | 2.41 |
| 180 | 194.7 | 379.6 | 353.2 | 845.9 | 1199.1 | 2.35 |
| 185 | 199.7 | 381.6 | 355.4 | 844.1 | 1199.5 | 2.30 |
| 190 | 204.7 | 383.7 | 357.6 | 842.2 | 1199.8 | 2.24 |
| 195 | 209.7 | 385.7 | 359.9 | 840.2 | 1200.1 | 2.18 |
| 200 | 214.7 | 387.7 | 362.0 | 838.4 | 1200.4 | 2.14 |
| 210 | 224.7 | 391.7 | 366.2 | 834.8 | 1201.0 | 2.04 |
| 220 | 234.7 | 395.5 | 370.3 | 831.2 | 1201.5 | 1.96 |
| 230 | 244.7 | 399.1 | 374.2 | 827.8 | 1202.0 | 1.88 |
| 240 | 254.7 | 402.7 | 378.0 | 824.5 | 1202.5 | 1.81 |
| 250 | 264.7 | 406.1 | 381.7 | 821.2 | 1202.9 | 1.74 |
| 260 | 274.7 | 409.3 | 385.3 | 817.9 | 1203.2 | 1.68 |
| 270 | 284.7 | 412.5 | 388.8 | 814.8 | 1203.6 | 1.62 |
| 280 | 294.7 | 415.8 | 392.3 | 811.6 | 1203.9 | 1.57 |
| 290 | 304.7 | 418.8 | 395.7 | 808.5 | 1204.2 | 1.52 |
| 300 | 314.7 | 421.7 | 398.9 | 805.5 | 1204.4 | 1.47 |
| 310 | 324.7 | 424.7 | 402.1 | 802.6 | 1204.7 | 1.43 |
| 320 | 334.7 | 427.5 | 405.2 | 799.7 | 1204.9 | 1.39 |

| | | | Heat Content§ | | | Specific Volume |
|---|---|---|---|---|---|---|
| **Gauge Pressure#** | **Absolute Pressure†** | **Temperature‡** | **Sensible $h_f$** | **Latent $h_{fg}$** | **Total $h_g$** | **Steam $v_g$ ‖** |
| 330 | 344.7 | 430.3 | 408.3 | 796.7 | 1205.0 | 1.35 |
| 340 | 354.7 | 433.0 | 411.3 | 793.8 | 1205.1 | 1.31 |
| 350 | 364.7 | 435.7 | 414.3 | 791.0 | 1205.3 | 1.27 |
| 360 | 374.7 | 438.3 | 417.2 | 788.2 | 1205.4 | 1.24 |
| 370 | 384.7 | 440.8 | 420.0 | 785.4 | 1205.4 | 1.21 |
| 380 | 394.7 | 443.3 | 422.8 | 782.7 | 1205.5 | 1.18 |
| 390 | 404.7 | 445.7 | 425.6 | 779.9 | 1205.5 | 1.15 |
| 400 | 414.7 | 448.1 | 428.2 | 777.4 | 1205.6 | 1.12 |
| 420 | 434.7 | 452.8 | 433.4 | 772.2 | 1205.6 | 1.07 |
| 440 | 454.7 | 457.3 | 438.5 | 767.1 | 1205.6 | 1.02 |
| 460 | 474.7 | 461.7 | 443.4 | 762.1 | 1205.5 | 0.98 |
| 480 | 494.7 | 465.9 | 448.8 | 757.1 | 1205.4 | 0.94 |
| 500 | 514.7 | 470.0 | 453.0 | 752.3 | 1205.3 | 0.902 |
| 520 | 534.7 | 474.0 | 457.6 | 747.5 | 1205.1 | 0.868 |
| 540 | 554.7 | 477.8 | 462.0 | 742.8 | 1204.8 | 0.835 |
| 560 | 574.7 | 481.6 | 466.4 | 738.1 | 1204.5 | 0.805 |
| 580 | 594.7 | 485.2 | 470.7 | 733.5 | 1204.2 | 0.776 |
| 600 | 614.7 | 488.8 | 474.8 | 729.1 | 1203.9 | 0.750 |
| 620 | 634.7 | 492.3 | 479.0 | 724.5 | 1203.5 | 0.726 |
| 640 | 654.7 | 495.7 | 483.0 | 720.1 | 1203.1 | 0.703 |
| 660 | 674.7 | 499.0 | 486.9 | 715.8 | 1202.7 | 0.681 |
| 680 | 694.7 | 502.2 | 490.7 | 711.5 | 1202.2 | 0.660 |
| 700 | 714.7 | 505.4 | 494.4 | 707.4 | 1201.8 | 0.641 |
| 720 | 734.7 | 508.5 | 498.2 | 703.1 | 1201.3 | 0.623 |
| 740 | 754.7 | 511.5 | 501.9 | 698.9 | 1200.8 | 0.605 |
| 760 | 774.7 | 514.5 | 505.5 | 694.7 | 1200.2 | 0.588 |
| 780 | 794.7 | 517.5 | 509.0 | 690.7 | 1199.7 | 0.572 |
| 800 | 814.7 | 520.3 | 512.5 | 686.6 | 1199.1 | 0.557 |

\* in Hg vac
† in psia
‡ in °F
§ in Btu/lb
‖ in cu ft/lb
# in psig

## STEAM TABLES (ABSOLUTE PRESSURE)...

| Absolute Pressure# p | Temperature† t | Specific Volume‡ | | Enthalpy§ | | | Entropy‖ | | |
|---|---|---|---|---|---|---|---|---|---|
| | | Saturated Liquid $v_f$ | Saturated Vapor $v_g$ | Saturated Liquid $h_f$ | Evap. $h_{fg}$ | Saturated Vapor $h_g$ | Saturated Liquid $s_f$ | Evap. $s_{fg}$ | Saturated Vapor $s_g$ |
| 1 | 101.74 | 0.01614 | 333.6 | 69.70 | 1036.3 | 1106.0 | 0.1326 | 1.8456 | 1.9782 |
| 2 | 126.08 | 0.01623 | 173.73 | 93.99 | 1022.2 | 1116.2 | 0.1749 | 1.7451 | 1.9200 |
| 3 | 141.48 | 0.01630 | 118.71 | 109.37 | 1013.2 | 1122.6 | 0.2008 | 1.6855 | 1.8863 |
| 4 | 152.97 | 0.01636 | 90.63 | 120.86 | 1006.4 | 1127.3 | 0.2198 | 1.6427 | 1.8625 |
| 5 | 164.24 | 0.01640 | 73.52 | 130.13 | 1001.0 | 1131.1 | 0.2347 | 1.6094 | 1.8441 |
| 6 | 170.06 | 0.01645 | 61.98 | 137.96 | 996.2 | 1134.2 | 0.2472 | 1.5820 | 1.8292 |
| 7 | 176.85 | 0.01649 | 53.64 | 144.76 | 992.1 | 1136.9 | 0.2581 | 1.5586 | 1.8167 |
| 8 | 182.86 | 0.01653 | 47.34 | 150.79 | 989.5 | 1139.3 | 0.2674 | 1.5383 | 1.8057 |
| 9 | 188.28 | 0.01656 | 42.40 | 156.22 | 985.2 | 1141.4 | 0.2759 | 1.5203 | 1.7962 |
| 10 | 193.21 | 0.01659 | 38.42 | 161.17 | 982.1 | 1143.3 | 0.2835 | 1.5041 | 1.7876 |
| 14.696 | 212.00 | 0.01672 | 26.80 | 180.07 | 970.3 | 1150.4 | 0.3120 | 1.4446 | 1.7566 |
| 15 | 213.03 | 0.01672 | 26.29 | 181.11 | 969.7 | 1150.8 | 0.3135 | 1.4415 | 1.7549 |
| 20 | 227.96 | 0.01683 | 20.089 | 196.16 | 960.1 | 1156.3 | 0.3356 | 1.3962 | 1.7319 |
| 25 | 240.07 | 0.01692 | 16.303 | 208.42 | 952.1 | 1160.6 | 0.3533 | 1.3606 | 1.7139 |
| 30 | 250.33 | 0.01701 | 13.746 | 218.82 | 945.3 | 1164.1 | 0.3680 | 1.3313 | 1.6993 |
| 35 | 259.28 | 0.01708 | 11.898 | 227.91 | 939.2 | 1167.1 | 0.3807 | 1.3063 | 1.6870 |
| 40 | 267.25 | 0.01715 | 10.498 | 236.03 | 933.7 | 1169.7 | 0.3919 | 1.2844 | 1.6763 |
| 45 | 274.44 | 0.01721 | 9.401 | 243.36 | 928.6 | 1172.0 | 0.4019 | 1.2650 | 1.6669 |
| 50 | 281.01 | 0.01727 | 8.515 | 250.09 | 924.0 | 1174.1 | 0.4110 | 1.2474 | 1.6585 |
| 55 | 287.07 | 0.01732 | 7.787 | 256.30 | 919.6 | 1175.9 | 0.4193 | 1.2316 | 1.6509 |
| 60 | 292.71 | 0.01738 | 7.175 | 262.09 | 915.5 | 1177.6 | 0.4270 | 1.2168 | 1.6438 |
| 65 | 297.97 | 0.01743 | 6.655 | 267.50 | 911.6 | 1179.1 | 0.4342 | 1.2032 | 1.6374 |
| 70 | 302.92 | 0.01748 | 6.206 | 272.61 | 907.9 | 1180.6 | 0.4409 | 1.1906 | 1.6315 |
| 75 | 307.60 | 0.01753 | 5.816 | 277.43 | 904.5 | 1181.9 | 0.4472 | 1.1787 | 1.6259 |
| 80 | 312.03 | 0.01757 | 5.472 | 282.02 | 901.1 | 1183.1 | 0.4531 | 1.1676 | 1.6207 |
| 85 | 316.25 | 0.01761 | 5.168 | 286.39 | 897.8 | 1184.2 | 0.4587 | 1.1571 | 1.6158 |
| 90 | 320.27 | 0.01766 | 4.896 | 290.56 | 894.7 | 1185.3 | 0.4641 | 1.1471 | 1.6112 |
| 95 | 324.12 | 0.01770 | 4.652 | 294.56 | 891.7 | 1186.2 | 0.4692 | 1.1376 | 1.6068 |
| 100 | 327.81 | 0.01774 | 4.432 | 298.40 | 888.8 | 1187.2 | 0.4740 | 1.1286 | 1.6026 |
| 110 | 334.77 | 0.01782 | 4.049 | 305.66 | 883.2 | 1188.9 | 0.4832 | 1.1117 | 1.5948 |
| 120 | 341.25 | 0.01789 | 3.728 | 312.44 | 887.9 | 1190.4 | 0.4916 | 1.0962 | 1.5878 |
| 130 | 347.32 | 0.01796 | 3.455 | 318.81 | 872.9 | 1191.7 | 0.4995 | 1.0817 | 1.5812 |
| 140 | 353.02 | 0.01802 | 3.220 | 324.82 | 868.2 | 1193.0 | 0.5069 | 1.0682 | 1.5751 |
| 150 | 358.42 | 0.01809 | 3.015 | 330.51 | 863.6 | 1194.1 | 0.5138 | 1.0556 | 1.5694 |
| 160 | 363.53 | 0.01815 | 2.834 | 335.93 | 859.2 | 1195.1 | 0.5204 | 1.0436 | 1.5640 |
| 170 | 368.41 | 0.01822 | 2.675 | 341.09 | 854.9 | 1196.0 | 0.5266 | 1.0324 | 1.5590 |
| 180 | 373.06 | 0.01827 | 2.532 | 346.03 | 850.8 | 1196.9 | 0.5325 | 1.0217 | 1.5542 |
| 190 | 377.51 | 0.01833 | 2.404 | 350.79 | 846.8 | 1197.6 | 0.5381 | 1.0116 | 1.5497 |
| 200 | 381.79 | 0.01839 | 2.288 | 355.36 | 843.0 | 1198.4 | 0.5435 | 1.0018 | 1.5453 |
| 250 | 400.95 | 0.01865 | 1.8438 | 376.00 | 825.1 | 1201.1 | 0.5675 | 0.9588 | 1.5263 |
| 300 | 417.33 | 0.01890 | 1.5433 | 393.84 | 809.0 | 1202.8 | 0.5879 | 0.9225 | 1.5104 |

| | | **...STEAM TABLES (ABSOLUTE PRESSURE)** | | | | | | | |
|---|---|---|---|---|---|---|---|---|---|
| **Absolute Pressure#** $p$ | **Temperature[†]** $t$ | **Specific Volume[‡]** | | **Enthalpy[§]** | | | **Entropy[ll]** | | |
| | | **Saturated Liquid $v_f$** | **Saturated Vapor $v_g$** | **Saturated Liquid $h_f$** | **Evap.** $h_{fg}$ | **Saturated Vapor $h_g$** | **Saturated Liquid $s_f$** | **Evap.** $s_{fg}$ | **Saturated Vapor $s_g$** |
| 350 | 431.72 | 0.01913 | 1.3260 | 409.69 | 794.2 | 1203.9 | 0.6056 | 0.8910 | 1.4966 |
| 400 | 444.59 | 0.0193 | 1.1613 | 424.0 | 780.5 | 1204.5 | 0.6214 | 0.8630 | 1.4844 |
| 450 | 456.28 | 0.0195 | 1.0320 | 437.2 | 767.4 | 1204.6 | 0.6356 | 0.8378 | 1.4734 |
| 500 | 467.01 | 0.0197 | 0.9278 | 449.4 | 755.0 | 1204.4 | 0.6487 | 0.8147 | 1.4634 |
| 550 | 476.94 | 0.0199 | 0.8424 | 460.8 | 743.1 | 1203.9 | 0.6608 | 0.7934 | 1.4542 |
| 600 | 486.21 | 0.0201 | 0.7698 | 471.6 | 731.6 | 1203.2 | 0.6720 | 0.7734 | 1.4454 |
| 650 | 494.90 | 0.0203 | 0.7083 | 481.8 | 720.5 | 1202.3 | 0.6826 | 0.7548 | 1.4374 |
| 700 | 503.10 | 0.0205 | 0.6554 | 491.5 | 709.7 | 1201.2 | 0.6925 | 00.7371 | 1.4296 |
| 750 | 510.86 | 0.0207 | 0.6092 | 500.8 | 699.2 | 1200.0 | 0.7019 | 0.7204 | 1.4223 |
| 800 | 518.23 | 0.0209 | 0.5687 | 509.7 | 688.9 | 1198.6 | 0.7108 | 0.7045 | 1.4153 |
| 850 | 525.26 | 0.0210 | 0.5327 | 518.3 | 678.8 | 1197.7 | 0.7194 | 0.6891 | 1.4085 |
| 900 | 531.98 | 0.0212 | 0.5006 | 526.6 | 668.8 | 1195.4 | 0.7275 | 0.6744 | 1.4020 |
| 950 | 538.43 | 0.0214 | 0.4717 | 534.6 | 659.1 | 1193.7 | 0.7355 | 0.6602 | 1.3957 |
| 1000 | 544.61 | 0.0216 | 0.4456 | 542.4 | 649.4 | 1191.8 | 0.7430 | 0.6467 | 1.3897 |
| 1100 | 556.31 | 0.0220 | 0.4001 | 557.4 | 630.4 | 1187.8 | 0.7575 | 0.6205 | 1.3780 |
| 1200 | 567.22 | 0.0223 | 0.3619 | 571.7 | 611.7 | 1183.4 | 0.7711 | 0.5956 | 1.3667 |
| 1300 | 577.46 | 0.0227 | 0.3293 | 585.4 | 593.2 | 1178.6 | 0.7840 | 0.5719 | 1.3559 |
| 1400 | 587.10 | 0.0231 | 0.3012 | 598.7 | 574.7 | 1173.4 | 0.7963 | 0.5491 | 1.3454 |
| 1500 | 596.23 | 0.0235 | 0.2765 | 611.6 | 556.3 | 1167.9 | 0.8082 | 0.5269 | 1.3351 |
| 2000 | 635.82 | 0.0257 | 0.1878 | 671.7 | 463.4 | 1135.1 | 0.8619 | 0.4230 | 1.2849 |
| 2500 | 668.13 | 0.0287 | 0.1307 | 730.6 | 360.5 | 1091.1 | 0.9126 | 0.3197 | 1.2322 |
| 3000 | 695.36 | 0.0346 | 0.0858 | 802.5 | 217.8 | 1020.3 | 0.9731 | 0.1885 | 1.1615 |
| 3206.2 | 705.40 | 0.0503 | 0.0503 | 902.7 | 0 | 902.7 | 1.0580 | 0 | 1.0580 |

* in psi
† in °F
‡ in cu ft/lb
§ in Btu/lb
ll in Btu/lb °F

| **COMMON ASME INTERNATIONAL BOILER CLASSIFICATIONS** | |
|---|---|
| **Name** | **Description** |
| Automatic boiler | Equipped with certain controls and limit devices per ASME code |
| Boiler | Closed vessel used for heating water or liquid, or for generating steam or vapor by direct application of hear |
| Boiler plant | One or more boilers, connecting piping, and vessels within the same premises |
| Hot water supply boiler | Low pressure hot water heating boiler having a volume exceeding 120 gal., a heat input exceeding 200,000 Btu/hr, or an operating temperature exceeding 200°F that provides hot water to be used externally |
| Low pressure hot water heating boiler | Boiler in which water is heated for the purpose of supplying heat at pressures not exceeding 160 psi or temperatures not exceeding 250°F |
| Low pressure steam heating boiler | Boiler operated at pressures not exceeding 15 psi for steam |
| Power hot water boiler | Boiler used for heating water or liquid to pressure exceeding 160 psi or a temperature exceeding 250°F |
| Power steam boiler | Boiler in which steam or vapor is generated at pressures exceeding 15 psi |
| Small power boiler | Boiler with pressures exceeding 15 psi but not exceeding 100 psi and having less than 440,000 btu/hr input |

# STEAM TABLES (SUPERHEATED STEAM)...
(*v* = specific volume, cu ft/lb; *h* = enthalpy, Btu/lb; *s* = entropy)

| Absolute Pressure* (sat. temp) | | Temperature† | | | | | | | | | |
|---|---|---|---|---|---|---|---|---|---|---|---|
| | | **400** | **500** | **600** | **700** | **800** | **900** | **1000** | **1100** | **1200** | **1400** |
| 1 (101.74) | *v* | 512.0 | 571.6 | 631.2 | 690.8 | 750.4 | 809.9 | 869.5 | 929.1 | 988.7 | 1107.8 |
| | *h* | 1241.7 | 1288.3 | 1335.8 | 1383.8 | 1432.8 | 1482.7 | 1533.5 | 1585.2 | 1637.7 | 1745.7 |
| | *s* | 2.1720 | 2.2233 | 2.2702 | 2.3137 | 2.3542 | 2.3932 | 2.4283 | 2.4625 | 2.4952 | 2.5566 |
| 5 (162.24) | *v* | 102.26 | 114.22 | 126.16 | 138.10 | 150.03 | 161.95 | 173.87 | 185.79 | 197.71 | 221.6 |
| | *h* | 1241.2 | 1288.0 | 1335.4 | 1383.6 | 1432.7 | 1482.6 | 1533.4 | 1585.1 | 1637.7 | 1745.7 |
| | *s* | 1.9942 | 2.0456 | 2.0927 | 2.1361 | 2.1767 | 2.2148 | 2.2509 | 2.2851 | 2.3178 | 2.3792 |
| 10 (193.21) | *v* | 51.04 | 57.05 | 63.03 | 69.01 | 74.98 | 80.95 | 86.92 | 92.88 | 98.84 | 110.77 |
| | *h* | 1240.6 | 1287.5 | 1335.1 | 1383.4 | 1432.5 | 1482.4 | 1533.2 | 1585.0 | 1637.6 | 1745.6 |
| | *s* | 1.9172 | 1.9689 | 2.0160 | 2.0596 | 2.1002 | 2.1383 | 2.1744 | 2.2086 | 2.2413 | 2.3028 |
| 14.696 (212.00) | *v* | 34.68 | 38.78 | 42.86 | 46.94 | 51.00 | 55.07 | 59.13 | 63.19 | 67.25 | 75.37 |
| | *h* | 1239.9 | 1287.1 | 1334.8 | 1383.2 | 1432.3 | 1482.3 | 1533.1 | 1584.8 | 1637.5 | 1745.5 |
| | *s* | 1.8743 | 1.9261 | 1.9734 | 2.0170 | 2.0576 | 2.0958 | 213.19 | 2.1662 | 2.1989 | 2.2603 |
| 20 (227.96) | *v* | 25.43 | 28.46 | 31.47 | 34.47 | 37.46 | 40.45 | 43.44 | 46.42 | 49.41 | 55.37 |
| | *h* | 1239.2 | 1286.6 | 1334.4 | 1382.9 | 1432.1 | 1482.1 | 1533.0 | 1584.7 | 1637.4 | 1745.4 |
| | *s* | 1.8396 | 1.8918 | 1.9392 | 1.9829 | 2.0235 | 2.0618 | 2.0978 | 2.1321 | 2.1648 | 2.2263 |
| 40 (267.25) | *v* | 12.628 | 14.168 | 15.688 | 17.198 | 18.702 | 20.20 | 21.70 | 23.20 | 24.69 | 27.68 |
| | *h* | 1236.5 | 1284.8 | 1333.1 | 1381.9 | 1431.3 | 1481.4 | 1532.4 | 1584.3 | 1637.0 | 1745.1 |
| | *s* | 1.7608 | 1.8140 | 1.8619 | 1.9058 | 1.9467 | 1.9850 | 2.0212 | 2.0555 | 2.0883 | 2.1498 |
| 60 (292.71) | *v* | 8.357 | 9.403 | 10.427 | 11.441 | 12.449 | 13.452 | 14.454 | 15.453 | 16.451 | 18.446 |
| | *h* | 1233.6 | 1283.0 | 1331.8 | 1380.9 | 1430.5 | 1480.8 | 1531.9 | 1583.8 | 1636.6 | 1744.8 |
| | *s* | 1.7135 | 1.7678 | 1.8162 | 1.8605 | 1.9015 | 1.9400 | 1.9762 | 2.0106 | 2.0434 | 2.1049 |
| 80 (312.03) | *v* | 6.220 | 7.020 | 7.797 | 8.562 | 9.322 | 10.077 | 10.830 | 11.582 | 12.332 | 13.830 |
| | *h* | 1230.7 | 1281.1 | 1330.5 | 1379.9 | 1429.7 | 1480.1 | 1531.3 | 1583.4 | 1636.2 | 1744.5 |
| | *s* | 1.6791 | 1.7346 | 1.7836 | 1.8281 | 1.8694 | 1.9079 | 1.9442 | 1.9787 | 2.0115 | 2.0731 |
| 100 (327.81) | *v* | 4.937 | 5.589 | 6.218 | 6.835 | 7.446 | 8.052 | 8.656 | 9.259 | 9.860 | 11.060 |
| | *h* | 1227.6 | 1279.1 | 1329.1 | 1378.9 | 1428.9 | 1479.5 | 1530.8 | 1582.9 | 1635.7 | 1744.2 |
| | *s* | 1.6518 | 1.7085 | 1.7581 | 1.8029 | 1.8443 | 1.8829 | 1.9193 | 1.9538 | 1.9867 | 2.0484 |
| 120 (341.25) | *v* | 4.081 | 4.636 | 5.165 | 5.683 | 6.195 | 6.702 | 7.207 | 7.710 | 8.212 | 9.214 |
| | *h* | 1224.4 | 1277.2 | 1327.7 | 1377.8 | 1428.1 | 1478.8 | 1530.2 | 1582.4 | 1635.3 | 1743.9 |
| | *s* | 1.6287 | 1.6869 | 1.7370 | 1.7822 | 1.8237 | 1.8625 | 1.8990 | 1.9335 | 1.9664 | 2.0281 |
| 140 (353.02) | *v* | 3.468 | 3.954 | 4.413 | 4.861 | 5.301 | 5.738 | 6.172 | 6.604 | 7.035 | 7.895 |
| | *h* | 1221.1 | 1275.2 | 1326.4 | 1376.8 | 1427.3 | 1478.2 | 1529.7 | 1581.9 | 1634.9 | 1743.5 |
| | *s* | 1.6087 | 1.6683 | 1.7190 | 1.7645 | 1.8063 | 1.8451 | 1.8817 | 1.9163 | 1.9493 | 2.0110 |
| 160 (363.53) | *v* | 3.008 | 3.443 | 3.849 | 4.244 | 4.631 | 5.015 | 5.396 | 5.775 | 6.152 | 6.906 |
| | *h* | 1217.6 | 1273.1 | 1325.0 | 1375.7 | 1426.4 | 1477.5 | 1529.1 | 1581.4 | 1634.5 | 1743.2 |
| | *s* | 1.5908 | 1.6519 | 1.7033 | 1.7491 | 1.7911 | 1.8301 | 1.8667 | 1.9014 | 1.9344 | 1.9962 |
| 180 (373.06) | *v* | 2.649 | 3.044 | 3.411 | 3.764 | 4.110 | 4.452 | 4.792 | 5.129 | 5.466 | 6.136 |
| | *h* | 1214.0 | 1271.0 | 1323.5 | 1374.7 | 1425.6 | 1476.8 | 1528.6 | 1581.0 | 1634.1 | 1742.9 |
| | *s* | 1.5745 | 1.6373 | 1.6894 | 1.7355 | 1.7776 | 1.8167 | 1.8534 | 1.8882 | 1.9212 | 1.9831 |
| 200 (381.79) | *v* | 2.361 | 2.726 | 3.060 | 3.380 | 3.693 | 4.002 | 4.309 | 4.613 | 4.917 | 5.521 |
| | *h* | 1210.3 | 1268.9 | 1322.1 | 1373.6 | 1424.8 | 1476.2 | 1528.0 | 1580.5 | 1633.7 | 1742.6 |
| | *s* | 1.5594 | 1.6240 | 1.6767 | 1.7232 | 1.7655 | 1.8048 | 1.8415 | 1.8763 | 1.9094 | 1.9713 |
| 220 (389.86 | *v* | 2.125 | 2.465 | 2.772 | 3.066 | 3.352 | 3.634 | 3.913 | 4.191 | 4.467 | 5.017 |
| | *h* | 1206.5 | 1266.7 | 1320.7 | 1372.6 | 1424.0 | 1475.5 | 1527.5 | 1580.0 | 1633.3 | 1742.3 |
| | *s* | 1.5453 | 1.6117 | 1.6652 | 1.7120 | 1.7545 | 1.7939 | 1.8308 | 1.8656 | 1.8987 | 1.9607 |
| 240 (397.37) | *v* | 1.9276 | 2.247 | 2.533 | 2.804 | 3.068 | 3.327 | 3.584 | 3.839 | 4.093 | 4.597 |
| | *h* | 1202.5 | 1264.5 | 1319.2 | 1371.5 | 1423.2 | 1474.8 | 1526.9 | 1579.6 | 1632.9 | 1742.0 |
| | *s* | 1.5319 | 1.6003 | 1.6546 | 1.7017 | 1.7444 | 1.7839 | 1.8209 | 1.8558 | 1.8889 | 1.9510 |
| 260 (404.42) | *v* | — | 2.063 | 2.330 | 2.582 | 2.827 | 3.067 | 3.305 | 3.541 | 3.776 | 4.242 |
| | *h* | — | 1262.3 | 1317.7 | 1370.4 | 1422.3 | 1474.2 | 1526.3 | 1579.1 | 1632.5 | 1741.7 |
| | *s* | — | 1.5897 | 1.6447 | 1.6922 | 1.7352 | 1.7748 | 1.8118 | 1.8467 | 1.8799 | 1.9420 |
| 280 (411.05) | *v* | — | 1.9047 | 2.156 | 2.392 | 2.621 | 2.845 | 3.066 | 3.286 | 3.504 | 3.938 |
| | *h* | — | 1260.0 | 1316.2 | 1369.4 | 1421.5 | 1473.5 | 1525.8 | 1578.6 | 1632.1 | 1741.4 |
| | *s* | — | 1.5796 | 1.6354 | 1.6834 | 1.7264 | 1.7662 | 1.8033 | 1.8383 | 1.8716 | 1.9337 |
| 300 (417.33) | *v* | — | 1.7675 | 2.005 | 2.227 | 2.442 | 2.652 | 2.859 | 3.065 | 3.269 | 3.674 |
| | *h* | — | 1257.6 | 1314.7 | 1368.3 | 1420.6 | 1472.8 | 1525.2 | 1578.1 | 1631.7 | 1741.0 |
| | *s* | — | 1.5701 | 1.6268 | 1.6751 | 1.7184 | 1.7582 | 1.7954 | 1.8305 | 1.8638 | 1.9260 |
| 350 (431.72) | *v* | — | 1.4923 | 1.7036 | 1.8980 | 2.084 | 2.266 | 2.445 | 2.622 | 2.798 | 3.147 |
| | *h* | — | 1251.5 | 1310.9 | 1365.5 | 1418.5 | 1471.1 | 1523.8 | 1577.0 | 1630.7 | 1740.3 |
| | *s* | — | 1.5481 | 1.6070 | 1.6563 | 1.7002 | 1.7403 | 1.7777 | 1.8130 | 1.8463 | 1.9086 |
| 400 (444.59) | *v* | — | 1.2851 | 1.4770 | 1.6508 | 1.8161 | 1.9767 | 2.134 | 2.290 | 2.445 | 2.751 |
| | *h* | — | 1245.1 | 1306.9 | 1362.7 | 1416.4 | 1469.4 | 1522.4 | 1575.8 | 1629.6 | 1739.5 |
| | *s* | — | 1.5281 | 1.5894 | 1.6398 | 1.6842 | 1.7247 | 1.7623 | 1.7977 | 1.8311 | 1.8936 |

## ...STEAM TABLES (SUPERHEATED STEAM)
(v = specific volume, cu ft/lb; h = enthalpy, Btu/lb; s = entropy)

| Absolute Pressure* (sat. temp) | | 500 | 600 | 620 | 640 | 660 | 680 | 700 | 800 | 900 | 1000 | 1200 | 1400 |
|---|---|---|---|---|---|---|---|---|---|---|---|---|---|
| 450 (456.28) | v | 1.1231 | 1.3005 | 1.3332 | 1.3652 | 1.3967 | 1.4278 | 1.4584 | 1.6074 | 1.7516 | 1.8928 | 2.170 | 2.443 |
| | h | 1238.4 | 1302.8 | 1314.6 | 1326.2 | 1337.5 | 1348.8 | 1359.9 | 1414.3 | 1467.7 | 1521.0 | 1628.6 | 1738.7 |
| | s | 1.5095 | 1.5735 | 1.5845 | 1.5951 | 1.6054 | 1.6153 | 1.6250 | 1.6699 | 1.7108 | 1.7486 | 1.8177 | 1.8803 |
| 500 (467.01) | v | 0.9927 | 1.1591 | 1.1893 | 1.2188 | 1.2478 | 1.2763 | 1.3044 | 1.4405 | 1.5715 | 1.6996 | 1.9504 | 2.197 |
| | h | 1231.3 | 1298.6 | 1310.7 | 1322.6 | 1334.2 | 1345.7 | 1357.0 | 1412.1 | 1466.0 | 1519.6 | 1627.6 | 1737.9 |
| | s | 1.4919 | 1.5588 | 1.5701 | 1.5810 | 1.5915 | 1.6016 | 1.6115 | 1.6571 | 1.6982 | 1.7363 | 1.8056 | 1.8683 |
| 550 (476.94) | v | 0.8852 | 1.0431 | 1.0714 | 1.0989 | 1.1259 | 1.1523 | 1.1783 | 1.3038 | 1.4241 | 1.5414 | 1.7706 | 1.9957 |
| | h | 1223.7 | 1294.3 | 1306.8 | 1318.9 | 1330.8 | 1342.5 | 1354.0 | 1409.9 | 1464.3 | 1518.2 | 1626.6 | 1737.1 |
| | s | 1.4751 | 1.5451 | 1.5568 | 1.5680 | 1.5787 | 1.5890 | 1.5991 | 1.6452 | 1.6868 | 1.7250 | 1.7946 | 1.8575 |
| 600 (486.21) | v | 0.7947 | 0.9463 | 0.9729 | 0.9988 | 1.0241 | 1.0489 | 1.0732 | 1.1899 | 1.3013 | 1.4096 | 1.6208 | 1.8279 |
| | h | 1215.7 | 1289.9 | 1302.7 | 1315.2 | 1327.4 | 1339.3 | 1351.1 | 1407.7 | 1462.5 | 1516.7 | 1625.5 | 1736.3 |
| | s | 1.4586 | 1.5323 | 1.5443 | 1.5558 | 1.5667 | 1.5773 | 1.5875 | 1.6343 | 1.6762 | 1.7147 | 1.7846 | 1.8476 |
| 700 (503.10) | v | — | 0.7934 | 0.8177 | 0.8411 | 0.8639 | 0.8860 | 0.9077 | 1.0108 | 1.1082 | 1.2024 | 1.3853 | 1.5641 |
| | h | — | 1280.6 | 1294.3 | 1307.5 | 1320.3 | 1332.8 | 1345.0 | 1403.2 | 1459.0 | 1513.9 | 1623.5 | 1734.8 |
| | s | — | 1.5084 | 1.5212 | 1.5333 | 1.5449 | 1.5559 | 1.5665 | 1.6147 | 1.6573 | 1.6963 | 1.7666 | 1.8299 |
| 800 (518.23) | v | — | 0.6779 | 0.7006 | 0.7223 | 0.7433 | 0.7635 | 0.7833 | 0.8763 | 0.9633 | 1.0470 | 1.2088 | 1.3662 |
| | h | — | 1270.7 | 1285.4 | 1299.4 | 1312.9 | 1325.9 | 1338.6 | 1398.6 | 1455.4 | 1511.0 | 1621.4 | 1733.2 |
| | s | — | 1.4863 | 1.5000 | 1.5129 | 1.5250 | 1.5366 | 1.5476 | 1.5972 | 1.6407 | 1.6801 | 1.7510 | 1.8146 |
| 900 (531.98) | v | — | 0.5873 | 0.6089 | 0.6294 | 0.6491 | 0.6680 | 0.6863 | 0.7716 | 0.8506 | 0.9262 | 1.0714 | 1.2124 |
| | h | — | 1260.1 | 1275.9 | 1290.9 | 1305.1 | 1318.8 | 1332.1 | 1393.9 | 1451.8 | 1508.1 | 1619.3 | 1731.6 |
| | s | — | 1.4653 | 1.4800 | 1.4938 | 1.5066 | 1.5187 | 1.5303 | 1.5814 | 1.6257 | 1.6656 | 1.7371 | 1.8009 |
| 1000 (544.61) | v | — | 0.5140 | 0.5350 | 0.5546 | 0.5733 | 0.5912 | 0.6084 | 0.6878 | 0.7604 | 0.8294 | 0.9615 | 1.0893 |
| | h | — | 1248.8 | 1265.9 | 1281.9 | 1297.0 | 1311.4 | 1325.3 | 1389.2 | 1448.2 | 1505.1 | 1617.3 | 1730.0 |
| | s | — | 1.4450 | 1.4610 | 1.4757 | 1.4893 | 1.5021 | 1.5141 | 1.5670 | 1.6121 | 1.6525 | 1.7245 | 1.7886 |
| 1100 (556.31) | v | — | 0.4532 | 0.4738 | 0.4929 | 0.5110 | 0.5281 | 0.5445 | 0.6191 | 0.6866 | 0.7503 | 0.8716 | 0.9885 |
| | h | — | 1236.7 | 1255.3 | 1272.4 | 1288.5 | 1303.7 | 1318.3 | 1384.3 | 1444.5 | 1502.2 | 1615.2 | 1728.4 |
| | s | — | 1.4251 | 1.4425 | 1.4583 | 1.4728 | 1.4862 | 1.4989 | 1.5535 | 1.5995 | 1.6405 | 1.7130 | 1.7775 |
| 1200 (567.22) | v | — | 0.4016 | 0.4222 | 0.4410 | 0.4586 | 0.4752 | 0.4909 | 0.5617 | 0.6250 | 0.6843 | 0.7967 | 0.9046 |
| | h | — | 1223.5 | 1243.9 | 1262.4 | 1279.6 | 1295.7 | 1311.0 | 1379.3 | 1440.7 | 1499.2 | 1613.1 | 1726.9 |
| | s | — | 1.4052 | 1.4243 | 1.4413 | 1.4568 | 1.4710 | 1.4843 | 1.5409 | 1.5879 | 1.6293 | 1.7025 | 1.7672 |
| 1400 (587.10) | v | — | 0.3174 | 0.3390 | 0.3580 | 0.3753 | 0.3912 | 0.4062 | 0.4714 | 0.5281 | 0.5805 | 0.6789 | 0.7727 |
| | h | — | 1193.0 | 1218.4 | 1240.4 | 1260.3 | 1278.5 | 1295.5 | 1369.1 | 1433.1 | 1493.2 | 1608.9 | 1723.7 |
| | s | — | 1.3639 | 1.3877 | 1.4079 | 1.4258 | 1.4419 | 1.4567 | 1.5177 | 1.5666 | 1.6093 | 1.6836 | 1.7489 |
| 1600 (604.90) | v | — | — | 0.2733 | 0.2936 | 0.3112 | 0.3271 | 0.3417 | 0.4034 | 0.4553 | 0.5027 | 0.5906 | 0.6738 |
| | h | — | — | 1187.8 | 1215.2 | 1238.7 | 1259.6 | 1278.7 | 1358.4 | 1425.3 | 1487.0 | 1604.6 | 1720.5 |
| | s | — | — | 1.3489 | 1.3741 | 1.3952 | 1.4137 | 1.4303 | 1.4964 | 1.5476 | 1.5914 | 1.6669 | 1.7328 |
| 1800 (621.03) | v | — | — | — | 0.2407 | 0.2597 | 0.2760 | 0.2907 | 0.3502 | 0.3986 | 0.4421 | 0.5218 | 0.5968 |
| | h | — | — | — | 1185.1 | 1214.0 | 1238.5 | 1260.3 | 1347.2 | 1417.4 | 1480.8 | 1600.4 | 1717.3 |
| | s | — | — | — | 1.3377 | 1.3638 | 1.3855 | 1.4044 | 1.4765 | 1.5301 | 1.5752 | 1.6520 | 1.7185 |
| 2000 (635.82) | v | — | — | — | 0.1936 | 0.2161 | 0.2337 | 0.2489 | 0.3074 | 0.3532 | 0.3935 | 0.4668 | 0.5352 |
| | h | — | — | — | 1145.6 | 1184.9 | 1214.8 | 1240.0 | 1335.5 | 1409.2 | 1474.5 | 1596.1 | 1714.1 |
| | s | — | — | — | 1.2945 | 1.3300 | 1.3564 | 1.3783 | 1.4576 | 1.5139 | 1.5603 | 1.6384 | 1.7055 |
| 2500 (668.13) | v | — | — | — | — | — | 0.1484 | 0.1686 | 0.2294 | 0.2710 | 0.3061 | 0.3678 | 0.4244 |
| | h | — | — | — | — | — | 1132.3 | 1176.8 | 1303.6 | 1387.8 | 1458.4 | 1585.3 | 1706.1 |
| | s | — | — | — | — | — | 1.2687 | 1.3073 | 1.4127 | 1.4772 | 1.5273 | 1.6088 | 1.6775 |
| 3000 (695.36) | v | — | — | — | — | — | — | 0.0984 | 0.1760 | 0.2159 | 0.2476 | 0.3018 | 0.3505 |
| | h | — | — | — | — | — | — | 1060.7 | 1267.2 | 1365.0 | 1441.8 | 1574.3 | 1698.0 |
| | s | — | — | — | — | — | — | 1.1966 | 1.3690 | 1.4439 | 1.4984 | 1.5837 | 1.6540 |
| 3206.2 (705.40) | v | — | — | — | — | — | — | — | 0.1583 | 0.1981 | 0.2288 | 0.2806 | 0.3267 |
| | h | — | — | — | — | — | — | — | 1250.5 | 1355.2 | 1434.7 | 1569.8 | 1694.6 |
| | s | — | — | — | — | — | — | — | 1.3508 | 1.4309 | 1.4874 | 1.5742 | 1.6452 |

* in PSI
† in °F

| | Measured or Initiating Variable | Modifier | Readout or Passive Function | Output Function | Modifier |
|---|---|---|---|---|---|
| **A** | Analysis | | Alarm | | |
| **B** | Burner Flame | | User's Choice | User's Choice | User's Choice |
| **C** | Conductivity (Electrical) | | | Control | |
| **D** | Density (Mass) or Specific Gravity | Differential | | | |
| **E** | Voltage (EMF) | | Primary Element | | |
| **F** | Flow Rate | Ratio (Fraction) | | | |
| **G** | Gaging (Dimensional) | | Glass | | |
| **H** | Hand (Manually Initiated) | | | | High |
| **I** | Current (Electrical) | | Indicate | | |
| **J** | Power | Scan | | | |
| **K** | Time or Time Schedule | | | Control Station | |
| **L** | Level | | Light (Pilot) | | Low |
| **M** | Moisture or Humidity | | | | Middle or Intermediate |
| **N** | User's Choice | | User's Choice | User's Choice | User's Choice |
| **O** | User's Choice | | Orifice (Restriction) | | |
| **P** | Pressure or Vacuum | | Point (Test Connection) | | |
| **Q** | Quantity or Event | Integrate or Totalize | | | |
| **R** | Radioactivity, Radiation | | Record or Print | | |
| **S** | Speed or Frequency | Safety | | Switch | |
| **T** | Temperature | | | Transmit | |
| **U** | Multivariable | | Multifunction | Multifunction | Multifunction |
| **V** | Viscosity, Vibration | | | Valve, Damper, or Louver | |
| **W** | Weight or Force | | Well | | |
| **X** | Unclassified | | Unclassified | Unclassified | Unclassified |
| **Y** | Event or State | | | Relay or Compute | |
| **Z** | Position | | | Drive, Actuate, or Unclassified Final Control Element | |

Table title: **INSTRUMENTATION TAG IDENTIFICATION**, with column group headers **First Letter** (Measured or Initiating Variable, Modifier) and **Second Letter** (Readout or Passive Function, Output Function, Modifier).

# SELECTED INSTRUMENTATION SYMBOLS

## General Instrument Symbols—Balloons

INSTRUMENT FOR SINGLE MEASURED VARIABLE (WITH ANY NUMBER OF FUNCTIONS)

INSTRUMENT FOR TWO MEASURED VARIABLES (OPTIONALLY, SINGLE-VARIABLE INSTRUMENT WITH MORE THAN ONE FUNCTION, ADDITIONAL TANGENT BALLOONS MY BE ADDED AS REQUIRED)

APPROXIMATELY $^7/_{16}''$ DIAMETER

LOCALLY MOUNTED

MOUNTED ON BOARD 1 (OR BOARD 2); BOARD 2 MAY ALTERNATIVELY BE DESIGNATED BY A DOUBLE HORIZONTAL LINE INSTEAD OF A SINGLE LINE, WITH DESIGNATION OUTSIDE BALLOON OMITTED.

MOUNTED BEHIND THE BOARD

LOCALLY MOUNTED INSTRUMENT WITH LONG TAG NUMBER (SIX IS OPTIONAL AND IS PLANT NUMBER); ALTERNATIVELY, A CLOSED CIRCLE MAY BE ENLARGED

LOCALLY MOUNTED

MOUNTED ON MAIN BOARD

## Control Valve Body Symbols

    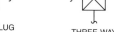 

GLOBE, GATE, OR OTHER IN-LINE TYPE NOT OTHERWISE IDENTIFIED

ANGLE

BUTTERFLY, DAMPER, OR LOUVER

ROTARY PLUG OR BALL

THREE-WAY

FOUR-WAY

## Actuator Symbols

DIAPHRAGM, SPRING-OPPOSED

DIAPHRAGM, SPRING-OPPOSED, WITH POSITIONER AND OVERRIDING PILOT VALVE THAT PRESSURIZES DIAPHRAGM WHEN ACTUATED

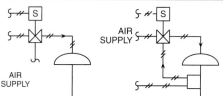

AIR SUPPLY

WITHOUT POSITIONER OR OTHER PILOT

PREFERRED FOR DIAPHRAGM THAT IS ASSEMBLED WITH PILOT SO THAT ASSEMBLY IS ACTUATED BY ONE CON-TROLLED INPUT (SHOWN TYPICALLY WITH ELECTRIC INPUT TO ASSEMBLY)

AIR SUPPLY

PREFERRED ALTERNATIVE

OPTIONAL ALTERNATIVE

DIAPHRAGM, PRESSURE-BALANCED

CYLINDER, WITHOUT POSITIONER OR OTHER PILOT

SINGLE-ACTING

DOUBLE-ACTING

PREFERRED FOR ANY CYLINDER THAT IS ASSEMBLED WITH PILOT SO THAT ASSEMBLY IS ACTUATED BY ONE CONTROLLED INPUT

ROTARY MOTOR (SHOWN TYPICALLY WITH ELECTRIC SIGNAL)

## Symbols for Self-Actuated Regulators, Valves, and Other Devices

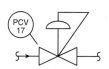 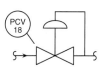  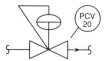 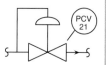

PRESSURE-REDUCING REGULATOR, SELF-CONTAINED

PRESSURE-REDUCING REGULATOR WITH EXTERNAL PRESSURE TAP

DIFFERENTIAL-PRESSURE-REDUCING REGULATOR WITH INTERNAL AND EXTERNAL PRESSURE TAPS

BACKPRESSURE REGULATOR, SELF-CONTAINED

BACKPRESSURE REGULATOR WITH EXTERNAL PRESSURE TAP

## Symbols for Actuator Action in Event of Actuator Power Failure

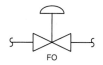   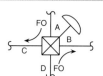 

TWO-WAY VALVE, FAIL OPEN

TWO-WAY VALVE, FAIL CLOSED

THREE-WAY VALVE, FAIL OPEN TO PATH A-C

FOUR-WAY VALVE, FAIL OPEN TO PATHS A-C AND D-B

ANY VALVE, FAIL LOCKED (POSITION DOES NOT CHANGE)

# GLOSSARY

## A

**aboveground storage tank (AST):** A tank and any underground piping connected to the tank that has less than 10% of its combined volume underground.

**absolute pressure (psia):** Gauge pressure plus atmospheric pressure.

**accessory:** A piece of equipment that is not directly attached to a boiler but is necessary for its safe and efficient operation.

**accumulation test:** A test used to establish the relief capacity of boiler safety valves.

**acid:** Any water with a pH less than 7.

**actuator:** A device that provides the power and motion needed to manipulate the moving parts of a valve or damper used to control fluid flow through a final element.

**air atomizing burner:** A fuel oil burner that sends compressed air and pressurized fuel oil through a nozzle into the furnace where the vaporized fuel mixes with air and is ignited.

**air heater:** A heat exchanger that is used to heat combustion air for a furnace and is located in the breeching between the boiler and the stack.

**alkaline:** Any water with a pH greater than 7.

**analyzer sampling system:** A system of piping, valves, and other equipment used to extract a sample from a process, condition it if necessary, and convey it to an analyzer.

**anion:** A negatively charged ion.

**annunciator:** An audible electronic alarm that alerts the operator of an operating condition that needs immediate attention.

**anthracite coal:** A geologically old coal that contains a high percentage of fixed carbon and a low percentage of volatile matter. Also known as hard coal.

**°API:** A value used by the American Petroleum Institute (API) to describe the density of fuel oil.

**area source:** A facility that emits less than 10 tons per year of any single HAP and less than 25 tons per year of any combination of air toxics.

**ash:** The solid material left behind in the process of burning coal.

**ASME Code:** A code written by the American Society of Mechanical Engineers (ASME) that governs and controls the types of material, methods of construction, and procedures used in the installation of boilers.

**automatic blowdown:** The process of automatically controlling the amount of boiler blowdown in order to maintain the level of total dissolved solids at the given setpoint.

## B

**baghouse:** A device that separates PM from the gases of combustion using cloth bags, similar to vacuum cleaner bags.

**balanced draft:** Draft produced from one or more forced draft fans located before the boiler and one or more induced draft fans after the boiler. Also known as combination draft.

**bent-tube boiler:** A watertube boiler with multiple drums connected by shaped tubes.

**best available control technology (BACT):** Technology that meets the environmental requirement for emissions from major new or modified sources in an attainment area.

**bimetallic steam trap:** A steam trap in which a temperature-sensitive bimetallic element controls a small discharge valve.

**bimetallic thermometer:** A thermometer that uses a strip of two metal alloys with different coefficients of thermal expansion that are fused together at the ends.

**biomass:** A biological material used as renewable energy fuel.

**bituminous coal:** A geologically young coal that contains a low percentage of fixed carbon and a high percentage of volatile gas. Also known as soft coal.

**blowdown:** The process of opening valves to blow water or steam through a fitting or from a boiler in order to remove any sludge, sediment, or other undesirable particles.

**blowdown heat recovery system:** Equipment that is installed to reclaim heat that is normally lost during continuous blowdown.

**blowdown separator:** A small tank in which makeup water is added to the boiler blowdown water after it flashes in order to reduce the discharge water temperature.

**blowdown tank:** A vented tank that cools blowdown water and steam to protect sewer lines from the high pressures and temperatures of blowdown water.

**boiler:** A closed vessel used for heating water to generate steam by direct application of heat from combustion fuels or electricity.

**boiler horsepower (BHP):** The amount of energy equal to the evaporation of 34.5 lb of water/hr from and at a feedwater temperature of 212°F.

**boiler vent:** A valve connected to the top of a boiler that allows air to be removed from the boiler when filling and heating and allows air to be drawn in when the pressure drops during cool-down or when draining the boiler.

**bottom ash:** Relatively large ash particles that are heavy enough to be collected and removed from the bottom of a furnace.

**bottom blowdown:** The process of removing water from the bottom of a boiler in order to remove impurities from the water.

**bottom blowdown valve:** A valve located at the lowest part of the water side of a firetube boiler or the mud drum of a watertube boiler so that sludge and sediment can be removed from the bottom of the boiler.

**Bourdon tube:** An oval metal tube inside a mechanical pressure gauge that is shaped like a question mark and has a tendency to straighten when pressurized.

**British thermal unit (Btu):** A measurement of the quantity of heat required to raise the temperature of 1 lb of water by 1°F.

**brown coal:** *See* lignite coal.

**bubbler:** A level-measuring instrument consisting of an air tube extending to the bottom of a tank, a pressure gauge and transmitter, a flow-control meter, and an air-pressure regulator.

**burner management system:** A system of control devices and control logic used to ensure safe burner operation.

## C

**carryover:** The act of small water particles and impurities being carried out of the boiler into the steam lines.

**cation:** A positively charged ion.

**caustic embrittlement:** A type of corrosion in which boiler metal becomes brittle because alkaline materials accumulate in cracks and crevices.

**centrifugal blower:** A blower that has a rotating impeller in a housing that throws the air to its outer edge, increasing the air's velocity and pressure.

**centrifugal feedwater pump:** A pump that uses the centrifugal force of a rotating element to pressurize water so it can be added to a boiler.

**chain grate stoker:** A stoker that feeds coal into a boiler on a traveling grate.

**chemical precipitation:** A process in which a chemical is added to raw makeup water to react with dissolved minerals, creating heavier particles that settle out of the solution.

**Clean Air Act (CAA):** A law that was passed by Congress in 1967 to allow for monitoring and controlling environmental air emissions.

**clinker:** The noncombustible components of coal that melt and fuse together during combustion, leaving chunks that interfere with ash handling. Also known as slag.

**closed feedwater heater:** A shell-and-tube feedwater heater used to capture heat from continuous blowdown water from a boiler or flash steam from a process.

**coagulation:** A process in which the chemicals added to raw water cause suspended solids to adhere to each other, making them larger and heavier and causing them to settle out of the solution.

**coal:** A solid black fossil fuel formed when organic material is hardened in the earth over millions of years.

**coal grade:** The size, heating value, and ash content of coal.

**coal rank:** The hardness of coal.

**cofiring furnace:** A furnace that mixes burnable material, such as biomass, with a fossil fuel for combustion.

**cogeneration:** The process of generating electricity and then using the waste heat from the generating process for heating buildings, providing process heat, or further electrical generation.

**combination draft:** *See* balanced draft.

**combustible material:** Any material that burns when it is exposed to oxygen and heat.

**combustion:** The rapid reaction of oxygen with a fuel that results in the release of heat.

**combustion control system:** An automatic boiler control system that regulates fuel supply, air supply, air-fuel ratio, and draft in a boiler in order to deliver the required amount of steam to a load.

**combustion efficiency:** A measure of the ability of burners to burn fuel efficiently.

**complete combustion:** Combustion that occurs when all the fuel is burned using a minimum amount of excess air.

**compressive stress:** Stress that occurs when two forces of equal intensity act from opposite directions and push toward the center of an object.

**condensate polisher:** An ion-exchange conditioner that removes impurities from condensate.

**condensate return tank:** An accessory that collects condensate returned from the point of use.

**condensing steam turbine:** A steam turbine that allows condensate to be reclaimed for use in the system.

**conductivity:** The ability of a material to allow the flow of electricity.

**conductivity meter:** An instrument that measures the electrical conductivity of a water sample to determine total dissolved solids present.

**conductivity probe:** A water level indicator that senses the boiler water level by detecting a change in current flow as water covers or uncovers the end of the probe.

**confined space:** A space large enough and so configured that an employee can physically enter and perform assigned work, has limited or restricted means for entry and exit, and is not designed for continuous employee occupancy.

**continuous blowdown:** The process of continuously draining a small, controlled amount of water from a boiler to control the quantity of impurities in the remaining water.

**continuous emissions monitoring system (CEMS):** A monitoring system used for continuous measurement of emissions from gases of combustion or from industrial processes.

**control device:** *See* final element.

**control element:** A device that takes an input signal, compares the signal to a setpoint, performs a computation, and sends an output signal to a final element. Also known as a controller.

**controller:** *See* control element.

**control loop:** A system that consists of a primary element and transmitter, a control element, and a final element.

**control system:** A system of measuring instruments and controllers that work together to control a process.

**control valve:** A final element that is used to modulate fluid flow in response to signals from a control element.

**convection:** A type of heat transfer that occurs due to the circulation of a fluid or gas.

**convection air heater:** An air heater that uses convection to transfer heat from the gases of combustion to combustion air.

**convection superheater:** A superheater that receives heat from convection currents in the gases of combustion.

**Coriolis flow meter:** A meter that measures the mass flow rate of a fluid.

**corrosion:** The deterioration of boiler metal caused by a chemical reaction with oxygen and carbon dioxide in water.

**critical pressure:** The pressure at which the density of water and steam are the same.

**cross-limited control:** A control strategy in which increases in airflow lead increases in fuel flow, and decreases in airflow lag behind decreases in fuel flow.

**cyclone separator:** A cylindrical device that separates water droplets from steam using centrifugal force.

## D

**damper:** An adjustable blade or set of blades used to control the flow of air.

**data plate:** A plate that is attached to a piece of equipment that provides important information about that equipment.

**deaerator:** A feedwater heater that operates under pressure and is used to separate oxygen and other gases from steam before releasing the gases to the atmosphere through a vent.

**dealkalizer:** An ion-exchange water conditioner that removes alkalinity from water.

**demineralizer:** An ion-exchange conditioner that consists of a strong cation-exchange resin and a strong anion-exchange resin and purifies water more than is possible with a zeolite softener.

**desuperheater:** An accessory used to remove heat from superheated steam to make it suitable for a process.

**diaphragm draft gauge:** A draft gauge that measures draft using a flexible diaphragm connected with a linkage to a pointer on a scale that is calibrated in inches or fractions of an inch of water column.

**digester gas:** Gas produced by the biological breakdown of organic matter in the absence of oxygen.

**digital control system:** An integrated system of control loops that use digital control signals to control a process.

**displacer:** A level-measuring instrument consisting of a cylinder, heavier than the liquid in which it is immersed, connected to a spring or torsion device that measures its weight. Also known as pneumatic level control.

**draft:** The flow of air or gases of combustion caused by a difference in pressure between two points.

**draft gauge:** A pressure gauge used to measure the difference in draft pressure between the atmosphere and a furnace, breeching, or stack.

**drum desuperheater:** A desuperheater that routes steam back through coils of submerged piping within a steam drum.

**drum internal:** *See* steam separator.

**dry-back boiler:** A firetube boiler with a refractory-lined door or chamber at the rear or top of the boiler that directs the gases of combustion from the furnace to the first pass of tubes or from one section of tubes to another.

**dry lay-up:** The storage of a boiler with all water drained.

**dry pipe separator:** A closed-end pipe that is perforated at the top, has drain holes on the bottom, and removes moisture from steam.

**dry-top boiler:** A vertical firetube boiler with an upper tube sheet that is dry.

## E

**economizer:** A feedwater heater that heats feedwater by passing it through a finned-tube heat exchanger placed in the path of the gases of combustion.

**electric boiler:** A boiler that produces heat using electrical resistance coils or electrodes instead of burning a fuel.

**electromagnetic induction:** The production of electricity as a result of a conductor passing through a magnetic field.

**electrostatic precipitator:** A device that removes finer PM from the gases of combustion using electric charge.

**energy audit:** An audit that identifies how energy is used in a facility and recommends ways to improve energy efficiency and reduce energy costs.

**excess air:** Air supplied to a boiler that is more than the theoretical amount of air needed for combustion.

**external boiler water treatment:** The treatment of boiler water before it enters a boiler.

**extraction steam:** Steam that is removed from the turbine at a controlled pressure after it has passed through some of the turbine stages.

**extra hazard area:** A location that includes wood-working shops, manufacturing plants using painting or dipping, and automotive repair shops.

## F

**feedwater:** Water supplied to a boiler.

**feedwater heater:** A device used to heat feedwater before it enters the boiler.

**feedwater line:** A pipeline that carries feedwater from a feedwater pump to a boiler.

**feedwater pump:** A pump that takes treated water from a tank and delivers it to a boiler at the proper pressure.

**feedwater regulator:** A device that maintains the water level in a boiler by controlling the amount of feedwater pumped to the boiler.

**field-erected boiler:** A boiler that is assembled at the final site because of size and complexity.

**filter:** A device that contains a porous substance through which fluid can pass but particulate matter cannot.

**filtration:** The process in which makeup water passes through a filter to remove sediment, particulates, and suspended solids.

**final element:** A device that controls the flow of liquid, gas or air, or electrical current. Also known as a control device.

**fire point:** The lowest temperature to which a fuel must be heated to burn continuously when exposed to an open flame.

**firetube steam boiler:** A boiler in which hot gases of combustion pass through tubes that are surrounded by water.

**fitting:** A component directly attached to a boiler that is required for the operation of the boiler.

**flame failure:** A situation where the pilot or main flame fails to light properly or goes out unintentionally during normal operation.

**flame rod:** A sensing device in a flame scanner that generates a small electric current when one end of the sensor is placed directly in a flame.

**flame scanner:** A safety device with a flame sensor that detects the presence of a flame and signals to the burner management system whether it is safe to operate the boiler.

**flame sensor:** A sensing device in a flame scanner that senses the pilot and main flame in the burner.

**flash point:** The lowest temperature at which the vapor of a fuel oil ignites when exposed to an open flame.

**flash steam:** The steam created when water at a high temperature experiences a sudden drop in pressure.

**flash tank:** A tank used with a continuous blowdown system to recover the flash steam from the water being removed during blowdown.

**flex-tube boiler:** A watertube boiler in which replaceable serpentine tubes are connected to the upper and lower drums and surround the firebox.

**float:** A level-measuring instrument with a hollow ball attached to it that floats on top of a liquid in a tank.

**float feedwater regulator:** A feedwater regulator that contains a steel or copper float ball connected to a switch by a linkage.

**float thermostatic steam trap:** A steam trap that contains a thermostatic bellows or another thermostatic element and also contains a steel ball float connected to a discharge valve by a linkage.

**flue:** The general term for the path used by the gases of combustion as they flow from the point of combustion to the point where they are released to the atmosphere.

**flue-gas recirculation (FGR):** An emissions control method for boilers in which moderate amounts of flue gas are captured from the exhaust and recirculated back through the burner along with the secondary air.

**fly ash:** 1. Relatively small ash particles that are light enough to be suspended in a combustible gas stream and carried though a boiler. 2. A type of PM consisting of noncombustible material found in the gases of combustion.

**foaming:** The rapid fluctuation of the water level that occurs when steam bubbles are trapped below a film of impurities on the surface of the boiler water.

**forced draft:** Mechanical draft produced by a fan supplying air to a furnace.

**fuel oil burner:** A device that delivers fuel oil to a furnace in a fine spray where it mixes with air to provide efficient combustion.

**fuel system:** A boiler system that provides fuel for combustion to produce the heat needed to evaporate water into steam.

**furnace:** A location where the combustion process takes place.

## G

**gasifier:** A device used to extract volatile gases from a solid fuel by heating the fuel in the absence of oxygen.

**gauge glass:** A water level indicator that consists of a glass column that indicates water level.

**gauge pressure (psig):** The pressure above atmospheric pressure.

**generator:** A device that converts mechanical energy, usually from a rotating shaft, into electrical energy by means of electromagnetic induction.

**guided-wave radar:** A water level indicator that uses a radar pulse that is sent down a waveguide that extends into the water to calculate the level based on the radar transit time.

## H

**hard coal:** *See* anthracite coal.

**hardness:** A measurement of scale-forming minerals dissolved in water.

**hard water:** Water that contains large quantities of dissolved scale-forming minerals, usually calcium and magnesium.

**hazardous air pollutant (HAP):** Any material from a list of approximately 200 pollutants that has significant health effects when it is emitted into the air.

**hazardous material:** A substance that can cause injury to personnel or damage to the environment.

**heating surface:** The part of a boiler with water on one side and heat and gases of combustion on the other.

**heating value:** The amount of heat that can be obtained from burning a fuel.

**heat recovery steam generator (HRSG):** A boiler that uses heat recovered from a hot gas stream, such as hot exhaust gases from a kiln or from a gas turbine.

**high-limit pressure control:** An ON/OFF control with a manual reset that sends a signal to the combustion control system to shut down the burner in the event of a high pressure condition that exceeds the operating pressure control cut-out setting.

**high/low/OFF control:** A combustion control strategy that senses the steam pressure and sends a control signal to start and stop the burner to maintain the steam pressure.

**high pressure steam boiler:** A boiler that operates at a maximum allowable working pressure (MAWP) of more than 15 psi.

**horizontal return tubular (HRT) boiler:** A firetube boiler that consists of a shell that contains tube sheets and fire tubes mounted over a firebox or furnace.

**huddling chamber:** The part on a safety valve that increases the area of the safety valve disc, thus increasing the total force, causing the valve to open quickly, or pop.

**hydrostatic pressure:** The pressure caused by the weight of a column of water.

## I

**incomplete combustion:** Combustion that occurs when all the fuel is not burned, resulting in the formation of soot and smoke.

**induced draft:** Draft produced by pulling air through the boiler furnace with a fan.

**infrared flame sensor:** A device in a flame scanner that generates a small amount of electric current when exposed to infrared light.

**instrumentation:** A group of measuring instruments and related devices that are part of a monitoring and control system.

**internal boiler water treatment:** The treatment of boiler water through the direct addition of chemicals into the boiler to prepare the boiler water for optimum operation.

**inverted bucket steam trap:** A steam trap that contains an inverted bucket connected to a discharge valve.

**ion-exchange softener:** A device that uses an ion-exchange resin, typically a zeolite, to exchange a sodium ion for an ion that causes hardness.

## L

**landfill gas:** Gas collected from a landfill that is made up of mainly methane gas and carbon dioxide with varying amounts of miscellaneous gases.

**lay-up:** The preparation of a boiler for out-of-service status for an extended period.

**light hazard area:** A building or room that is used as a church, office, classroom, or assembly hall.

**lignite coal:** A geologically young coal that contains a low percentage of fixed carbon and a very high percentage of volatile gas and moisture. Also known as brown coal.

**lime-soda softener:** A water softener that uses lime and soda ash to remove hardness from water.

**line desuperheater:** A desuperheater that injects feedwater into a superheated steam line to reduce the steam temperature.

**lockout:** The use of locks, chains, or other physical restraints to positively prevent the operation of specific equipment.

**lowest achievable emission rate (LAER):** A rate for emissions from a new source in nonattainment areas that meets the environmental requirement.

**low pressure steam boiler:** A boiler that operates at a maximum allowable working pressure (MAWP) of not more than 15 psi.

**low-water fuel cutoff:** A boiler control that secures the burner in the event of a low-water condition.

## M

**magnetic float indicator:** A water level indicator that consists of a tube chamber and an internal float.

**magnetostrictive sensor:** Part of a level-measuring system for boiler water consisting of an electronics module, a waveguide, and a float containing a magnet that is free to move up and down a brass rod pipe inserted into a vessel from the top.

**major source:** A facility that emits 10 or more tons per year of any single HAP or 25 tons or more per year of any combination of HAPs.

**manometer:** A draft gauge that measures pressure with a liquid-filled tube.

**MATT:** An acronym that stands for "mixture, atomization, temperature, and time."

**maximum achievable control technology (MACT):** Technology that meets a level of control the EPA uses to regulate HAP emissions based on the level of emissions control for the best-performing technology used in similar major sources.

**maximum allowable working pressure (MAWP):** The operating pressure of a boiler as determined by the design and construction of the boiler in conformance with the ASME Code.

**mechanical draft:** Draft produced by power-driven fans.

**mechanical dust collector:** A device that separates PM from the gases of combustion using centrifugal force to spin the PM out of the gas stream.

**membrane boiler:** A watertube boiler that uses strips of metal alloy welded between the tubes to form a seal.

**metering control system:** A combustion control system in which the flow of fuel and air are measured and adjusted in correct proportions based on steam pressure.

**modulating control system:** A combustion control system that is used to adjust the firing rate to a degree that is proportional to the steam load through the use of a PLC or digital control system. Also known as positioning control system.

**modulating pressure control device:** A pressure control device that provides local control of the firing rate proportional to the steam pressure.

**moisture separator:** A device placed in a wet steam line used to remove entrained water droplets.

**municipal solid waste (MSW):** The trash or garbage from residential and commercial users that is collected for disposal.

## N

**natural draft:** Draft produced by the natural action resulting from temperature differences between air and gases of combustion.

**natural gas:** A colorless and odorless combustible gas that consists mainly of methane with small quantities of ethylene, hydrogen, and other gases.

**New Source Performance Standards (NSPS):** Health-based environmental standards for emissions from major new sources of emissions.

**nitrogen oxides ($NO_x$):** A general term used to include all possible forms of molecules containing nitrogen and oxygen that result from combustion at high temperatures.

**noncondensing steam turbine:** A steam turbine that exhausts at atmospheric pressure or above.

**normal operating water level (NOWL):** The water level designated by the manufacturer as being the proper water level for safe boiler operation.

## O

**one-element feedwater regulator:** A feedwater regulator that regulates the amount of feedwater by measuring only the actual water level in the boiler.

**ON/OFF control:** A combustion control strategy that is used to start and stop a burner without any modulation of the flame.

**open feedwater heater:** A feedwater heater in which steam and feedwater mix with each other at atmospheric pressure to raise the temperature of the water.

**operating pressure control:** An ON/OFF control with an adjustable differential that regulates the operating range of the boiler between the cut-in pressure and the cut-out pressure by opening and closing electrical contacts.

**ordinary hazard area:** A building or room that is used as a shop and related storage facility, light manufacturing plant, automobile showroom, or parking garage.

**orifice plate:** A flow restriction device consisting of a thin circular metal plate with a sharp-edged hole in it and a tab that protrudes from a flange.

**Orsat analyzer:** An old method of measuring the carbon dioxide, carbon monoxide, and oxygen in the gases of combustion by selectively absorbing the gases and measuring the volume removed from the gases of combustion.

**oxygen trim:** A control strategy used to adjust the air-fuel ratio of a burner according to the results from a flue gas oxygen analyzer.

**P**

**package boiler:** A boiler that comes completely assembled with its own pressure vessel, burner, draft fans, and fuel train.

**parallel positioning:** A modulating control strategy that uses steam pressure as the input signal and outputs one signal to a fuel valve to modulate the fuel flow and another signal to a variable-speed blower motor or to an air damper to modulate the airflow.

**perfect combustion:** Combustion that occurs when all the fuel is burned using only the theoretical amount of air.

**personal protective equipment (PPE):** Any device worn by a boiler operator to prevent injury.

**pH:** A measurement representing whether the substance is acidic, neutral, or alkaline.

**pH meter:** A water analyzer used to monitor the acidity or alkalinity of a boiler water.

**photocell:** A device in a flame scanner that generates a small amount of electric current when exposed to visible light.

**pilot-operated pressure regulator:** A regulator that uses upstream fluid as a pressure source to power the diaphragm of a larger valve.

**plant master:** A master controller that calculates and distributes steam production requirements across several boilers to maintain the steam pressure at the setpoint.

**pneumatic level control:** *See* displacer.

**popping pressure:** The predetermined pressure at which a safety valve opens and remains open until the pressure drops.

**pop test:** A safety valve test performed to determine if a safety valve opens at the correct pressure.

**positioning control system:** *See* modulating control system.

**positive-displacement flow meter:** A flow meter that admits fluid into a chamber of known volume and then discharges it.

**postpurge:** A purge that occurs after burner shutdown.

**pour point:** The lowest temperature at which a liquid will flow from one container to another.

**pre-purge:** A purge that occurs before a burner is allowed to fire.

**pressure atomizing burner:** A fuel oil burner that uses pressure to force fuel through a nozzle or sprayer and into the furnace where the vaporized fuel mixes with air and is ignited.

**pressure regulator:** An adjustable valve that is designed to automatically control the pressure downstream.

**pressurized furnace:** A furnace that operates at slightly above atmospheric pressure.

**primary air:** The air supplied to a burner to atomize fuel oil or convey pulverized coal and control the rate of combustion, thus determining the amount of fuel that can be burned.

**primary element:** An instrument that measures a process variable and produces a usable output in the form of mechanical movement, electrical output, or instrument air-pressure output. Also known as a sensor.

**priming:** The act of large slugs of water and impurities being carried out of the boiler into the steam lines.

**programming device:** A device used to control the actions of a PLC, DDC system, or computer.

**proving the main flame:** A process in which a flame scanner observes the main flame to verify that it is lit.

**proving the pilot:** A process in which a flame scanner is used to observe the pilot to verify that it is lit.

**proximate analysis:** An analysis used to determine the amount of moisture, volatile gas, fixed carbon, and ash in a coal specimen.

**pulverized coal:** Coal that is ground into a fine powder and is then blown into the combustion chamber where it is burned in suspension.

**purge cycle:** A process during which air is blown through a furnace for a predetermined period of time to remove any combustible vapors.

## R

**radiant superheater:** A superheater that is directly exposed to the radiant heat of the boiler.

**reagent:** A chemical used in a chemical test to indicate the presence of a specific substance.

**reasonably available control technology (RACT):** Technology that meets the environmental requirement for emissions from an existing source in a nonattainment area.

**reciprocating feedwater pump:** A positive-displacement pump that uses steam to apply pressure to a large piston in one cylinder connected to a small piston in another cylinder that pumps the water.

**refuse boiler:** A boiler that uses the municipal solid waste that would normally go to landfills.

**regenerative air heater:** An air heater that is divided into zones: the zone containing the gases of combustion, the sealing zones, and the air zone containing the air for combustion.

**register:** A formed sheet-metal plate with slots where the metal is bent on an angle.

**regulator:** A self-operating control valve used for pressure and temperature control.

**resistance temperature detector (RTD):** An electronic thermometer consisting of a high-precision resistor with resistance that varies with temperature.

**reverse osmosis (RO):** A filtration process in which the water to be treated is pressurized and applied against the surface of a semipermeable membrane in order to demineralize it and remove its impurities.

**rotameter:** A variable-area flow meter consisting of a tapered tube and a float with a fixed diameter.

**rotary cup burner:** A fuel oil burner consisting of a quickly spinning cup that discharges the fuel oil with high-velocity air into the boiler, resulting in a finely atomized fuel that mixes with air and ignites.

## S

**safety data sheet (SDS):** Printed material used to relay chemical hazard information from the manufacturer, importer, or distributor to the employer.

**safety valve:** An automatic, full-open, pop-action valve that is opened by an overpressure in a boiler and used to relieve the overpressure before damage occurs.

**safety valve blowback:** *See* safety valve blowdown.

**safety valve blowdown:** A drop in pressure between popping pressure and reseating pressure as a safety valve relieves boiler overpressure. Also known as safety valve blowback.

**safety valve capacity:** The amount of steam that can be relieved by a safety valve, measured in the number of pounds of steam per hour it is capable of discharging under a given pressure.

**saturated steam:** Steam that is in equilibrium with water at the same temperature and pressure.

**saturation pressure:** The pressure at which water and steam are at the same temperature.

**saturation temperature:** The temperature at which water and steam are at the same pressure.

**scale:** An accumulation of compounds such as calcium carbonate and magnesium carbonate on the water side of the heating surfaces of a boiler.

**scotch marine boiler:** A firetube boiler with an internal furnace.

**secondary air:** The air supplied to the furnace by draft fans to control combustion efficiency by controlling how completely the fuel is burned.

**sediment:** The hard, sandy particles of foreign matter that have precipitated out of water.

**selective catalytic reduction (SCR):** An emissions control method for boilers in which ammonia gas is introduced over a catalyst located in a module that is installed in the boiler exhaust stack.

**sensor:** *See* primary element.

**shear stress:** Stress that occurs when two forces of equal intensity act parallel to each other but in opposite directions.

**single-point positioning:** A modulating control strategy that uses steam pressure as the input signal and outputs a signal to a modulating motor that turns a jackshaft to modulate the air and fuel flow.

**slag:** *See* clinker.

**sludge:** The soft, muddy residue produced from impurities in water that accumulates in low spots in a boiler.

**soft coal:** *See* bituminous coal.

**solenoid valve:** An electrically operated control valve that quickly adjusts a valve to an open or closed position.

**soot:** The carbon deposits resulting from incomplete combustion.

**soot blower:** A device used to remove soot deposits from around tubes and permit better heat transfer in the boiler.

**specific gravity:** The ratio of the weight of a given volume of a substance to the weight of the same volume of water at a standard temperature of 60°F.

**spontaneous combustion:** The process where a material can self-generate heat until its ignition point is reached.

**spreader stoker:** A stoker that feeds the coal into the boiler in suspension and on the grate.

**squirrel cage blower:** A blower with a wheel that has blades attached at the rim and rotates in a housing.

**steam:** the vapor that forms when water is heated to its boiling point.

**steam atomizing burner:** A fuel oil burner that sends steam and pressurized fuel oil through a nozzle into the furnace where the vaporized fuel mixes with air and is ignited.

**steam impingement:** The condition where steam strikes a metal surface, causing erosion of that surface.

**steam pressure gauge:** A boiler fitting that displays the amount of pressure inside a boiler, steam line, or other pressure vessel.

**steam quality:** The ratio of dry steam to the total amount of water evaporated.

**steam separator:** A device that is located in the steam drum of a boiler and used to increase the quality of steam. Also known as a drum internal.

**steam trap:** An accessory that removes air and noncondensable gases and condensate from steam lines and heat exchangers without a loss of steam.

**steam turbine:** A rotary mechanical device used to drive rotating equipment, such as a generator, by extracting thermal energy from pressurized steam.

**stoker:** A mechanical device used for feeding coal into a boiler.

**stop valve:** A valve that is opened or closed by the operator and typically used to isolate the feedwater to the boiler or the steam discharge from the boiler.

**straight-tube boiler:** A watertube boiler with box headers connected together by straight, inclined water tubes.

**stress:** Pressure or tension applied to an object.

**superheated steam:** Steam that has been heated above the saturation temperature.

**superheater:** A bank of tubes through which steam passes after leaving the boiler where additional heat is added to the steam.

**surface blowdown:** The process of removing water from a boiler near the NOWL to control the quantity of impurities in the remaining water or to remove a film of impurities on the surface of the water.

**swell:** A process where the water level in a boiler momentarily rises with an increase in steam demand.

# T

**tagout:** The process of attaching a danger tag to the source of power to indicate that the equipment may not be operated until the tag is removed.

**tensile stress:** Stress that occurs when two forces of equal intensity act on an object, which pulls the object in opposite directions.

**therm:** A unit used to measure the heat content of natural gas and is equal to 100,000 Btu.

**thermal efficiency:** The ratio of the heat absorbed by a boiler (output) to the heat available in the fuel (input) including radiation and convection losses.

**thermocouple:** A thermometer consisting of two dissimilar metals that are joined together to form a circuit that generates a voltage proportional to the difference in temperature between the hot and cold ends of the wires.

**thermodynamic steam trap:** A steam trap that has a single movable disc that rises to allow the discharge of air and cool condensate.

**thermoexpansion feedwater regulator:** A feedwater regulator with a thermostat that expands and contracts with exposure to steam and moves a linkage that modulates a regulator valve.

**thermohydraulic feedwater regulator:** A feedwater regulator that has a regulating valve, a bellows, a generator, and stop valves and varies feedwater flow in direct response to changes in the boiler water level by using changes in temperature to create changes in hydraulic pressure.

**thermometer:** The general term for an instrument used to measure temperature.

**thermostatic steam trap:** A steam trap that contains a temperature-operated element, such as a corrugated bellows.

**three-element feedwater regulator:** A feedwater regulator that regulates the amount of feedwater by measuring the steam flow from the boiler and the feedwater flow into the boiler in addition to the water level.

**titration test:** A water treatment test in which a reagent is added to a sample to determine the concentration of a specific dissolved substance.

**total dissolved solids:** A measurement of the concentration of dissolved impurities in boiler water.

**transducer:** A device that receives a signal representing a variable, such as temperature, pressure, or flow, and converts that signal into an electrical or digital signal that is compatible with a digital control system.

**transmitter:** A device that conditions a low-energy signal from a sensor and produces a suitable signal for transmission to other components or devices.

**try cock:** A valve located on the water column of a boiler that is used to determine the boiler water level if the gauge glass is not functioning.

**turbine:** A machine that converts the energy from the expansion of high-pressure steam into the mechanical energy of rotation.

**turbine feedwater pump:** A rotary positive-displacement pump that uses a flat impeller with small, flat, perpendicular fins machined into the impeller rim to discharge feedwater into the boiler.

**turndown ratio:** The ratio of the maximum firing rate to the minimum firing rate.

**two-element feedwater regulator:** A feedwater regulator that regulates the amount of feedwater by measuring the steam flow from the boiler in addition to the water level.

## U

**ultimate analysis:** An analysis used to determine the elements present in a coal specimen.

**ultrasonic sensor:** A level-measuring instrument that uses an ultrasonic signal to measure level.

**ultraviolet flame sensor:** A device in a flame scanner that generates a small amount of electric current when exposed to ultraviolet light.

**underfeed stoker:** A stoker that feeds coal into a boiler from under the fire.

**underground storage tank (UST):** A tank and any underground piping connected to the tank that has at least 10% of its combined volume underground.

## V

**vacuum gauge:** A pressure gauge used to measure pressure less than atmospheric pressure.

**variable-area flow meter:** A meter that maintains a constant differential pressure and allows the flow area to change with the flow rate.

**variable-speed drive (VSD):** A motor controller used to vary the frequency of the electrical signal supplied to an AC motor in order to control its rotational speed.

**velocity compounding:** A method in which the energy from steam is extracted using multiple stages in a turbine to reduce the steam velocity and pressure in a short time.

**vent condenser:** An internal element in some types of deaerators that condenses steam and uses the heat produced to warm the feedwater.

**venturi tube:** A flow restriction device consisting of a fabricated pipe section with a converging inlet section, a straight throat, and a diverging outlet section.

**vertical firetube boiler:** A one-pass firetube boiler that has tubes in a vertical position.

**viscosity:** The measurement of the internal resistance of a fluid to flow.

**volatile gas:** The gas given off when coal burns.

## W

**waste boiler:** A boiler that uses fuel from an industrial process that would normally be wasted.

**water column:** A boiler fitting that reduces the turbulence of boiler water to provide an accurate water level in the gauge glass.

**water softener:** A boiler water pretreatment device used to reduce hardness.

**watertube steam boiler:** A boiler that has water inside the tubes with heat and gases of combustion around the tubes.

**waterwall:** A set of tubes that is placed in the furnace area of watertube boilers and used to increase the heating surface of boilers and the service life of refractory.

**wet-back boiler:** A firetube boiler with three tube sheets and a water-cooled turnaround chamber, with a water leg formed between the rear tube sheet and the chamber.

**wet lay-up:** The storage of a boiler filled with warm, chemically treated water.

**windbox:** A chamber surrounding a burner assembly or coal stoker that allows pressurized air from a forced draft fan to enter the burner.

# Z

**zeolite:** One of a group of minerals including silicates of aluminum, sodium, and calcium that are used in ion-exchange softeners.

# INDEX

*Page numbers in italic refer to figures.*